"In the past thirty years, modeling and simulation has undergone a revolution that has transformed this specialty into a vital discipline. Visionaries saw the potential to develop virtual environments to provide real-life training in fields such as defense and health care. And they encouraged new paths to help solve complex problems that could not be unraveled using traditional methods."

"Dr. Tolk's book chronicles this revolution and captures the essence of how and why it happened. It also acknowledges Old Dominion University's role in expanding modeling and simulation with our academic programs and our research in new methods to apply this technology. We are proud to have helped refine a field that will play a critical role in the twenty-first century, and I appreciate Dr. Tolk's recognition of this crucial discipline."

John R. Broderick
President
Old Dominion University

In this volume, nearly three dozen experts in Modeling and Simulation (M&S) come together to make a compelling case for the recognition of M&S as a profession. They do so by citing evidence of the unique body of knowledge possessed by those with M&S skills, the broad and indispensable contributions they make to science and industry, the status of M&S in other parts of the world, and the strong economic, educational, and ethical foundations that exist for this work. This book is important reading for anyone seeking to elevate the standing of this vital field.

Alfred (Al) Grasso
President & CEO
The MITRE Corporation

The Profession of
Modeling and Simulation

Wiley Series in Modeling and Simulation

The Wiley Series in Modeling and Simulation provides an interdisciplinary and global approach to the numerous real-world applications of modeling and simulation (M&S) that are vital to business professionals, researchers, policymakers, program managers, and academics alike. Written by recognized international experts in the field, the books present the best practices in the applications of M&S as well as bridge the gap between innovative and scientifically sound approaches to solving real-world problems and the underlying technical language of M&S research. The series successfully expands the way readers view and approach problem solving in addition to the design, implementation, and evaluation of interventions to change behavior. Featuring broad coverage of theory, concepts, and approaches along with clear, intuitive, and insightful illustrations of the applications, the Series contains books within five main topical areas: Public and Population Health; Training and Education; Operations Research, Logistics, Supply Chains, and Transportation; Homeland Security, Emergency Management, and Risk Analysis; and Interoperability, Composability, and Formalism.

Founding Series Editors:

Joshua G. Behr, Old Dominion University

Rafael Diaz, MIT Global Scale
Advisory Editors:
Homeland Security, Emergency Management, and Risk Analysis
Interoperability, Composability, and Formalism

Saikou Y. Diallo, Old Dominion University

Mikel Petty, University of Alabama

Operations Research, Logistics, Supply Chains, and Transportation

Loo Hay Lee, National University of Singapore

Public and Population Health

Peter S. Hovmand, Washington University in St. Louis

Bruce Y. Lee, University of Pittsburgh

Training and Education

Thiago Brito, University of Sao Paolo

Spatial Agent-Based Simulation Modeling in Public Health: Design, Implementation, and Applications for Malaria Epidemiology
by *S. M. Niaz Arifin, Gregory R. Madey, Frank H. Collins*

The Digital Patient: Advancing Healthcare, Research, and Education
by *C. D. Combs (Editor), John A. Sokolowski (Editor), Catherine M. Banks (Editor)*

The Profession of Modeling and Simulation

Discipline, Ethics, Education, Vocation,
Societies, and Economics

1st Edition

Edited By
Andreas Tolk
The MITRE Corporation, Hampton, VA, USA

Tuncer Ören
University of Ottawa, Ottawa, Ontario, Canada

This edition first published 2017
© 2017 John Wiley & Sons, Inc

The right of Andreas Tolk, Tuncer Ören to be identified as the author(s) of the editorial material in this work has been asserted in accordance with law.

Registered Office
John Wiley & Sons, Inc., 111 River Street, Hoboken, NJ 07030, USA

Editorial Office
111 River Street, Hoboken, NJ 07030, USA

For details of our global editorial offices, customer services, and more information about Wiley products visit us at www.wiley.com.

Wiley also publishes its books in a variety of electronic formats and by print-on-demand. Some content that appears in standard print versions of this book may not be available in other formats.

Library of Congress Cataloging-in-Publication Data
Names: Tolk, Andreas, editor. | Ören, Tuncer I., editor.
Title: The profession of modeling and simulation : discipline, ethics, education, vocation, societies, and economics / edited By Andreas Tolk, Tuncer Ören.
Description: 1st edition. | Hoboken, NJ : John Wiley & Sons, 2017. | Series: Wiley series in modeling and simulation | Includes bibliographical references and index. |
Identifiers: LCCN 2017011678 (print) | LCCN 2017019085 (ebook) | ISBN 9781119288107 (pdf) | ISBN 9781119288220 (epub) | ISBN 9781119288084 (cloth)
Subjects: LCSH: Mathematics–Vocational guidance. | Mathematical models. | Computer simulation.
Classification: LCC QA10.5 (ebook) | LCC QA10.5 .P74 2017 (print) | DDC 331.7/615118–dc23
LC record available at https://lccn.loc.gov/2017011678

Hardback 9781119288084

Cover image: © kentoh/Gettyimages
Cover design by Wiley

Set in 10.5/12.5pt WarnockPro-Regular by Thomson Digital, Noida, India

10 9 8 7 6 5 4 3 2 1

This text is dedicated to Mr. **William (Bill) F. Waite** *(1946–2015), who has been a true Professional of Modeling and Simulation (M&S). He dedicated his life to contribute to the establishment of the M&S Body of Knowledge, was active as a servant leader in many professional M&S Societies, and was a pioneer in the Economics of M&S. He touched the lives of many scholars, students, and practitioners. He truly will be missed by all his friends and colleagues.*

Contents

Foreword

In memoriam Bill Waite

I first met Bill in the 1990s at an annual Interservice/Industry Training, Simulation and Education Conference (I/ITSEC). Our meeting was unplanned and casual, but Bill's influence on and respect by the Modeling and Simulation (M&S) community were clearly evident throughout that I/ITSEC and those that followed. Although Bill, as a founder and principal in Aegis Technologies, was an industry icon, he had a deep affinity for the academic community that I represented. Bill was a powerful proponent of standards for M&S as well as for developing a Body of Knowledge (BoK) for the field. Bill saw that the lack of an agreed-upon BoK meant that no one "owned" M&S and that many disciplines had and would continue to use M&S in parallel, with little coordination and sharing of best practices.

Bill had many remarkable qualities. Those that stood out to me were vision, persistence, and a sense of community. He was one of the few to truly develop and share a vision for M&S as a true academic discipline in addition to its established role as an indispensable part of system design, manufacturing, and training/education. Bill and I – yes, Bill and most of the M&S world – would engage in passionate debates about an array of related topics, from the need for and nature of the M&S BoK to the value of professional recognition for M&S practitioners to creating the standards that would allow the maturation of the M&S discipline.

This book pulls together all the elements that Bill championed as an M&S advocate. Part I, Foundation, after Dr. Tolk's introduction to the M&S profession, deals directly with two of Bill's greatest passions – a Body of Knowledge and a Code of Ethics. These truly are the basis for the entire profession, and every M&S professional must embrace both the need for a dynamic Body of Knowledge and the necessary ethical underpinnings of our profession. To build professionals, we must prepare them. Part II, Education, draws upon the work of many of Bill's friends from the academic community to make the case that M&S is an academic discipline and that there are educational programs that directly produce the professionals needed by the field. The section also documents the professional certification program that Bill was instrumental in developing through the National Training Systems Association.

Part III, Society, introduces the professional organizations that are essential to the development of the practitioner and to the dissemination of new knowledge as it is discovered. The ubiquity of M&S in the military community demands that the uniformed services' approach to professional development and recognition be fully described. Recognizing that M&S is truly an international profession, it is appropriate that the way in which other nations approach developing the discipline and its professionals be included. Finally, one of Bill's focuses – enterprise M&S – is explored in terms of how to draw into professional development the influences of the multitude of domains in which M&S is routinely used.

Part IV, Application, presents two examples – complexity/innovation dynamics and cyber environments – of areas where M&S has more recently begun to make major impacts and calls on all of us to accelerate the creation of M&S professionals to address these new challenges. Of course, no professional can be sustained and continue to evolve without financial success. In Part V, Economics, Dr. Tolk has engaged key thinkers to produce three chapters on the funding of academic M&S research as well as documenting the return on M&S investment, including a deep dive into how M&S supports cost avoidance. Finally, in Part VI, Policy, the book describes past and current efforts to build a national M&S coalition in the United States – an essential pathway to sustaining government support for research, education, and application development.

There is no doubt that this volume will serve as one of the tangible products to celebrate Bill's many and varied contributions to M&S. On behalf of a grateful M&S community, I extend our deepest appreciation to Dr. Andreas Tolk for conceiving this book and bringing it to life.

Bill Waite will be remembered as a visionary leader who, with persistence, shepherded our developing field to adulthood, creating enduring friendships along the way. To him goes the age-old salute: *Per aspera ad astra.*[1] Bill labored long and hard to see M&S arrive at its rightful place among the great disciplines of science and engineering. His legacy will endure and inspire all M&S professionals to reach for the stars.

President *Emeritus*, Texas A&M University R. Bowen Loftin
Former Chancellor, University of Missouri
January 8, 2017

---

1 Through hardships to the stars.

Preface

Looking through the various chapters of this book, I am surprised that all this information on the profession of modeling and simulation (M&S) has not been compiled in a single volume like this one before. Although I have been working for many years as an M&S professional, this is the first time I have fully considered what it means to be an M&S professional, the many facets of our professional domain, and how all these aspects influence scholars and practitioners.

These insights did not surprise me, as my friend and colleague William (Bill) F. Waite led the way in many of these contributing fields, sometimes long before others recognized the value and contribution of his work. He was a pioneer, visionary, leading advocate, and champion of M&S as a technology, industry, marketplace, and profession for many, many years. The industry lost its most ardent supporter, and outspoken promoter, and many of us lost a dear friend when Bill unexpectedly passed away in July 2015. To honor his work and memory, as well as to assemble the contributions and the resulting work of others, this book has been compiled by many of Bill's closest friends and colleagues from the M&S Industry. It is a tribute to his passion, accomplishments, and enduring legacy to the profession, and a source of exceptional information for scholars, students, and practitioners.

Bill passionately believed in M&S as a calling, as such he dedicated several decades of his life to leading the charge to ensure M&S was indeed a recognized profession. He was a true pioneer in M&S, as his 45-year career spans critical decades in the coming-of-age of M&S. Bill spent a majority of the last 25 years raising national and international awareness of the power of M&S technology. He was instrumental in advocating, supporting, or personally leading many initiatives in this technology's advancement: establishing academic foundations, a code of ethics, and professional certification; encouraging M&S application in diverse industries; studying and advocating for the economic values of M&S; and creating multiple organizations and societies to further the industry.

Bill worked diligently to transform the industry's ability to collaborate and communicate through the establishment of industry forums such as the Alabama Modeling and Simulation Council (AMSC) and the Virginia Modeling and Simulation Partnership (VMSP). He took this concept to a national and international scale, birthing the idea and supporting the full implementation of both the National Modeling and Simulation Coalition (NM&SSC) and the international *SimSummit* Round Table. He twice led the Society for Modeling and Simulation International (SCSI) as Chairman of the Board of Directors, and he served twice on the Executive Committee of the Simulation Interoperability Standards Organization (SISO). Bill was also active in the National Training and Simulation Association (NTSA), served the National Defense Industrial Association (NDIA) as a member of its Board of Trustees, and was on the Economic Development Partnership of Alabama (EDPA) Foundation Board of Directors.

Bill was instrumental in supporting Congressman Randy Forbes and the Congressional M&S Caucus in establishing House Resolution 487, recognizing M&S as a National Critical Technology. When an initiative involving many people and organizations to establish an M&S NAICS code failed initially, Bill rolled up his sleeves, doubled his efforts, and led the second attempt to establish a much needed federal procurement designation for the M&S industry.

M&S education was a priority to Bill. He served as an influential, active member of the Modeling and Simulation Professional Certification Commission (M&SPCC), which established the concept, protocols, testing, and training for the Certified M&S Professional (CMSP) program. This is our industry's highest certification, comparable to becoming a CPA in the accounting field. Bill also worked diligently in advocacy and support to establish Master's Degree and Ph.D. programs in M&S at several major universities: the University of Alabama Huntsville, University of Central Florida, Old Dominion University, and the Naval Post Graduate School, to name but a few.

Bill passionately and selflessly believed that a "rising tide raises all ships" and that through increased awareness and application of M&S skills, tools, and technologies, the Aerospace/Defense industry, DoD and other government agencies, commercial industries, and ultimately mankind would greatly benefit, not only saving time, money, and resources, but also accomplishing many things that previously, simply could not be done.

Finally, of course, Bill was extremely passionate and committed to building AEgis Technologies – always striving for our company to be a true international leader in M&S technology, services, products, and training. AEgis Technologies and its people are clearly the legacy for which he was proudest.

By all accounts and to everyone who knew him in the Aerospace/Defense industry, on the national and international scale, and in a corporate or personal manner – Bill's exceptionalism was evident. His vibrant, passionate personality leaves a legacy in many hearts and minds. The relationship I shared with Bill as we worked side by side in the development of AEgis Technologies, each of us with differing gifts and abilities but a common goal, is one that significantly influenced me personally. Bill has been profoundly missed by the AEgis Technologies family and the many, many people with whom he worked, mentored, and collaborated over several decades of success. Most of all, for his loving wife Katie, and children Elliott, Portia, and Emily, Bill's absence leaves a void that is not easily or ever truly filled.

A special thanks to all of the authors and Andreas Tolk, who led this exceptional group of M&S colleagues in compiling this text. Clearly, Bill's M&S legacy will have a lasting impact for decades more to come, as will the professional commitments of each of you who contributed to this text. Whoever works in our domain as an M&S professional should be aware of the many facets of the profession of M&S, as compiled by these leading experts, inspired by and in honor of Bill Waite, a true M&S professional by all accounts.

President, CEO, and Co-Founder with   *Steve Hill*
Bill Waite of AEgis Technologies

List of Contributors

Robert K. Armstrong
Eastern Virginia Medical School, Norfolk, VA, USA

Curtis L. Blais
Naval Postgraduate School, Monterey, CA, USA

Tim Cooley
DynamX Consulting, Castle Rock, CO, USA

Rudolph P. Darken
Naval Postgraduate School, Monterey, CA, USA

Saikou Y. Diallo
Old Dominion University, Norfolk, VA, USA

Umut Durak
German Aerospace Center (DLR), Braunschweig, Germany

Richard Fujimoto
Georgia Institute of Technology, Washington, DC, USA

Randall B. Garrett
National M&S Coalition, Washington, DC, USA

Steven Gordon
Georgia Tech Research Institute, Orlando, FL, USA

Robert M. Gravitz
AEgis Technologies, Orlando, FL, USA

Gary M. Lightner
AEgis Technologies, Orlando, FL, USA

Margaret L. Loper
Georgia Tech Research Institute, Atlanta, GA, USA

Christopher J. Lynch
Old Dominion University, Norfolk, VA, USA

Roland R. Mielke
Old Dominion University, Norfolk, VA, USA

Saurabh Mittal
The MITRE Corporation, McLean, VA, USA

Navonil Mustafee
University of Exeter, Exeter, UK

Tuncer Ören
University of Ottawa, Ottawa, Ontario, Canada

Ivar Oswalt
Simulation U Analytics, LLC, Fredericksburg, VA, USA

Mikel D. Petty
University of Alabama in Huntsville, Huntsville, AL, USA

Gregory S. Reed
Torch Technologies, Huntsville AL, USA

James A. Robb
National Training & Simulation Association, Arlington, VA, USA

Richard J. Severinghaus
CRTN Solutions, LLC, Washington, DC, USA

John A. Sokolowski
Old Dominion University, Norfolk, VA, USA

Steve Swenson
AEgis Technologies, Newport, RI, USA

Simon J.E. Taylor
Brunel University London, London, UK

Andreas Tolk
The MITRE Corporation, Hampton, VA, USA

William V. Tucker
Simulationist.US, Huntsville, AL, USA

Charles D. Turnitsa
Georgia Tech Research Institute, Atlanta, GA, USA

Yingnian Wu
Beijing Information Science & Technology University, Beijing, P. R. China

Gengjiao Yang
Beihang University; Engineering Research Center of Complex Product Advanced Manufacturing Systems, Ministry of Education, Beijing, P. R. China

Levent Yilmaz
Auburn University, Auburn, AL, USA

Bernard P. Zeigler
University of Arizona, Tucson, AZ; RTSync Corporation, Rockville, MD, USA

Lin Zhang
Beihang University; Engineering Research Center of Complex Product Advanced Manufacturing Systems, Ministry of Education, Beijing, P. R. China

Notes on Contributors

Robert K. Armstrong serves in multiple roles at Eastern Virginia Medical School (EVMS) in Norfolk, Virginia. He is Director of the Sentara Center for Simulation and Immersive Learning, leading a team of 140+ personnel providing healthcare training and education to the EVMS School of Medicine and School of Health Professions, various local Graduate Medical Education programs, regional emergency medical providers, and over 60 external healthcare clients. He is Director of the National Center for Collaboration in Medical Modeling and Simulation, helping commercial businesses to diversify their modeling and simulation products into the commercial healthcare sector. He is the Director of Corporate Relations, providing a link between EVMS researchers and regional commercial and entrepreneurial opportunities. He joined EVMS from Booz Allen Hamilton, where he provided modeling and simulation support and analysis, developed M&S training tools, and drafted M&S policy for U.S. Navy, Marine Corps, and Air Force clients. Prior to his time at Booz Allen, he was Director of Technology at the Virginia Modeling, Analysis and Simulation Center at Old Dominion University in Norfolk, Virginia. Bob holds a M.S. in Computer Science from the U.S. Naval Postgraduate School in Monterey, California, and a B.S. in General Engineering from the U.S. Naval Academy in Annapolis, Maryland. He is on the Board of Directors of the Society for Simulation in Healthcare. He is a U.S. Marine Corps veteran.

Curtis L. Blais is a member of the Naval Postgraduate School (NPS) research faculty in the Modeling, Virtual Environments, and Simulation (MOVES) Institute. He has over 42 years of experience in the specification, design, development, and employment of military simulation systems, including the past 17 years with NPS as a principal investigator, technical expert, and instructor in modeling and simulation projects and education. He began his career as an Operations Research Analyst at the Naval Electronics Laboratory Center (now called the Space and Naval Warfare

Systems Center, Pacific) developing analytical simulation models of command and control systems and tactical communications systems and combat simulation models of Marine Corps operations for command staff training, as well as conducting evaluations of artificial intelligence techniques and natural language processing systems for military applications. Over a period of 20 years working in industry, Blais continued in software development for follow-on Marine Corps training systems, while holding progressively higher positions in technical management. Blais is active in several U.S. military and international modeling and simulation standards and professional education organizations. Blais earned B.S. and M.S. degrees in Mathematics from the University of Notre Dame.

Tim Cooley is President and Founder of DynamX Consulting, a veteran-owned consulting firm located in Castle Rock, Colorado. He is also an adjunct faculty member at Colorado College. He spent 16 years on the U.S. Air Force Academy faculty, both in uniform and as a civilian, holding numerous positions, including the DMSO Modeling and Simulation Chair, Deputy Department Head, and Senior Researcher. His previous military assignments included Squadron Commander of the 1987th Communications Squadron at Lowry AFB and Land Mobile Radio Manager for CINCNORAD. Cooley has completed a number of innovative mathematical and cost analyses for the USMC, USAF, and OSD and performed counterterrorism research for Joint Staff. He developed and taught the USAF Combat Analysis Class and has provided decision support tools and research for the FAA. He is coauthor of the 2011 Defense Acquisition University Research Paper of the Year and 2011 Hirsch Prize recipient. Cooley received his Ph.D. in Computer Science/Biomedical Engineering from Rutgers University in 1996.

Rudolph P. Darken is Professor of Computer Science at the Naval Postgraduate School in Monterey, California. He is a former Director of the Modeling, Virtual Environments, and Simulation (MOVES) Institute. He has served on advisory boards for the National Institutes of Health (NIH), the NASA Ames Research Center, the National Science Foundation, the Engineering and Physical Sciences Research Council (UK) as well as several technology companies. He was an Associate Editor of Presence Journal (MIT Press). He received his D.Sc. and M.S. degrees in Computer Science from the George Washington University and his B.S. in Computer Science Engineering from the University of Illinois at Chicago. He also holds a J.D. from the Monterey College of Law.

Saikou Y. Diallo is Research Associate Professor at the Virginia Modeling, Analysis and Simulation Center and adjunct Professor of Modeling,

Simulation, and Visualization Engineering at Old Dominion University. He received his M.S and Ph.D. in Modeling and Simulation from Old Dominion University. His research focuses on the theory and practice of interoperability and the advancement of M&S. Diallo has authored over 100 publications, including a number of awarded papers and articles in conferences, journals, and book chapters. He is a member of SCS, IEEE, and ACM.

Umut Durak is a Research Scientist in the Flight Dynamics and Simulation Department of the German Aerospace Center (DLR) Institute of Flight Systems. His research interests include model-based simulation engineering, simulation-based systems engineering, and ontologies in simulation. He is teaching graduate-level courses on simulation at the Clausthal University of Technology, Department of Informatics as an adjunct lecturer. He received his B.S., M.S., and Ph.D. degrees in Mechanical Engineering from Middle East Technical University (METU) in Turkey. In the last 15 years, he contributed various research and development projects and published more than 50 papers in national and international conferences, workshops, and journals. He is a member of the Society for Computer Simulation International (SCS), Arbeitsgemeinschaft Simulation (ASIM), and the American Institute of Aeronautics and Astronautics (AIAA). He is a member of the AIAA Modeling and Simulation Technical Committee and editorial teams of the *Simulation: Transactions of the Society for Modeling and Simulation International* and the *International Journal of Modeling, Simulation, and Scientific Computing.*

Richard Fujimoto is a Regents' Professor in the School of Computational Science and Engineering at the Georgia Institute of Technology. He received the M.S. and Ph.D. degrees from the University of California at Berkeley in 1980 and 1983, respectively, in Computer Science and Electrical Engineering. He did his undergraduate work at the University of Illinois at Urbana-Champaign where he received B.S. degrees in Computer Science and Computer Engineering in 1977 and 1978, respectively. He was the founding chair of the School of Computational Science and Engineering (CSE) at Georgia Tech, an academic unit focused on computational models of natural and engineered systems. In this role he led in the creation of M.S. and Ph.D. degree programs in CSE as well as two undergraduate minor programs. He has been an active researcher in the parallel and distributed simulation field since 1985 and has published over 200 papers in this area. He is author or coauthor of three books. He has received several best paper awards for his research as well as the ACM SIGSIM Distinguished Contributions in Simulation Award. He led the definition of the time management services for the High Level Architecture (IEEE Standard

1516). Fujimoto has served as Co-Editor-in-Chief of the journal *Simulation: Transactions of the Society for Modeling and Simulation International* and was a founding area editor for the *ACM Transactions on Modeling and Computer Simulation* journal. He has also served on the organizing committees for several leading conferences in the parallel and distributed simulation field.

Randall B. Garrett remains actively involved throughout M&S and Research and Development (R&D) communities. This includes successful application of science and engineering for full-spectrum Live, Virtual, Constructive (LVC) training and immersive game applications. Garrett leads various R&D teams with the responsibility for practical evaluation of emerging technologies and identification of next-generation architectures supporting scientific and operational needs. His interests include the application of M&S principles for computational dynamics, analytics, and autonomous system controls. His interests also include the effective use of Artificial Intelligence (AI) for immersive games and robotics. Garrett holds a B.A. from the University of Arkansas and M.S. from Marshall University; he earned his Ph.D. from Old Dominion University with a degree in M&S. He served in the military as a Naval Officer where he held many leadership positions including that of Commanding Officer. His industry experience includes positions as Chief Scientist, Principal Investigator (PI), and Technical Director for large corporations. Garrett serves on regional, state, and national boards to include task forces addressing information technology standards and policies. Garrett actively participates in the National M&S Coalition (NMSC) as Vice Chair of the Policy Committee.

Steven Gordon is the Orlando Field Office Manager and a Principal Research Engineer for Georgia Tech Research Institute. He is the Director of the Georgia Tech Test and Evaluation Research and Education Center. He served 26 years in the U.S. Air Force with tours as an F-111 Weapons Systems Officer, Instructor, and Wing Electronic Warfare Officer; Air Force Board Structure scheduler; Air Staff Modeling and Simulation Division Chief; 13th Air Force Director of Operations and Air Operations Center Director throughout PACOM; and Air Force Academy Department of Mathematics Professor and Head. He also served as the first Technical Director of the Air Force Agency for Modeling and Simulation. Gordon has a Bachelor's Degree in Mathematics (Marymount College); Master's Degrees in Education (Peabody College of Vanderbilt), Industrial Engineering/Operations Research (Purdue), and in Business (University of Florida, 2009); and a Ph.D. in Aero and Astro Engineering (Purdue, 1991).

Robert M. Gravitz is a Principal Member of the Technical Staff at the AEgis Technologies Group, Inc. He has an M.S., Systems Management, from the Florida Institute of Technology; and B.S., from the University of Florida. He is also a graduate of the Materiel Acquisition Management Course at the U.S. Army Logistics Management Center, and the Combined Arms and Services Staff School at the U.S. Army's Command and General Staff College. He has over 30 years of experience in the fields of research, development, test, and evaluation (RDT&E) of major Department of Defense (DoD) weapon programs using simulation-based engineering, test, and analysis methodologies. He has directed the execution of model and simulation (M&S) verification, validation, and accreditation (VV&A) tasks for several Major Defense Acquisition Programs (MDAPs), including the Airborne Laser (ABL) Program, the Ground-Based Midcourse Defense (GMD) Program, the Theater High-Altitude Area Defense (THAAD) Program, and the Ballistic Missile Defense System (BMDS). He also serves as the corporation's Quality Management Lead in executing its ISO 9001 quality program, including training, internal audits, and external certification audits. Notably. Gravitz supported the genesis of NASA's Technical Standard, NASA-STD-7009 Standard for Models and Simulations.

Gary M. Lightner only recently retired from his position as Chief Engineer and Program Manager for the AEgis Technologies Group. He received a B.S. in Computer Science from Arizona State University, and the M.S. in Computer Science from the U.S. Air Force Institute of Technology. As Director of the Simulation Technologies and Advanced Research Group, he established, communicated, and fostered the vision, goals, and agenda for researching, developing, and advancing new and advanced M&S technologies, techniques, and standards as well as for providing comprehensive M&S training and education courses. He was pivotal in introducing the simulation interoperability standards IEEE 1516 High Level Architecture to industry and government. He has been affiliated with AEgis for more than 20 years. He is a U.S. Air Force veteran who served for 24 years in various lead programming and project manager roles. He was an active member of the Simulation Interoperability Standards Organization (SISO), the Society for Computer Simulation (SCS), the Institute of Electrical and Electronics Engineers (IEEE), the National Defense Industrial Association (NDIA), the National Training Systems Association (NTSA), and the American Society for Training and Development (ASTD).

Margaret L. Loper is the Chief Scientist for the Information & Communications Laboratory at the Georgia Tech Research Institute. She holds a Ph.D. in Computer Science from the Georgia Institute of Technology, a M.S. in

Computer Engineering from the University of Central Florida, and a B.S. in Electrical Engineering from Clemson University. Margaret has worked in M&S since 1985, focusing primarily on distributed simulation. She is a founding member of the Simulation Interoperability Standards Organization (SISO) and received service awards for her work with the Distributed Interactive Simulation (DIS) and High-Level Architecture (HLA) standards and the DIS/SISO transition. Margaret is a founding member of Georgia Tech's Professional Masters in Applied Systems Engineering degree program, where she teaches the core M&S course. In 2006 she started the M&S professional certificate through the GT School of Professional Education, and more than 50 certificates have been awarded to date. For 7 years, Margaret also led a STEM event that introduced K-12 teachers and administrators to M&S.

Christopher J. Lynch is a Senior Project Scientist and member of the Modeling and Simulation Science lab at the Virginia Modeling, Analysis and Simulation Center (VMASC). He received his M.S. in Modeling and Simulation from Old Dominion University in 2012 and a B.S. in Electrical Engineering from Old Dominion University in 2011. He is currently pursuing a Ph.D. in M&S from Old Dominion University. His research interests include multiparadigm modeling, conceptual modeling, and verification of simulation models. Currently, he is applying these skills to the Modeling Religion Project in collaboration with the Institute for the Bio-Cultural Study of Religion (IBCSR). He has authored or coauthored over 20 publications in conferences, journals, and book chapters and has conducted several guest lectures and a tutorial session on multiparadigm modeling. He is a member of the Society for Modeling and Simulation (SCS).

Roland R. Mielke is a Professor in the Department of Modeling, Simulation and Visualization Engineering (MSVE) at Old Dominion University and holds the designation University Professor. Mielke received the B.S., M.S., and Ph.D. degrees from the University of Wisconsin – Madison, all in Electrical Engineering. At Old Dominion University, he served as Chair of the Department of Electrical and Computer Engineering from 1981 to 1996, Technical Director of the Virginia Modeling, Analysis and Simulation Center (VMASC) from 1996 to 2006, and Director of M&S Graduate Programs from 2006 to 2010. He was the Founding Chair of the MSVE Department from 2010 to 2013 and played a key role in the development and implementation of the undergraduate program in Modeling and Simulation Engineering. Mielke's research interests include mathematical systems theory, composability theory, and applications of discrete event simulation. He also has interests in developing and expanding modeling and simulation educational opportunities.

Saurabh Mittal is Lead Systems Engineer at MITRE, Founder and President of Dunip Technologies, LLC, and Vice President-Memberships for Society of Computer Simulation (SCS) International. He is also currently serving as General Chair for Spring Simulation Multi-Conference 2017 and Associate Editor-in-Chief for *Transactions of SCS*. Previously, he was a full-time scientist and architect at National Renewable Energy Laboratory, Department of Energy at Golden, Colorado, where he contributed to complex energy systems. He also worked at L3 Link Simulation & Training as a contractor to U.S. Air Force Research Lab at Wright-Patterson Air Force Base, Ohio, where he integrated artificial agents and various cognitive architectures in Live, Virtual, and Constructive (LVC) environments using formal systems theoretical principles. He was a Research Assistant Professor at the Department of Electrical and Computer Engineering at the University of Arizona. Mittal served as General Chair of SummerSim'15, Vice General Chair for SpringSim'16 and SummerSim'14, and Program Chair for Spring-Sim'15. He has coauthored nearly 60 articles in various international conferences and journals, including one book.

Navonil Mustafee is a Senior Lecturer (Management Studies) and Program Director for BA B&M at University of Exeter. He has research interests in Hybrid Systems Modelling (application of simulation with the wider Operational Research techniques like problem structuring methods, game theory, and forecasting), Multi-Methodology/Hybrid Simulation, and Serious Games. He uses techniques like Discrete-Event, Agent-Based and Spreadsheet/Monte Carlo simulation, and has applied them in the context of healthcare operations management, supply chains, and engineering asset maintenance. His healthcare research is with stakeholders from the UK National Health Service (NHS); he is an honorary researcher at Torbay & South Devon NHS Foundation Trust. Mustafee has a B.A. degree from Calcutta University, an M.Sc. with distinction in Distributed Systems, and a Ph.D. in Information Systems and Computing, both from Brunel University. His M.Sc./Ph.D. research focused on the application of distributed systems, including Desktop Grid Computing, to execute large and complex simulations in healthcare, banking, and the auto industry. Some of this work involved interoperability and the reuse of simulation models through Parallel and Distributed Simulation (PADS) principles, standards (IEEE 1516 High-Level Architecture), and software (Run-Time Infrastructure).

Previous experience includes working as a research fellow in Warwick Business School, Brunel (in an EU-funded project on e-Science and e-Infrastructures – BELIEF-II) and a Lectureship in Swansea University.

Tuncer Ören is an emeritus professor of computer science at the University of Ottawa, Canada. His Ph.D. is in systems engineering from the University of Arizona. He has been active in simulation since 1965. His research interests include (i) advancing modeling and simulation methodologies; (ii) agent-directed simulation; (iii) agents for cognitive and emotive simulations; (iv) reliability, quality assurance, failure avoidance, and ethics; as well as (v) body of knowledge and (vi) multi-lingual terminology of modeling and simulation. Ören has over 500 publications, including 40 books and proceedings. He has contributed to over 500 conferences and seminars (about half as invited/honorary) in about 40 countries. He was inducted to SCS Modeling and Simulation Hall of Fame –Lifetime Achievement Award. He is also a distinguished lecturer, an SCS fellow, and an AVP for ethics. He is recognized, by IBM Canada, as a pioneer of computing in Canada. A book about him was edited in 2015 by Prof. L. Yilmaz: *Concepts and Methodologies for Modeling and Simulation: A Tribute to Tuncer Ören*, Springer. Other distinctions include invitations from United Nations and plaques and certificates of appreciation from organizations including ACM, Atomic Energy of Canada, and NATO.

Ivar Oswalt is the President of Simulation U Analytics, LLC, which focuses on the analysis of M&S and on using M&S in support of mission applications. These projects employ his academic training in multiattribute utility theory and decision analysis, operations research experience assessing M&S applications, and significant expertise in measures and metrics development. He defines M&S requirements, assesses their value, proposes design and development concepts, and evaluates their application. He developed, under the sponsorship of the Office of Naval Research, a unique approach to using defined parameters to assess the effectiveness of M&S support to gaming. He also contributed to a Marine Corps Systems Command effort to develop a Model-Based System Engineering Education and Training Road Map. He has defined measures of merit that reflect M&S effectiveness, and has coauthored "Using Data Types and Scales for Analysis and Decision Making" for the *Acquisition Review Quarterly*. Currently, he supports the Navy's M&S Office and the Naval Research Lab in policy, planning, analysis, and M&S verification, validation, and accreditation. He received a Ph.D. in Political Science from Claremont Graduate School in 1989.

Mikel D. Petty is currently a Senior Scientist for Modeling and Simulation at the University of Alabama in Huntsville's Information Technology and Systems Center and an Associate Professor of Computer Science. Prior to joining UAH, he was Chief Scientist at Old Dominion University's Virginia Modeling, Analysis, and Simulation Center and Assistant Director at the University of Central Florida's Institute for Simulation and Training. He received a Ph.D. in Computer Science from the University of Central Florida in 1997. Petty has worked in modeling and simulation research and education since 1990 in areas that include simulation interoperability and composability, verification and validation methods, and human behavior modeling. He has published over 200 research papers and has been awarded over $16 million in research funding. He served on National Research Council and National Science Foundation committees on modeling and simulation, is a Certified Modeling and Simulation Professional, and is Editor-in-Chief of the journal *SIMULATION: Transactions of the Society for Modeling and Simulation International.* He has served as dissertation advisor to six graduated Ph.D. students in three different academic disciplines (Modeling and Simulation, Computer Science, and Industrial and Systems Engineering); his students include the first two students to receive Ph.D. in Modeling and Simulation at Old Dominion University and the first two students to receive a Ph.D. in Modeling and Simulation at UAH.

Gregory S. Reed is a Senior Member of the Technical Staff at Torch Technologies, Inc. He also periodically serves as a researcher, professor, and/or graduate committee member at the University of Alabama in Huntsville (UAH). His primary research areas include modeling and simulation (M&S), decision support, visualization, systems engineering, and human factors and ergonomics. He has also contributed to or led user interfaces and decision support tools for a wide variety of users, including intelligence analysts at DARPA, NASA risk assessment engineers, U.S. Army decision-makers, and hospital nurses. He also manages the CAVE visualization system housed at UAH, which allows researchers and designers to view and interact with data, environments, and designs in full 3D. He additionally led the development of the Certified Modeling and Simulation Professional (CMSP) website's underlying infrastructure, user interface, and deployment platform. He earned a Ph.D. in Modeling and Simulation from UAH in 2013.

James A. Robb is President of the National Training & Simulation Association. RADM USN (retd.), Robb is a native of Corpus Christi, Texas. Raised in northern Virginia, he graduated from Rensselaer Polytechnic Institute in 1972, earned a Master of Science degree from the University of West Florida in 1973, and upon commissioning in the U.S. Navy, was designated a Naval Aviator in 1974. A veteran Navy combat pilot, RADM Robb served, among numerous other assignments, as officer in command of TOPGUN, the Navy's Fighter Weapons School. Navy Staff Flag officer assignments included service as the Director of Aviation Plans and Requirements (N880), and as Director of Fleet Readiness (N43). His final active duty assignment was as Director, Fleet Readiness Division (N43), Office of the Chief of Naval Operations, Washington, DC. His civilian career has included working as an independent consultant, specializing in strategic planning, joint operations, defense acquisition reform, and global political/military affairs. Robb assumed the role of President of NTSA in 2012.

Richard J. Severinghaus is President and Director, CRTN Solutions, LLC. Education includes a B.S. degree in Economics, U.S. Naval Academy, a M.S. in Systems Management, University of Southern California, and graduation from the Navy's advanced nuclear engineering and nuclear propulsion technology training programs. He has been into naval service for 23 years, with submarine command and post-command experience. He is member of the Simulation Interoperability Standards Organization (an international 501(c)3), past Chair, SISO Executive Committee (3 years), and past Chair, SISO Conference Committee (3 years). He is a member of the Society for Modeling & Simulation International, and of the Society for Simulation in Healthcare (SSH). Working with NASA, he has serviced as Program Coordinator for the Simulation Exploration Experience, a unique program promoting STEM education and practical experience in distributed simulation leading to postgraduation employment. His research over the past 18 years has included technology and human performance study, use of M&S for training and assessment, and the application of human factors to design and training implementation, consulting in the areas of human performance improvement, training, education, and M&S. He has authored/contributed to over 20 articles, papers, and technical reports addressing training, human performance, and return on investment.

John A. Sokolowski is the Executive Director of Old Dominion University's (ODU), Virginia Modeling, Analysis & Simulation Center (VMASC), and Associate Professor of Modeling and Simulation Engineering at ODU. He holds a Bachelor of Science degree in Computer Science from Purdue University, a Master of Engineering Management from Old Dominion University (ODU), and a Ph.D. in Engineering with a concentration in the Modeling and Simulation from ODU. His research interests include human behavior modeling, multiagent system simulation, and simulation techniques for representing social systems. He edited multiple books that were produced particularly for M&S education on the graduate level.

Steve Swenson is Vice President at AEgis Technologies, leading a high-performance team of scientists and engineers for M&S and Human Systems Integration in various domains. He is responsible for the operations of the AEgis offices in Rhode Island, Virginia, Florida and Ohio. Swenson has 20 years of experience as a civil servant for the U.S. Navy occupying several critical roles in DoD's Modeling and Simulation core. He was the Deputy Director, Strategic Initiatives for the Defense Modeling and Simulation Office, as well as Chief Engineer for the Undersea Engineering Project Office where he helped establish the Congressionally mandated Submarine Design Improvement Process. He worked in the office of the Chief of Naval Operations and was one of the principal developers of the Navy's premiere hardware-in-the-loop weapon simulator. He has served as President and Conference Committee Chair of the Simulation Interoperability Standards Organization (SISO) and the President and Founder of the New England Modeling and Simulation Consortium. He is a member of the Institute for Electrical and Electronics Engineers (IEEE), the IEEE Standards Agency, and the Simulation Interoperability Standards Organization (SISO). He is a recipient of the Meritorious Civilian Service Award.

Simon J.E. Taylor is a Reader in the Department of Computer Science, Brunel University London, and leads the Modelling & Simulation Group. He was the Editor-in-Chief of the *Journal of Simulation* for 10 years and continues to serve as the special issues editor. He is a former ACM SIGSIM chair, and a member of the SIGSIM and the UK ORS Simulation Special Interest Group Steering Committee. His recent work has focused on how advanced ICT such as high-performance computing and cloud computing can benefit industry and open science (particularly in Africa). He has published over 150 publications on a wide variety of modeling and simulation topics and his research has made over £1 M of cost savings in industry.

Andreas Tolk is Technology Integrator in the Modeling, Simulation, Experimentation, and Analytics (MSEA) Technical Center of the MITRE Corporation. He is also Adjunct Full Professor of Engineering Management and Systems Engineering and Modeling, Simulation, and Visualization Engineering at Old Dominion University in Norfolk, Virginia. He holds an M.S. and a Ph.D. in Computer Science from the University of the Federal Armed Forces in Munich, Germany. He published more than 200 contributions to journals, book chapters, and conference proceedings and edited several books on Modeling & Simulation and Systems Engineering. He received the Excellence in Research Award from the Frank Batten College of Engineering and Technology in 2008, the Technical Merit Award from the Simulation Interoperability Standards Organization (SISO) in 2010, and the Outstanding Professional Contributions Award from the Society for Modeling and Simulation (SCS) in 2012, and the Distinguished Achievement Award from SCS in 2014. He is on the Board of Directors of the Society for Modeling and Simulation (SCS) as well as of the Association for Computing Machinery (ACM) Special Interest Group Simulation (SIGSIM). He is a senior member of ACM, IEEE, and SCS, and a fellow of SCS.

William V. Tucker, now retired, applied systems and software engineering, engineering management, and program management to modeling and simulation technology and applications across aerospace during his 34-year tenure with Boeing. His publications relate to modeling and simulation architecture, standards, education, and professional development. He has long been active in modeling and simulation industry groups, including the Simulation Interoperability Standards Organization, the Society for Modeling and Simulation International, the Modeling and Simulation Professional Certification Council, the National Training and Simulation Association, the Alabama Modeling and Simulation Council, and the National Defense Industrial Association. He completed a B.S. in Electrical Engineering from Wichita State University and served as an officer in the U.S. Army.

Charles D. Turnitsa is a senior research scientist for GTRI, working in the Information and Communications Laboratory. He has previously been involved in teaching and researching a variety of different topics related to information systems, system's engineering, computational science, and modeling and simulation. He has worked on research efforts related to scientific modeling, data visualization, systems interoperability, and knowledge-based modeling. He has a Bachelor's of Science degree in Computer Science, a Master's Degree in Electrical and Computer Engineering, and a Ph.D. in Modeling and Simulation from Old Dominion University. He is currently an instructor in the Professional Master's in Applied Systems Engineering for GTRI, teaches modeling and simulation (occasionally) as an adjunct assistant professor, and has dozens of published research works on a variety of different topics.

Yingnian Wu is an associate professor at Beijing Information Science & Technology University. He got the doctor's degree from Beihang University. He is the Secretary-General of the Committee for Intelligent IOT Systems

Modeling and Simulation of China Simulation Federation (CSF) and an expert of Beijing Association of Internet of Things. His research interests include system simulation, modeling theory and technology research, high-performance computing in complex environment simulation applications, Internet of things and network control, robot and human–computer interaction, and brain–computer interaction. He teaches the courses "microcomputer theory and interface technology," "intelligent thing and perception technology" for undergraduates and "cloud control technology" for graduates. He has presided over and participated in several national- and ministerial-level projects, such as the Beijing Municipal Education Commission of Talented People Project, the National Natural Science Foundation Project, and others. He published more than 10 papers in cloud computing and high-performance simulation, and IoT control. His papers have been cited more than 170 times.

Gengjiao Yang received the B.Sc. degree in applied mathematics from Yuncheng University, Yuncheng, China, and the M.Sc. degree in applied mathematics from the Liaoning University of Technology, Jinzhou, China, in 2007 and 2010, respectively. Since 2015, she has become a Ph.D. student of School of Automation Science and Electrical Engineering at Beihang University, Beijing, China. Currently, her research focuses on simulation model credibility evaluation, verification, validation, and accreditation (VV&A), and model engineering for simulation.

Levent Yilmaz is Professor of Computer Science and Software Engineering at Auburn University with a joint appointment in Industrial and Systems Engineering. He holds M.S. and Ph.D. degrees from Virginia Tech. His research interests are in agent-directed simulation, cognitive computing, and model-driven science and engineering. He is the former Editor-in-Chief of *Simulation: Transactions of the SCS* and founding organizer and general chair of the Agent-Directed Simulation Conference series.

Bernard P. Zeigler is Chief Scientist at RTSync Corp., Professor Emeritus of Electrical and Computer Engineering at the University of Arizona (UA), and Affiliated Research Professor in the C4I Center at George Mason University. He is internationally known for his seminal contributions in modeling and simulation theory. He has published several books, including *Theory of Modeling and Simulation* and *Modeling & Simulation-Based Data Engineering: Introducing Pragmatics into Ontologies for Net-Centric Information Exchange*. He is the originator of the Discrete Event System Specification (DEVS) formalism in 1976, which has spurred the development of a worldwide research community, with its own conference meetings, and research/technology awards. Due to its combination of modeling rigor and simulation power, the DEVS formalism is now being widely used in advanced information systems that incorporate simulation modeling components in an integral manner. In 1995, Zeigler was named Fellow of the IEEE in recognition of his contributions to the theory of discrete event simulation.

Lin Zhang is a full professor of Beihang University. He received the B.S. degree in 1986 from the Department of Computer and System Science at Nankai University, China. He received the M.S. degree and the Ph.D. degree in 1989 and 1992, respectively, from the Department of Automation at Tsinghua University, China. From 2002 to 2005, he worked at the U.S. Naval Postgraduate School as a senior research associate of the U.S. National Research Council. His research interests include service-oriented modeling and simulation, agent-based control and simulation, cloud manufacturing, and model engineering for simulation. Currently, he serves as the Past President of the Society for Modeling & Simulation International (SCS), a Fellow of the Federation of Asian Simulation Societies (ASIASIM), the executive vice president of China Simulation Federation (CSF), an IEEE senior member, a chief scientist of 863 key projects, and associate Editor-in-Chief and associate editors of six peer-reviewed journals. He authored and coauthored more than 200 papers, 5 books, and chapters. He received the National Award for Excellent Science and Technology Books in 1999, the Outstanding Individual Award of China High-Tech R&D Program in 2001, the 2nd prize of Ministry of Education Science and Technology Progress Award in 2013, and the National Excellent Scientific and Technological Workers Awards in 2014.

Part I

Foundation

1

An Introduction to the Facets of the Profession of Modeling and Simulation

Andreas Tolk

The MITRE Corporation, Hampton, VA, USA

1.1 Profession, Professionals, and Professionalism

Modeling and simulation (M&S) seems to be applied everywhere. Within the M&S community, we are currently celebrating many 50 years' anniversaries; so M&S seems to be often applied and has a large group of supporters, researchers, and practitioners. Within the academic realm, the annually conducted *Winter Simulation Conference*, organized by a group of simulation organization and societies, started as the "Conference on the Applications of Simulation Using GPSS" in November 1967 in New York, NY, and evolved into one of the biggest simulation conferences worldwide. It celebrates its 50th anniversary in December 2017. The *Annual Simulation Symposium* is the oldest international symposium continuously operating conference dedicated to simulation under the lead of the Society of M&S, it celebrated its 50th anniversary in April 2017. Practitioners in the field have also celebrated their 50th anniversary of the *Interservice/Industry Training, Simulation and Education Conference* (I/ITSEC) in November 2016. These conferences are all well attended and address a variety of application domains, covering all simulation paradigms and methods. In addition, many specialty conferences feature simulation track, in particular in the medical and healthcare community, but also in not so well-known fields, such as religion. Undoubtedly, simulation is ubiquitous, and many people apply simulation. *But is M&S a profession, and if not, should it be?*

To answer this question, some definitions are needed first. The Professional Standards Councils of Australia (http://www.psc.gov.au (accessed December 2016)) compiled a set of key definitions based on international literature research. Furthermore, they provided some different views on

The Profession of Modeling and Simulation: Discipline, Ethics, Education, Vocation, Societies, and Economics, First Edition. Andreas Tolk and Tuncer Ören.
© 2017 John Wiley & Sons, Inc. Published 2017 by John Wiley & Sons, Inc.

professions and professionals from various perspectives of academia, regulators, and more. Their recommended definitions are as follows:

- A *profession* is a disciplined group of individuals who adhere to ethical standards. This group positions itself as possessing special knowledge and skills in a widely recognized body of learning derived from research, education, and training at a high level, and is recognized by the public as such. A profession is also prepared to apply this knowledge and exercise these skills in the interest of others.
- A *professional* is a member of a profession. Professionals are governed by codes of ethics, and profess commitment to competence, integrity and morality, altruism, and the promotion of the public good within their expert domain. Professionals are accountable to those served and to society.
- *Professionalism* comprises the personally held beliefs about one's own conduct as a professional. It is often linked to the upholding of the principles, laws, ethics, and conventions of a profession as a way of practice.
- *Professionalization* is the pattern of how a profession develops, as well as the process of how becoming a profession.

A practical definition often referenced is that professionals use their *special skills to make money*. These skills may require certain characteristics by an individual or a group of individuals, such as professional athletes, but in any case they involve prolonged training and a formal qualification, such as for teachers or engineers.

In his groundbreaking work on the attributes of profession, Greenwood (1957) identified the following topics to be pivotal for any profession, independent of its specialty:

- Each profession needs systematic body of theory or *body of knowledge* that codifies the skills and expertise needed.
- A profession provides *authority and credibility* in the field. This includes examinations and certifications documenting the expertise of professionals.
- A profession is guided by ethical behavior. A professional *code of ethics* usually builds a moral contract and the foundation for a professional culture, or a culture of values, norms, and symbols. It also provides the basis for sanctions or regulation and control of its members, ensuring that ethics, authority, and credibility are sustained.

Ethical behavior of professionals is a common factor in the various definitions: professionals behave professionally, and this excludes lying, cheating, or deceiving. As important as general ethics, education in the special domain to master the skill sets plays an outstanding role for professions. A profession shall clearly communicate what the entry-level qualifications are, what the supported curriculums of academia need to cover, and how professional continuous education needs to look like.

Related disciplines are often described regarding their *theoretical contributions* (science), their *derived methods* that make the research results applicable to solve daily life problems (engineering), and often a practical branch that mainly focuses on the *application* of such methods (practice). Examinations to formally test the knowledge of professionals do belong to this category as well, and are often referenced as unique certificates testifying the professionalism of those who pass it.

Professional societies are another important pillar for each profession. Societies publish journals, raise public awareness, and bestow awards on outstanding members. Through their work, they help to define, publish, and archive the continuously growing body of knowledge. They set the standards for their professional fields, and they promote high standards of quality and ethics. Societies also provide access to the experts to support work with subject matter expertise, or to conduct peer reviews for research and applications.

Finally, the economics of the profession are an important factor. Professions allow for improved access of the community to their special services and skills, provided by the professional organizations in the best form possible. But does it matter? Are there enough people in the community requiring a professional level of such services? What is the return of investment? Only if there are solid economic foundations, a profession will survive over time.

Sarjoughian and Zeigler (2001) summarized and proposed similar concepts for M&S as a discipline while contributing to the development of simulation curriculums at academic institutions.

Figure 1.1 visualizes the main concepts and their relations that make up a profession and may guide research as well as contributions to future discussions. This figure is neither attempting to be complete nor exclusive, but provides a framework that can be used to be populated with additional concepts and structures over time.

1.2 Contributions of the Chapters

Based on these definitions, experiences, and observations, we can ask again: *Is there a profession of M&S?* We asked the chapter authors, all of them being experts in M&S, to help address the various facets identified in this introduction. We could use many activities of the late Bill Waite, who

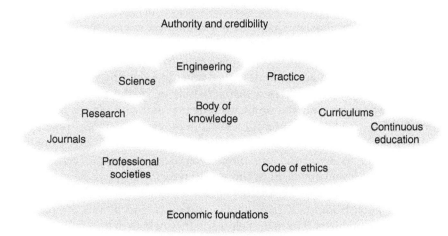

Figure 1.1 Main concepts and relations describing a profession.

can truly be recognized as one of the first true professionals of M&S. He recognized the need to create a body of knowledge and was active in this field. He was among the first M&S experts recognizing the need for strong economic foundations for M&S, so he organized workshops and tracks to address this topic. He supported partners in academia, industry, and government. And he was active in various leadership positions in M&S societies. All chapter authors knew and worked with him over the recent years, and all value his tireless contributions.

The individual contributing chapters follow the structure provided in this introduction. Chapter 2, "An Index to the Body of Knowledge of Simulation Systems Engineering" by Umut Durak, Tuncer Ören, and Andreas Tolk, provides an approach to understand what skills and knowledge make M&S unique as a discipline, and what defines an M&S professional.

Chapter 3 provides information about the "Code of Ethics" for M&S professionals. Andreas Tolk provides not only the code of professional ethics for simulationist, which has been adapted by many simulation organizations, but also puts the code into a broader context and gives some examples on how to apply the ideas in the life of a professional.

The academic foundations are described and explained in Chapter 4, "M&S as a Profession and an Academic Discipline: A Contemporary View" by John Sokolowski and Roland Mielke. Over the last decades, curriculums have been developed on the undergraduate, graduate, and postgraduate levels, allowing receiving all academic levels of recognition.

A slightly different approach is captured in Chapter 5, "Academic Education Supporting the Professional Landscape" by Margaret Loper. Her focus is on the support of the workforce in the M&S field, not only by programs leading to an academic degree but also including continuous education after graduation.

Like many other professional domains, the M&S community has an examination and certification program, "The Certified M&S Professional Certification and Examination," described in Chapter 6 by Mikel D. Petty, Gregory S. Reed, and William V. Tucker. They not only describe the current version of this professional certification effort but also give insights into the historical developments, as it actually underwent several changes from its original inception to the current day, being a testament to the dynamics of changes within the M&S community.

In Chapter 7, Bob Armstrong and Simon Taylor provide an overview of "M&S Societies Shaping the Profession." Such an overview can neither be complete nor exclusive, but by bringing experts from different continents with international experience on board, this overview may become a good starting point for further research. Although the number of professional societies is still fluctuating, the principles behind these societies are stabilizing, and as several of the featured professional societies are already several decades old, it can hence be assumed to be pretty stable and established.

Among the largest customers for M&S services, in particular for their training needs, but also for analysis of alternatives and procurement decisions, is the defense and security domain. Not only does the efficient use of M&S services require experts producing these services, but also the application side require experts. Chapter 8, "The Uniformed Military M&S Professional" by Rudolph P. Darken and Curtis L. Blais, describes how the United States Armed Forces ensure that soldiers have the necessary education to use M&S in support of their various tasks.

One of the fastest growing markets for M&S services is the Asian-Pacific realm. This growth is not driven externally, but local universities and societies are driving these efforts. Chapter 9, "M&S as a Profession and Discipline in China" by Lin Zhang, Yingnian Wu, and Gengjiao Yang, describes these developments, and how these activities are related to other efforts described in this book, and provides insight into their philosophy of simulation, application domains, education, and industry.

Chapter 10, "M&S for the Enterprise: Integrating application domains for the M&S Professional" by Steve Swenson, Mike Lightner, and Bob Gravitz, provides insights that are not often found in textbooks, namely, how M&S is used within industry. Their chapter allows gaining some ideas not only about what the needs and expectations are, but also how industry as an active partner is driving the necessary developments. It also provides some industry insights not necessarily found in academic papers.

The academic contribution is described in Chapter 11, "A Complexity and Creative Innovation Dynamics Perspective to Sustaining the Growth and Vitality of the M&S Profession" by Levent Yilmaz. Unlike many other fields, M&S is defined by its support for other disciplines. It supports many engineering disciplines and allows computational science. Yilmaz therefore evaluates the critical building blocks of M&S research to gain better insight into how to keep it successful and creative.

Chapter 12, "Theory and Unified Process for the Practice of M&S in Cyber Environments" by Saurabh Mittal and Bernard P. Zeigler, provides an example of how M&S methods and skills can be used to address one of the main current challenges, better understanding and managing cyber environments. The described research steps and processes are applicable to other research topics as well and not limited to cyber challenges. However, as cyber environments will remain a focus area of research, some particular insights are provided for this special application domain as well.

Chapter 13, "Funding an Academic Simulation Project: The Economics of M&S" by Saikou Diallo, Christopher J. Lynch, and Navonil Mustafee, addresses the need to better understand the monetary issues behind research The authors describe the results of content analysis on three sets of abstracts from funded projects that use M&S as one of the methods of investigation and compare and contrast the funding across geographical and disciplinary dimensions, providing empirical evidence to allow a better understanding of currently supported research needs.

In Chapter 14, Steven Gordon, Tim Cooley, and Ivar Oswalt ask the provocative question: "Why Spend One More Dollar for M&S? Observations on the Return of Investment." They address the need to measure the positive effect of M&S applications, as they are often perceived to be expensive and time consuming without an immediately noticeable difference. The chapter documents several approaches to show the positive return of investment in M&S services.

In Chapter 15, "Does M&S Help? Operationalizing Cost Avoidance and Proficiency Evaluations," the same group of authors looks at alternative means to evaluate the contributions of M&S that are not necessarily connected with the traditional return of investment.

Finally, in Chapter 16, Randy Garrett, RADM James Robb, Richard Severinghaus, and Richard Fujimoto document the approach of "Building a National M&S Coalition" within the United States. This coalition brings key partners from government, industry, and academia together to discuss the various research needs and application opportunities, but also funding and regulation constraints.

After studying these chapters, we hope that scholars, students, and practitioners of M&S will agree that there truly exists a profession of M&S, defined by a unique body of knowledge, well-defined curriculums and professional education programs, distinct contributions to science and engineering, supporting professional societies, and a solid economic foundation.

References

Greenwood, E. (1957) Attributes of a profession. *Social Work*, 2, 44–55.
Sarjoughian, H.S. and Zeigler, B.P. (2001) Towards making modeling & simulation into a discipline. *Simulation Series*, 33 (2), 130–135.

2

An Index to the Body of Knowledge of Simulation Systems Engineering

Umut Durak,[1] Tuncer Ören,[2] and Andreas Tolk[3]

[1]*German Aerospace Center (DLR), Braunschweig, Germany*
[2]*University of Ottawa, Ottawa, Ontario, Canada*
[3]*The MITRE Corporation, Hampton, VA, USA*

2.1 Introduction

The rise of the simulation discipline has been phenomenal. It is progressing not only as a discipline but also as a profession. It is extensively practiced in many conventional and unconventional areas to perform experimentations, to gain experience or for entertainment. As the need for a professional education for simulation is growing exponentially, the urgency for a Body of Knowledge (BoK) is becoming crystal clear. BoK is the knowledge that the members of a discipline use in practicing their profession (Ören, 2005). It represents the common understanding of relevant professionals and professional organizations to explain a professional domain by a comprehensive and concise description of concepts, terms, and activities needed (Tolk, 2010).

Simulation systems engineering (SSE) is regarded as a discipline that defines the contemporary practice of modeling and simulation (M&S) studies that emerged through the evolution of modeling and simulation. The Simulation Systems Engineering BoK is in this sense a part of the M&S BoK.

The early studies about the M&S BoK date back to initiatives of the IEEE Computer Society Technical Committee on Simulation (Ören, 2005). In the late 1990s and early 2000s, a couple of workshops were conducted toward the M&S Body of Knowledge (Rogers, 1997; Szczerbicka et al., 2000; Waite, 2001; Birta, 2003a). In 2003, Birta (2003b) published his M&S BoK proposal. Later Ören (2006) introduced his recommendations for developing an index for the M&S BoK. The M&S Body of Knowledge Index has then been

The Profession of Modeling and Simulation: Discipline, Ethics, Education, Vocation, Societies, and Economics, First Edition. Andreas Tolk and Tuncer Ören.
© 2017 John Wiley & Sons, Inc. Published 2017 by John Wiley & Sons, Inc.

developed following these recommendations (Ören, 2016). Ören with his colleagues published various articles that introduce the M&S Body of Knowledge Index; some of the important ones are Ören and Waite (2010), Lacy et al. (2010), and Ören (2011, 2014).

Recent scientometric studies conducted by Mustafee et al. (2014) concluded that the modeling and simulation discipline is self-organizing into meaningful clusters. Aligned with these studies, in 2015, Diallo et al. (2015a) applied content analysis on research articles to identify the key modeling and simulation concepts and their relations using computational means in order to come up with prominent topics and themes for the M&S BoK. On the other hand, Diallo et al. (2015b) propose that the diversity of the application areas and the excessive number of independently developed modeling and simulation theories, frameworks, and tools constitute a challenge toward realizing a universally recognized M&S BoK.

This chapter particularly focuses on the SSE BoK. It will first present the foundations and applications of SSE. In the core of the chapter, an initial proposal for the knowledge areas of SSE will be revealed. Later, an ontology development effort for SSE will be visited, before concluding the chapter.

2.2 Foundations of Simulation Systems Engineering

Simulation as a term exists in English since the fourteenth century. Its meaning has been imitation, pretence, or fake. Nowadays we define simulation as the use of dynamic models to perform experiments, to gain experience, or for entertainment. Here, a model is an abstract representation of a certain aspect of reality that we are concerned with for a particular purpose. The part of reality that is subjected to modeling is known as simuland.

Ören (1984a) states that "simulation is like a gem; it is multifaceted." Figure 2.1 presents a graphical interpretation of Ören's statement with faces that correspond to various aspects of simulation. Accordingly, he introduces an extensive review of definitions and types of simulation in Ören (2011b). While the gem analogy is very useful for identifying facets of simulation, the body of knowledge requires a top-down decomposition. It is the act of building the simulation tree. Regarding simulation as a multistem tree, the facets of the simulation gem can be described as the stems.

As simulation is conceived from an abstract knowledge processing viewpoint (Ören, 2009), it can be categorized as presented in Figure 2.2. In one facet, simulation is a knowledge generation activity, whereas in the other one it is knowledge processing. It can also be classified as a model-based

Figure 2.1 The simulation gem.

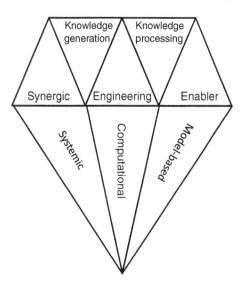

activity, or a computational activity as well as a systemic activity. In this chapter, we are more interested in simulation as an engineering activity.

The evolution of simulation can be presented through a sequence of aspects of simulation (Figure 2.3). At the beginning (aspect 1), simulation was noncomputational; thought experiments (Brown, 2017), scale models, and sandbox models were used.

Computerized simulation is the second aspect that was reached by the use of computers. While analog and later hybrid computers were used at the beginning, the advent of digital computers, programming, and later software engineering contributed to simulation becoming a versatile tool. Nowadays high-performance computing offers simulationists tremendous

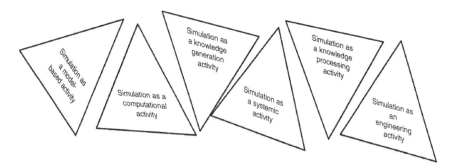

Figure 2.2 Simulation as a multifaceted activity.

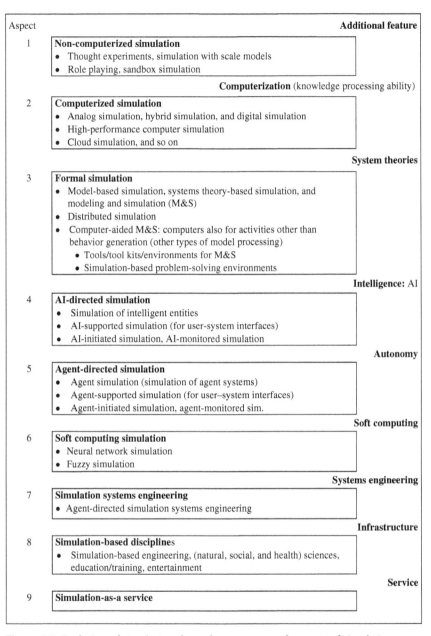

Figure 2.3 Evolution of simulation through a sequence of aspects of simulation. (Adapted from Ören (2009).)

opportunities. The emerging ubiquitousness of computers as well as cloud computing may open new vistas for simulationists too.

In the 1960s, the third aspect, theory-based simulation, started a new era of formal simulation with important implications such as model-based simulation that promotes utilizing formal models (which also started model-based approaches in other disciplines such as software engineering). The contributions of systems theory made simulation even more powerful.

Intelligent (or cybernetic) simulation can be introduced as the fourth aspect that the synergies of simulation with artificial (or machine) intelligence and later with software agents made it possible (Yilmaz and Ören, 2011).

The contribution of soft computing to simulation is the fifth aspect, namely, soft computing simulation. It involves neural network simulation, fuzzy simulation (for nonnumerical computation in simulation), and swarm simulation.

The increasing maturity and mutual contributions of both simulation and systems engineering (SE) offer the opportunity of their synergistic relations. The simulation systems engineering that results from the contribution of systems engineering to simulation is named as the sixth aspect.

Systems engineering is presented as a methodological and disciplined approach for the design, realization, technical management, operation, and retirement of a system, where system is defined as a construct or collection of different elements that together produce results not obtainable by the elements alone (NASA, 2007). The system architecture describes the structure of the system components, relationships, and rules governing their design and evolution over time (Tolk and Hughes, 2014). Simulation is an execution of the system architecture model. The models of the system architecture are placed at the core of the engineering process with the Model-Based Systems Engineering (MBSE) approach (Estefan, 2008). Simulation-based systems engineering further promotes executable models. Simulation is proposed as the native mechanisms to address measures of performance and measures of effectiveness throughout the conceptual design, development, and later life cycle phases.

As the system complexity increases, the executable system architecture (i.e., the simulation model) also becomes complex. It becomes necessary to assess it not only as a valid support of SE processes but also as an objective of SE efforts. Executing an interdisciplinary systems engineering process for developing, maintaining, and employing simulations, which enable systems engineers to experiment and gain insights about the systems of interest, is referred to as simulation systems engineering (Durak and Ören, 2016).

2.3 Applications of SSE

Simulation is the execution of dynamic models. Its application areas extend beyond the limits of engineering and physical and life sciences to social and political sciences. A recent publication of Ören (2011b) documents the richness of simulation and catalogs almost 400 types of simulation. Its application areas can be categorized into experimentation, gaining experience and entertainment. While the experimentation is extensively used in decision support, understanding, and education, simulation is also applied for gaining experience, that is, for training. Finally, execution of dynamic models of reality is utilized for entertainment, such as in video games. Figure 2.4 summarizes this taxonomy of application areas.

Simulation systems engineering can be introduced as an engineering approach for developing, maintaining, and employing simulations and is applicable where a simulation product is to be developed, maintained, or employed for any of the purposes mentioned above. It is defined by the simulation systems engineering process and the activities associated with it.

Figure 2.5 depicts the proposed set of activities that are associated with the application of SSE. These activities are simulation conceptual analysis,

Experimentation	Decision support
	Understanding
Experience	Live
	Virtual
	Constructive
Entertainment	Computer games
	Augmented reality
	Physics-based animation

Figure 2.4 Utilization of simulation.

Application of SSE	Simulation conceptual analysis
	Simulation design
	Simulation development
	Simulation integration
	Simulation V&V
	Simulation deployment and operation
	Simulation maintenance
	Simulation quality
	Simulation management

Figure 2.5 Application of SSE.

design, development, integration, verification and validation, deployment and operation, maintenance, quality, and management. These keywords will again be visited in the following sections as parts of the SSE knowledge areas.

2.4 SSE Knowledge Areas

2.4.1 An Index of SSE Knowledge Areas

The material in BoK studies is traditionally organized in knowledge areas (KAs). Each KA can be organized as subareas and eventually to topics. Generally two to three levels of breakdown provide an effective clustering. The aim is to have a list of topics that are grouped according to the shared understanding of academia, research, and industry. This chapter proposes a three-layered structure for KAs of SSE, as depicted in Figure 2.6.

The SSE BoK is categorized into three KAs at the first level, namely, *Foundations, Engineering,* and *Practice. Fundamentals* and *Science* are the subareas of *Foundations. Methodology and Infrastructures* and *Process* are the subareas of *Engineering* and, finally, *Ethics* and *Credibility and Reliability* subareas belong to *Practice.* This chapter intends to propose this initial index of KAs. They are briefly introduced in the following sections, including some key references. A full-fledged SSE BoK needs to extend this index with an adequate explanation of topics and with an extensive list of reference material. Figure 2.7 depicts the resulting areas and subareas.

Area

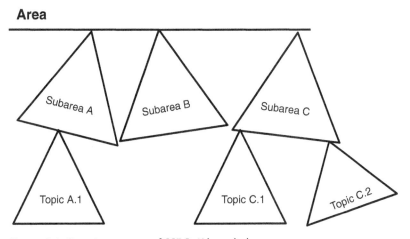

Figure 2.6 Generic structure of SSE BoK knowledge areas.

Foundations

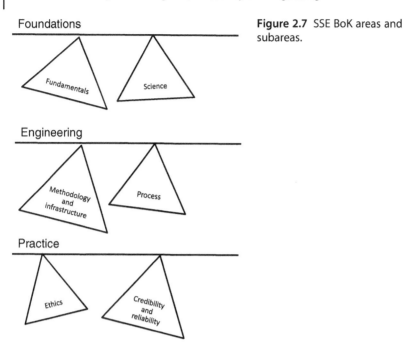

Figure 2.7 SSE BoK areas and subareas.

Engineering

Practice

2.4.2 Foundations

The *Foundations* area corresponds to the foundational knowledge regarding the simulation systems engineering and is shown in Figure 2.8. The subareas are *Fundamentals* of simulation and the simulation *Science*. With *Fundamentals*, the shared definition and categorization of the discipline, namely, the *Definitions of Simulation* and the *Types of Simulation* are addressed. The

Foundations

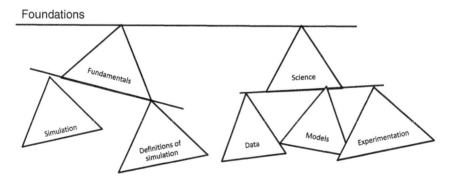

Figure 2.8 Foundations knowledge area.

works of Ören (2011a,b, 2014, 2016), Ören and Waite (2010), and Birta (2003a,b) provide valuable references for these topics.

Simulation *Science* refers to the basic theories of the discipline and defines the academic foundations of simulation (Padilla et al., 2011). It can be categorized under the topics *Data, Models,* and *Experimentation. Data* correspond to scientific fundamentals of input data modeling and output analysis that have their roots in statistics and probability theory. Among various textbooks, the one from Law and Kelton (1991) is one of the most referred texts that extensively explains these topics. Recent directions in simulation data modeling that are introduced by Tolk (2015), namely, big data and deep learning, can also be categorized under this topic.

Simulation has long been perceived as a model-based activity (Zeigler et al., 1979; Ören, 1984b). *Models* as the core elements of simulation require special attention as a topic. Under this topic, academic foundations of model building (modeling formalisms and languages), model base management, and model processing have been studied. *Theory of Modeling and Simulation,* the book by Zeigler et al. (2000), is a major reference for modeling formalisms, particularly the Discrete Event Systems Specification (DEVS). A typical example of a simulation language is Modelica. It is introduced extensively by Fritzson (2010). Model base management facilitates the management of model specifications. We mention the System Entity Structures and Model Base (SES/MB) framework as breakthrough in this field (Zeigler et al., 1991). It enables efficiency, reusability, and interoperability. Model processing, on the other hand, refers to foundations of model analysis, model transformations, and behavior generation and representation. Model analysis can have various objectives, some of which are comparison, checking, or verification. A recent example is the MATLAB Simulink Model Analysis and Transformation Environment (MATE) that has been developed to check modeling guidelines in MATLAB Simulink models (Legros et al., 2010).

Model transformations refer to model reduction, model construction, model pruning, and model elaboration. As an example, de Lara et al. (2003) used model transformation in multiparadigm modeling for transforming models between formalisms. Durak (2016) employed it for model elaboration and refactoring.

Behavior generation and representation refers to simulators that execute the models, the execution methodologies, and the visualization to represent the results. A simulator can be defined as simulation engine that executes the simulation models and compute their behavior. A good example would be a DEVS simulator that executes a DEVS model (Zeigler, 2003). Execution strategies like real-time simulation (Popovici and Mosterman, 2012) or Monte Carlo simulation (Mooney, 1997) are also a part of this topic.

Visualization refers to online or offline representation of the simulation results. Shen's chapter in the book by Sokolowski and Banks (2010) on visualization can be referred here as an introductory guide for simulation visualization.

2.4.3 Engineering

Engineering proposes applicable solution patterns based on simulation science (Padilla et al., 2011). The subareas that we propose under *Engineering* are *Methodology and Infrastructure* and *Process.*

As shown in Figure 2.9, the *Methodology and Infrastructure* subarea can be categorized into four topics, namely, *Life Cycles, Standards, Architectures,* and *Tools.* The life cycle process of modeling and simulation studies was initially defined by Balci in 1990 (Balci, 1990). Later, a process for federation development, particularly for distributed simulations that utilize high-level architecture (HAL), was developed and published as an IEEE standard, namely, IEEE Std 1516.3-2003, IEEE Recommended Practice for High Level Architecture Federation Development and Execution Process (FEDEP) (IEEE, 2003). After being well accepted as a starting framework for tailoring an end-to-end process for the development and execution of HLA federations, the Simulation Interoperability Standards Organization (SISO) FEDEP Product Development Group (PDG) generalized FEDEP in order to come up with an engineering process for all types of distributed simulation development and execution, namely, the IEEE Std 1730-2010 IEEE Recommended Practice for Distributed Simulation Engineering and Execution Process (DSEEP) (IEEE, 2010a).

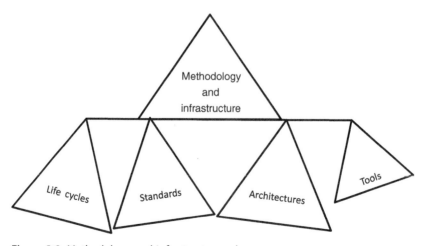

Figure 2.9 Methodology and infrastructure subarea.

There have been various standardization efforts in simulation, which target particular aspects or domains. They include terms (IEEE, 1989), language specifications (SISO-STD-011-2014, 2014), object model specifications (SISO-STD-003-2006, 2006; SISO-STD-001.1-2015, 2015), architectures (IEEE, 2010b, 2010c; IEEE, 2010d), process specifications (IEEE, 2010a), simulation data (SISO-STD-008-01-2012, 2012), distributed simulation (IEEE, 2012), and model interfaces (MODELISAR Consortium, 2016; ECSS-E-TM-40-07,2011). Table 2.1 presents a selected list of simulation standards. Tolk (2012) further provides a comprehensive review of simulation standards for distributed simulation.

Simulation architectures are important elements of SSE BoK. They provide a structured and well-formed approach for building simulation systems. Well-known and widely employed simulation architectures include HLA (IEEE, 2010b, 2010c, 2010d) and Test and Training Enabling Architecture (TENA) (Noseworthy, 2008).

Table 2.1 A set of simulation standards.

Number	Name
IEEE 1278	IEEE Standard for Distributed Interactive Simulation (IEEE, 2012)
IEEE 1516	IEEE Standard for Modeling and Simulation (M&S) High Level Architecture (HLA) (IEEE, 2010b, 2010c, 2010d)
IEEE 1730	IEEE Draft Recommended Practice for Distributed Simulation Engineering and Execution Process (DSEEP) (IEEE, 2010a)
IEEE 610	IEEE Standard Glossary of Modeling and Simulation Terminology (IEEE, 1989). Withdrawn Inactive Date: 6-3-2000
SISO-STD-011-2014	Standard for Coalition Battle Management Language (C-BML) (SISO-STD-011-2014, 2014)
SISO-STD-008-01-2012	Standard for Core Manufacturing Simulation Data (SISO-STD-008-01-2012, 2012)
SISO-STD-003-2006	Base Object Model (BOM) Template Specification (SISO-STD-003-2006, 2006)
SISO-STD-001.1-2015	Standard for Real-Time Platform Reference Federation Object Model (SISO-STD-001.1-2015, 2015)
FMI 2.0	FMI for Model Exchange and Co-Simulation (MODELISAR Consortium, 2016)
E-TM-40-07	Space Engineering: Simulation Modelling Platform (ECSS-E-TM-40-07, 2011)

As stated by Bergson (1911) in *Creative Evolution*, tool making is an essential characteristic of humans. Accordingly, simulationists have developed various tools. There can be various categorizations of simulation tools. For instance, we make a classification mainly based on execution strategy, simulation mechanics, and application domain. Parallel, distributed, or real-time simulation tools are sample categories based on execution strategy. With such thinking, Gotschlich et al. (2014) categorize their 2Simulate framework as a distributed real-time simulation tool. Discrete-event, continuous, hybrid, or agent-based simulation tools are examples of tool classification based on the simulation mechanics. In that manner, Liu et al. (2001) categorize their tool Ptolemy II as simulation environment for hybrid systems. Similarly, in their article, Railsback et al. (2006) provide a review of agent-based simulation tools. Classification of simulations based on their application domains is also possible. Here the examples can be supply chain simulation tools (Kleijnen, 2001) or multibody simulation tools (Schiehlen, 2013).

DSEEP is an invaluable source for a baseline, shared, and well-accepted definition of an SSE process. It can further be augmented for systems engineering process areas using ISO/IEC/IEEE 15288:2015, Systems and Software Engineering – System Life Cycle Processes (ISO/IEC/IEEE, 2015). D'Ambrogio and Durak (2016) propose an integrative approach for DSEEP and ISO/IEC/IEEE 15288:2015 for SSE. The proposed topics for the Process subarea are process areas of SSE, namely, *Conceptual Analysis, Design, Development, Integration, Verification and Validation, Deployment and Operation, Maintenance, Quality, and Management*. They are depicted in Figure 2.10.

Conceptual modeling and scenario development are two important activities of *Conceptual Analysis*. One of the key references for simulation conceptual model development (Pace, 2000) introduces it as a modeling activity that addresses the context of simulation. Scenario development is a

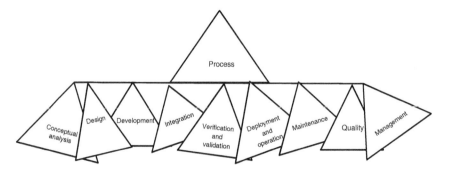

Figure 2.10 Process subarea.

comprehensive activity that begins with the stakeholders' descriptions of the scenario and finishes with the generation of the corresponding executable specifications (Durak et al., 2014). The NATO Science and Technology Organization (STO) Technical Report "Guideline on Scenario Development for (Distributed) Simulation Environments" is a recent and extensive reference on scenario development (NATO, 2014).

Referring to DSEEP, the topics *Simulation Design* and *Simulation Development* cover the design of the simulation environment, the design of simulation applications, the development of a simulation data exchange models, simulation environment agreements, and the implementation (IEEE, 2010a). The emerging practice in simulation design and development is tending more toward utilizing model-based, model-driven methodologies that are adopted from engineering of software-intensive systems (Topçu et al., 2016).

Integration refers to bringing all of the simulation applications into a unifying operating environment (IEEE, 2010a). Integration comes with one of the great challenges of simulation, namely, interoperability. It exceeds the technical implementation; a meaningful interoperation of simulation applications requires composability of their underlying conceptual models (Tolk and Muguira, 2003).

In his often-referenced paper, Sargent (2005) defines *Verification and Validation* as the efforts in a simulation life cycle that target ensuring that the simulations are correct. Validation aims at ensuring that the simulation and its associated data accurately represent the real world regarding the intended use, whereas verification ensures simulation implementation and the associated data conform to the conceptual specification of the developer (US DoD, 2009).

Deployment refers to the activities between the development and the release of the simulation for operation (Durak et al., 2016). *Operation* consists of executing the simulation, collecting results, and analyzing and evaluating them (IEEE, 2010a).

Due to constant change in technology and user requirements, the evolution of simulations during their operational use is inevitable. *Maintenance* refers to activities that ensure that simulation satisfies changing user requirements. Modernization is an evolutionary maintenance effort that comprehends and transforms legacy assets. Topçu et al. (2016) present a model-based approach for simulation modernization in the section "Simulation Evolution and Modernization" of their book.

Quality and *Management* are the topics that are proposed based on technical management processes of ISO/IEC/IEEE 15288:2015 System life cycle processes. They include activities such as project management, risk management, configuration management, information management, and quality assurance. While these technical management topics are well studied in the systems engineering literature and apparent in the Systems Engineering Body of Knowledge (BKCASE Editorial Board, 2016), the particulars of these topics in SSE have been relatively overlooked.

2.4.4 Practice

The *Practice* subarea is composed of the topics *Ethics* and *Credibility and Reliability*. Ören et al. (2002) propose that the professional ethics for simulationists lies on personal development and the profession, professional competence, trustworthiness, property rights and due credit, and compliance. Chapter 3 will provide further elaboration on code of ethics in modeling and simulation.

Credibility and Reliability are major factors toward the acceptability of a simulation study (Ören, 1981). The components of acceptability can be listed as acceptability of simulation results, data, models and parameters, experiment specifications, programs, and methodologies.

2.5 The Ontology of SSE

2.5.1 Ontologies and Simulation

Ontology is a term in philosophy that stands for systematic explanation of existence. Neches et al. (1991) introduced ontology in computer science as the basic terms and relations comprising the vocabulary of a topic area as well as the rules for combining terms and relations to define extensions to the vocabulary. Later, Gruber's definition "Ontology is explicit specification of conceptualization" (Gruber, 1995) was also well accepted. The merits of ontologies have been itemized as (a) common vocabulary, (b) explication of what has often been left implicit, (c) systematization of knowledge, (d) standardization, and (e) meta-model functionality (Mizoguchi, 2001).

In the 1990s, we started to see the first applications of ontologies in engineering, such as PhysSys (Borst et al., 1995). It was developed to formally define how design engineers or the end users of computer-aided engineering (CAE) systems understand their domain. It aimed to provide a foundation for the conceptual schema for data structuring in engineering databases, libraries, and other CAE information systems. PhysSys is regarded as a baseline for the development of a library of reusable models for engineering and design.

Miller et al. (2004) started the discussion on how ontologies can be helpful in modeling and simulation in 2004. They suggested ontologies for concept browsing, querying and navigating, simulation service discovery, simulation component repository development, hypothesis testing, platform- or machine-independent simulation specification, and shared conceptual framework development. Thereafter, many efforts have utilized ontologies for modeling and simulation.

Tolk (2013) in his recent book *Ontology, Epistemology and Teleology for Modeling and Simulation* categorizes the modeling and simulation ontologies into two categories: methodological ones that capture "how we model" or referential ones that capture "what we model". Following Ören's ideas for utilizing ontologies in BoK studies (Ören, 2011a) in order to provide an ontology-based dictionary for the terms to show also their logical relationships, we will present an ontological representation of the *Process* area under the *Engineering* topic. This is a methodological ontology that aims at providing a common vocabulary and systematization of knowledge.

2.5.2 Simulation Systems Engineering Ontology

The ontology of the simulation systems engineering process that is presented in this section is adopted from the recent work of Durak and Ören (2016). While ontology deals with what exists, epistemology deals with how do we come to know (Tolk, 2013). Although there can be other approaches to reveal the body of knowledge about SSE, such as scientometrics (Mustafee et al., 2014; Diallo et al., 2015a), the knowledge that is presented in this ontology of the simulation systems engineering process is based on DSEEP. Protégé (Knublauch et al., 2005) is used as the ontology development environment by virtue of its popularity. The ontology is developed using the Web Ontology Language (OWL) (Bechhofer et al., 2004).

Figure 2.11 presents the top-level entities of the SSE ontology: *Activity, Data_Store, Information, Product, Role, Step* and *Task*. The entities are then further inherited to their child entities. The reader can see the seven steps of simulation systems engineering under the *Step* entity. The structure of DSEEP proposes that steps are composed of activities and activities have a number of recommended tasks. The ontology is structured accordingly. The data stores captured in the ontology are authoritative resources, data dictionaries, M&S repositories, scenario databases, simulation data exchange models, and other resources. They are captured as the child entities of *Data_Store*.

DSEEP describes roles that act in the SSE process. They are captured as seven entities in the ontology, as depicted in Figure 2.12. These roles are

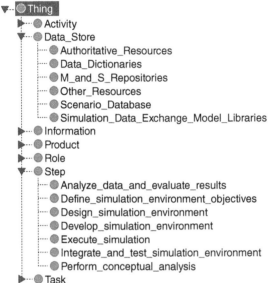

Figure 2.11 Hierarchy of SSE ontology. (Adapted from Durak and Ören (2016).)

user, sponsor, simulation environment manager, development team, integration team, verification and validation agent, and accreditation and acceptance agent

The entities that inherit from *Product* and *Information* are depicted in Figure 2.13. The items that are input to the overall process are named as *Information* and the items that are created with the activities are named as *Product*. Example *Information* entities are existing conceptual models or existing scenarios, whereas example *Product* entities are needs statement or new member applications.

Such an ontology development effort for the simulation systems engineering process is regarded as a step toward creating a common shared conceptualization of products, information exchange, data stores, roles, steps, activities, and tasks of the process. On the one hand, it may enable interoperability and cooperation within the simulation systems engineering

Figure 2.12 Roles in SSE process. (Adapted from Durak and Ören (2016).)

Figure 2.13 Product and information entities. (Adapted from Durak and Ören (2016).)

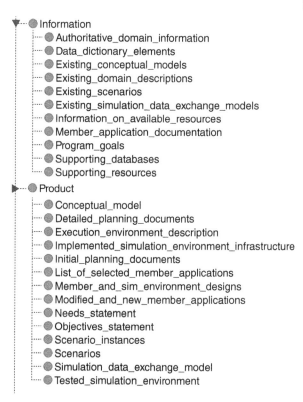

process with machine-readable shared vocabulary; on the other hand, it may be presented using ontology visualization tools in order to make it accessible for the nonontology experts.

2.6 Conclusion and Outlook

The Simulation Systems Engineering Body of Knowledge is regarded as a part of the overall Modeling and Simulation Body of Knowledge. Accordingly, this chapter first introduces the body of knowledge efforts in modeling and simulation and then brings out the simulation systems engineering as one of the latest steps in the evolution of modeling and simulation. As such, it completes the discussions on the need for theoretical foundations by M&S Science, the development of applicable methods based on these foundations, augmented by heuristics by M&S Engineering, and the use of such methods to solve real-world challenges by practitioners by M&S Applications (Tolk et al., 2015). Bill Waite was a driving force from the industrial site who tirelessly collaborated with academia to ensure solid foundations and bridged this knowledge to the practitioners of the field to ensure applicability of the resulting methods and heuristics.

In this chapter, an index is proposed for the Simulation Systems Engineering Body of Knowledge. Short definitions of areas, subareas, and topics are provided referring the key references of the field. Certainly, a comprehensive elaboration is required with an extensive list of references for a mature body of knowledge. But we believe that the content provided in this chapter is a good starting point.

Later in the chapter, we provided a short discussion about ontologies and utilized them to formalize the body of knowledge. An ontology is also presented to exemplify the idea of body of knowledge as an ontology.

An established body of knowledge is one of the pillars of an established discipline. All body of knowledge efforts including this very chapter are steps toward a well-understood, well-practiced profession of modeling and simulation.

References

Balci, O. (1990) Guidelines for successful simulation studies. *Proceedings of the 22nd Winter Simulation Conference, New Orleans, LA.*

Bechhofer, S., van Harmelen, F., Hendler, J., Horrocks, I., McGuinness, D.L., Patel-Schneider, P.F., and Stein, L.A. (2004) W3C Recommendation. *OWL Web Ontology Language Reference* (eds M. Dean and G. Schreiber). Available at http://www.w3.org/TR/owl-ref/ (accessed August 20, 2016).

Bergson, H. (1911) *Creative Evolution*, Henry Holt and Company, New York, NY.

Birta, L.G. (2003a) The Quest for the Modelling and Simulation Body of Knowledge. *Proceedings of the 6th Conference on Computer Simulation and Industry Applications, Tijuana, Mexico.*

Birta, L.G. (2003b) A perspective of the modeling and simulation body of knowledge. *SCS M&S Magazine*, 2 (1).

BKCASE Editorial Board (2016) *The Guide to the Systems Engineering Body of Knowledge (SEBoK)*, v. 1.6. R.D. Adcock, Hoboken, NJ. Available at www.sebokwiki.org. (accessed August 20, 2016).

Borst, R., Akkermans, J., Pos, A., and Top, J. (1995) The PhysSys ontology for physical systems. *9th International Workshop on Qualitative Reasoning, Amsterdam, The Netherlands.*

Brown, J.R. (2015) *Thought experiments*. Stanford Encyclopedia of Philosophy, Pittsburgh, PE. Available at http://plato.stanford.edu/entries/thought-experiment/ (accessed May 1, 2017).

D'Ambrogio, A. and Durak, U. (2016) Setting systems and simulation life cycle processes side by side. *IEEE International Symposium on Systems Engineering, Edinburgh, Scotland.*

De Lara, J., Guerra, E., and Vangheluwe, H. (2003) Meta-modelling, graph transformation and model checking for the analysis of hybrid systems, in *International Workshop on Applications of Graph Transformations with Industrial Relevance*, Springer, Berlin.

Diallo, S.Y., Gore, R.J., Padilla, J.J., and Lynch, C.J. (2015a) An overview of modeling and simulation using content analysis. *Scientometrics*, 103 (3), 977–1002.

Diallo, S., Mustafee, N., and Zacharewicz, G. (2015b) Towards an encyclopedia of modeling and simulation methodology. *Proceedings of the 2015 Winter Simulation Conference, Huntington Beach, CA.*

Durak, U. (2016) Pragmatic model transformations for refactoring in Scilab/Xcos. *International Journal of Modeling, Simulation, and Scientific Computing*, 7 (1), 154100.

Durak, U. and Ören, T. (2016) Towards an Ontology of Simulation Systems Engineering. *Proceedings of the 49th Annual Simulation Symposium (ANSS'16), Pasadena, CA.*

Durak, U., Öztürk, A., and Katircioglu, M. (2016) Simulation deployment blockset for MATLAB/Simulink. *Proceedings of the Symposium on Theory of Modeling & Simulation (TMS/DEVS'16), Pasadena, CA.*

Durak, U., Topçu, O., Siegfried, R., and Oğuztüzün, H. (2014) Scenario development: a model-driven engineering perspective. *2014 International Conference on Simulation and Modeling Methodologies, Technologies and Applications (SIMULTECH), Wien, Austria.*

ECSS-E-TM-40-07 (2011) Volumes 1A to 5A, Space Engineering Simulation Modeling Platform.

Estefan, J.A. (2008) *Survey of model based systems engineering methodologies.* International Council of Systems Engineering, San Diego, CA.

Fritzson, P. (2010) *Principles of Object-Oriented Modeling and Simulation with Modelica 2.1*, John Wiley & Sons, Inc., Hoboken, NJ.

Gotschlich, J., Gerlach, T., and Durak, U. (2014) *2Simulate: a distributed real-time simulation framework.* Workshop der ASIM/GI-Fachgruppen STS und GMMS. Reutlingen, Germany.

Gruber, T. (1995) Toward principles for the design of ontologies used for knowledge sharing. *International Journal of Human-Computer Studies*, 43, 907–928.

IEEE (1989) IEEE SA 610.3–1989, IEEE Standard Glossary of Modeling and Simulation Terminology, New York, NY. Withdrawn Inactive Date: 6-3-2000.

IEEE (2003) IEEE STD 1516.3–2003, IEEE Recommended Practice for High Level Architecture (HLA) Federation Development and Execution Process (FEDEP).

IEEE (2010a) IEEE STD 1730TM-2010, IEEE Recommended Practice for Distributed Simulation Engineering and Execution Process (DSEEP).

IEEE (2010b) IEEE STD 1516–2010, IEEE Standard for Modeling and Simulation High Level Architecture (HLA): Framework and Rules. New York, NY.

IEEE (2010c) IEEE STD 1516.1–2010, IEEE Standard for Modeling and Simulation (M&S) High Level Architecture (HLA): Federate Interface Specification, New York, NY.

IEEE (2010d) IEEE STD 1516.2–2010, IEEE Standard for Modeling and Simulation High Level Architecture (HLA): Object Model Template (OMT) Specification, New York, NY.

IEEE (2012) IEEE STD 1278.1–2012, Protocols for Distributed Interactive Simulation Applications: Entity Information and Interaction, New York, NY.

ISO/IEC/IEEE (2015) 15288:2015, ISO/IEC/IEEE International Standard: Systems and Software Engineering – System Life Cycle Processes.

Kleijnen, J.P. (2001) Supply chain simulation tools and techniques: a survey. *International Journal of Simulation and Process Modelling*, 1 (1–2), 82–89.

Knublauch, H., Horridge, M., Musen, M., Rector, A., Stevens, R., Drummond, N., Lord, P., Noy, N., Seidenberg, J., and Wangl, H. (2005) *The Protégé OWL experience*. OWL: Experiences and Directions Workshop (OWLED), Galway, Ireland.

Lacy, L.W., Gross, D.C., Ören, T., and Waite, B. (2010) A realistic roadmap for developing a Modeling and Simulation Body of Knowledge Index. *Proceedings of Fall Simulation Interoperability Workshop, Orlando, FL.*

Law, A.M. and Kelton, W.D. (1991) *Simulation Modeling and Analysis*, McGraw-Hill, New York, NY.

Legros, E., Schäfer, W., Schürr, A., and Stürmer, I. (2010) MATE: a model analysis and transformation environment for MATLAB Simulink, in *Model-Based Engineering of Embedded Real-Time Systems*, Springer, Berlin.

Liu, H., Liu, X., and Lee, E.A. (2001) Modeling distributed hybrid systems in Ptolemy II. *Proceedings of the 2001 American Control Conference, Arlington, VA.*

Miller, J.A., Baramidze, G.T., Sheth, A.P., and Fishwick, P.A. (2004) Investigating ontologies for simulation modeling. *37th Annual Simulation Symposium, Arlington, VA.*

Mizoguchi, R. (2001) Ontological engineering: foundations of the next generation knowledge processing. *Web Intelligence 2001, Maebashi City, Japan.*

MODELISAR Consortium (2016) *Functional mock-up interface for model exchange and co-simulation*. Available at https://svn.modelica.org/fmi/branches/public/specifications/v2.0/FMI_for_ModelExchange_and_CoSimulation_v2.0.pdf (accessed August 20, 2016).

Mooney, C.Z. (1997) *Monte Carlo Simulation*, Vol. 116, Sage Publications, Thousand Oaks, CA.

Mustafee, N., Katsaliaki, K., and Fishwick, P. (2014) Exploring the modelling and simulation knowledge base through journal co-citation analysis. *Scientometrics*, 98 (3), 2145–2159.

National Aeronautics and Space Administration (NASA) (2007) *Systems Engineering Handbook*, NASA/SP-2007-6105 Rev 1, NASA Headquarters, Washington, DC.

NATO (2014) MSG-086 Guideline on Scenario Development for (Distributed) Simulation Environments. NATO STO Technical Report, Brussels, Belgium.

Neches, R., Fikes, R.E., Finin, T., Gruber, T.R., Senator, T., and Swartout, W.R. (1991) Enabling technology for knowledge sharing. *AI Magazine*, 12 (3), 36–56.

Noseworthy, J.R. (2008) The Test and Training Enabling Architecture (TENA) supporting the decentralized development of distributed applications and LVC simulations. *12th IEEE/ACM International Symposium on Distributed Simulation and Real-Time Applications, British Columbia, Canada.*

Ören, T.I. (1981) Concepts and criteria to assess acceptability of simulation studies: a frame of reference. *Communications of the ACM*, 24 (4), 180–189.

Ören, T. (1984a) Foreword to the book: *Multifaceted Modelling and Discrete Event Simulation* (ed. B.P. Zeigler), Academic Press, London, Foreword.

Ören, T.I. (1984b) Model-based activities: a paradigm shift, in *Simulation and Model-Based Methodologies: An Integrative View* (eds T.I. Ören, B.P. Zeigler, and M.S. Elzas), Springer, Heidelberg, Germany.

Ören, T.I. (2005) Toward the body of knowledge of modeling and simulation. *Interservice/Industry Training, Simulation and Education Conference (I/ITSEC), Orlando, FL.*

Ören, T.I. (2006) Body of knowledge of modeling and simulation (M&SBOK): pragmatic aspects. *Proceedings of the 2nd European Modeling and Simulation Symposium, Barcelona, Spain.*

Ören, T.I. (2009) Modeling and simulation: a comprehensive and integrative view. *Agent-Directed Simulation and Systems Engineering*, Wiley-VCH Verlag GmbH, Weinheim, Germany.

Ören, T.I. (2011a) A Basis for a Modeling and Simulation Body of Knowledge Index: professionalism, stakeholders, Big Picture, and other BoKs. *SCS M&S Magazine*, 2 (1), 40–48.

Ören, T. (2011b) A critical review of definitions and about 400 types of modeling and simulation. *SCS M&S Magazine*, 2 (3), 142–151.

Ören, T.I. (2014) The richness of modeling and simulation and an index of its body of knowledge, in *Simulation and Modeling Methodologies, Technologies and Applications*, Advances in Intelligent Systems and Computing (eds M.S. Obaidat, J. Filipe, J. Kacprzyk, and N. Pina), Vol. 256, Springer.

Ören, T.I. (2016) *Modeling and Simulation Body of Knowledge (M&S BoK): Index.* Available at http://www.site.uottawa.ca/~Ören/MSBOK/MSBOK-index.pdf. (accessed July 17, 2016).

Ören, T.I. and Waite, B. (2010) Modeling and Simulation Body of Knowledge Index: an invitation for the final phases of its preparation. *SCS M&S Magazine*, 1 (4).

Ören, T.I., Elzas, M.S., Smit, I., and Birta, L.G. (2002) A code of professional ethics for simulationists. *Proceedings of the 2002 Summer Computer Simulation Conference, San Diego, CA.*

Pace, D.K. (2000) Ideas about simulation conceptual model development. *Johns Hopkins APL Technical Digest*, 21 (3), 327–336.

Padilla, J.J., Diallo, S.Y., and Tolk, A. (2011) Do we need M&S science? *SCS M&S Magazine*, 8, 161–166.

Popovici, K. and Mosterman, P.J. (eds) (2012) *Real-Time Simulation Technologies: Principles, Methodologies, and Applications*, CRC Press, Boca Raton, FL.

Railsback, S.F., Lytinen, S.L., and Jackson, S.K. (2006) Agent-based simulation platforms: review and development recommendations. *Simulation*, 82 (9), 609–623.

Rogers, R. (1997) What makes a modeling and simulation professional? The consensus view from one workshop. *Winter Simulation Conference (WCS), Atlanta, GA.*

Sargent, R.G. (2005) Verification and validation of simulation models. *Proceedings of the 37th Winter Simulation Conference, Orlando, FL.*

Schiehlen, W. (ed.) (2013) *Advanced Multibody System Dynamics: Simulation and Software Tools*, Springer Science+Business Media.

SISO-STD-001.1-2015 (2015) Standard for Real-time Platform Reference Federation Object Model Version 2.0. Orlando, FL.

SISO-STD-003-2006 (2006) Base Object Model (BOM) Template Specification. Orlando, FL.

SISO-STD-008-01-2012 (2012) Standard for Core Manufacturing Simulation Data: XML Representation 1, Orlando, FL.

SISO-STD-011-2014 (2014) Standard for Coalition Battle Management Language (C-BML) Phase 1, Orlando, FL.

Sokolowski, J.A. and Banks, C.M. (2010) *Modeling and Simulation Fundamentals: Theoretical Underpinnings and Practical Domains*, John Wiley & Sons, Inc., Hoboken, NJ.

Szczerbicka, H., Banks, J., Ören, T.I., Rogers, R.V., Sarjoughian, H.S., and Zeigler, B.P. (2000) Conception of curriculum for simulation education (panel). *Winter Simulation Conference (WCS), Orlando, FL.*

Tolk, A. (2010) M&S Body of Knowledge: progress report and look ahead. *SCS M&S Magazine*, 1 (4).

Tolk, A. (2012) Standards for distributed simulation. *Engineering Principles of Combat Modeling and Distributed Simulation*, John Wiley & Sons, Inc., Hoboken, NJ.

Tolk, A. (2013) *Ontology, Epistemology, and Teleology for Modeling and Simulation*, Springer, Heidelberg, Germany.

Tolk, A. (2015) The next generation of Modeling & Simulation: integrating big data and deep learning. *Proceedings of the Summer Computer Simulation Conference, Chicago, IL.*

Tolk, A. and Hughes, T.K. (2014) Systems engineering, architecture, and simulation, in *Modeling and Simulation-Based Systems Engineering Handbook* (eds D. Gianni, A. D'Ambrogio, and A. Tolk), CRC Press, Boca Raton, FL.

Tolk, A. and Muguira, J.A. (2003) The levels of conceptual interoperability model. *Proceedings of the 2003 Fall Simulation Interoperability Workshop, Orlando, FL.*

Tolk, A., Balci, O., Combs, C.D., Fujimoto, R., Macal, C.M., Nelson, B.L., and Zimmerman, P. (2015) Do we need a national research agenda for modeling and simulation? *Proceedings of the 2015 Winter Simulation Conference, IEEE Press*, pp. 2571–2585.

Topçu, O., Durak, U., Oğuztüzün, H., and Yilmaz, L. (2016) *Distributed Simulation: A Model Driven Engineering Approach*, Springer International Publishing, Cham, Germany.

US DoD (2009) DoD Modeling and Simulation (M&S) Verification, Validation, and Accreditation (VV&A), DoD Instruction 5000.61.

Waite, B. (2001) M&S Professional Body of Knowledge/Code of Ethics. *Summer Computer Simulation Conference (SCSC), Orlando, FL.*

Yilmaz, L. and Ören, T.I. (eds) (2011) *Agent-Directed Simulation and Systems Engineering*, Wiley-VCH Verlag GmbH, Weinheim, Germany.

Zeigler, B.P. and Sarjoughian, H.S. (2003) Introduction to DEVS Modeling and Simulation with Java: Developing Component-Based Simulation Models, Technical Document, Arizona Center for Integrative Modeling and Simulation, University of Arizona, Tucson, AZ.

Zeigler, B.P., Elzas, M.S., Klir, J.G., and Ören, T.I. (1979) *Methodology in System Modelling & Simulation*, North-Holland, Amsterdam.

Zeigler, B.P., Luh, C.J., and Kim, T.G. (1991) Model base management for multifacetted systems. *ACM Transactions on Modeling and Computer Simulation*, 1 (3), 195–218.

Zeigler, B.P., Praehofer, H., and Kim, T.G. (2000) *Theory of Modeling and Simulation: Integrating Discrete Event and Continuous Complex Dynamic Systems*, Academic Press, San Diego, CA.

3

Code of Ethics

Andreas Tolk

The MITRE Corporation, Hampton, VA, USA

3.1 Introduction

This chapter introduces the reader to professional codes of conduct and ethics as they are used in building a strong ethically sound foundation for engineering professions. From a general introduction into the topic, followed by the discussion to what degree professional ethics can be coded, and under what constraints this can be done, the professional code of ethics for simulationists is introduced. This code was developed for the Society of Modeling and Simulation (SCS), translated into several languages, and has been adapted not only by various professional organizations but also by several industry partners. The chapter is summarized by several examples how the code can support morally sound decision making for simulationists in various situations.

3.2 Ethics in Technology and Engineering

It is an expectation of the society that a professional should know what is right or wrong. But that is easier said than done, as what is considered right or wrong can be relative to cultural backgrounds, personal experiences, current laws and regulations, or simply conscious or unconscious bias with a group. Morality and society are interdependent things. Even in the normative sense, morality refers to a code of conduct that would be accepted by anyone who meets certain intellectual and volitional conditions, almost always including the condition of being rational (Gert and Gert, 2016). However, morality is tightly related to ethics. The difference between ethics

The Profession of Modeling and Simulation: Discipline, Ethics, Education, Vocation, Societies, and Economics, First Edition. Andreas Tolk and Tuncer Ören.

and moral is often formulated as that *ethics are the science of morals*, and *morals are the practice of ethics*. Moral is the collection of values and standards, ethics is the formal study and encoding of these standards. While it is difficult to address morality, ethics codifies agreements on moral foundations, making it tangible. Morality assumes a common code of conduct, ethics captures and formulates the code of conduct.

But why is this important for an engineer in general, and for simulationists in particular?

Many engineering students, including simulationists, may have to take an ethics class within the requirements of the general education, but that is more often seen as a necessity to pass the class than as a foundation for the future profession. Nonetheless, the *obligations of an engineer* are part of what students have to commit to when they are inducted into the order of the engineer at graduation. These obligations focus on an engineer being a trusted servant of the public good. Table 3.1 shows one of the popular versions, but slightly different versions have been adapted as well.

In order to fulfill these obligations, students have to be morally sensitive. Within their engineering education, they learn how to identify the users' needs and derive the problems that need to be solved. Once the problem is identified, several alternative solutions and designs are often created that have to be compared to identify the best feasible solution. In a structure decision process, the design is selected and implemented. All these processes, assumptions, experiments, constraints, and results are documented to ensure reproducibility of results and decisions made.

Engineers have to develop the same sensitivity for moral issues as they have for technical challenges. The processes proposed by van de Poel and

Table 3.1 Order of the engineer: the obligations of an engineer.

I am an Engineer. In my profession I take deep pride. To it, I owe solemn obligations.

Since the Stone Age, human progress has been spurred by the engineering genius. Engineers have made usable nature's vast resources of material and energy for Humanity's benefit. Engineers have vitalized and turned to practical use the principles of science and the means of technology. Were it not for this heritage of accumulated experience, my efforts would be feeble.

As an Engineer, I pledge to practice integrity and fair dealing, tolerance and respect, and to uphold devotion to the standards and the dignity of my profession, conscious always that my skill carries with it the obligation to serve humanity by making the best use of Earth's precious wealth.

As an Engineer, I shall participate in none but honest enterprises. When needed, my skill and knowledge shall be given without reservation for the public good. In the performance of duty and in fidelity to my profession, I shall give the utmost.

Royakkers (2011) can be easily mapped to the engineering cycle, as every stage is connected with certain moral skills.

In the problem definition phase, which usually ends with a high-level understanding on how a solution may look like, the engineer also needs to apply *moral sensibility* to identify the moral needs and possible challenges. Are there any moral concerns regarding the solution or components thereof? Are certain risks already obvious that may endanger live or health of humans? Any challenges for the environment perceived that need to be addressed? The engineer, as an agent for the public good, is responsible to address such issues as much as technical challenges in the problem identification phase.

When the solution is discussed in detail, and the implementation and production is technically specified, the engineer must continue to conduct a *moral analysis*. All systems engineering phases can contain morally challenging decisions that often are diametric in metrics: An inexpensive solution may create risks for life, health, and environment that is not justifiable. A stronger material may not be easily disposed at the end of its life cycle. A fast numerical approximation may make the code faster, but the precision is insufficient for critical phases.

Just like more than one design need to be established, *moral creativity* establishes and evaluates moral alternatives. Particularly in multicultural and diverse environments, the moral viewpoints on the same solution may significantly differ. The engineer must be sensitive to such issues and creative enough to recognize these challenges before they endanger the project.

Just like the engineers need technical competence for a thorough evaluation of the system, the *moral evaluation* is as important. What is the best solution? Are there still risks? Can these risks be mitigated? Is the technical solution also morally feasible? Are their cultural or other biases that have to be addressed before a solution can be implemented?

One of the most obvious processes requiring best judgment is the *moral decision making*. Are the ideals of practicing integrity and fair dealing upheld? Is the burden sharing appropriate? How are responsibilities and rewards distributed? Is the public good really optimized with the decision?

Last but not the least, the moral decisions need to be understood by all team members as well as by possible external evaluators, hence *moral argumentation* is a skill required by engineers as well.

van de Poel and Royakkers (2011) introduced the ethical cycle in their book as captured in Figure 3.1. It comprises five phases with possible feedback loops, as shown in Figure 3.1.

Each ethical cycle starts with a moral problem statement. This can be the potential violation of a moral norm, but it can also be facing a dilemma in

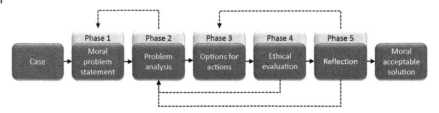

Figure 3.1 The ethical cycle as introduced by van de Poel and Royakkers (2011).

which two or more moral norms or values cannot be fully supported, as they are not aligned with each other or the overall case objectives. The objective of phase 1 is to unambiguously identify the problem and – even more important – who has to make a decision.

During the following problem analysis, additional challenges may emerge that trigger a new problem statement process in a feedback loop. Very often, the facts and supporting data may still be vague, incomplete, and even contradictive. The analysis shall help clarify the situation. Another aspect is that the stakeholders and their interests are not well understood. Only if the context and supporting facts are made tangible, the identification of the moral values at stake can be captured.

Phase 3 utilizes the moral creativity of engineers to identify options for actions. One option is always "to do nothing," nonaction is an action by itself that may be the most immoral decision. Another option can be "whistle blowing," in particular when stakeholders, managers, or other important members of team are not willing to choose a morally superior course of action. Hopefully, however, the result will be collaborative strategies and creative middle-way solutions.

The ethical evaluation of these options builds the fourth phase. How this evaluation is conducted is topic of several moral books. It is possible to use intuition, but that makes justification and reproducibility challenging. It is better to use methods that can be documented, that may even be based on utility theory. Another option is the use of reference cases. Often, additional questions are emerging that require another problem analysis process. The result is a choice for an option or, in case of uncertainty, a strategy to be followed.

Finally, the reflection phase reevaluates the decision in the context of the ethical framework used in order to improve this over time. Did the framework provide the right methods? Did it help to ask the necessary questions and find appropriate answers? Is the underlying ethical theory applicable, or does it need to be revisited? All these widespread reflection may actually lead to the reentry into phases 2 or 3, as new problems may be identified, or new options have to be evaluated.

Moral sensibility is predominantly needed in the earlier phase, while moral argumentation is needed in the later processes, including documenting the decision processes along with its assumptions and constraints. The ethical cycle provides a framework to not only guide the decisions of engineers but also help identify morally challenging constellations. What if the ethical framework and moral values within the team or with the stakeholders are not aligned? What if the understanding of the "public good" is interpreted quite differently? While the ethical cycle does not provide answers, it provides structure to identify problem areas and may provide the foundation to overcome these challenges.

3.3 Codifying Ethics – A Discussion

If engineers in general, and simulationists in particular, have a common moral foundation to work for the public good, why is it necessary to codify them explicitly? What forms of codes are used? Can we codify moral behavior anyhow? Is it not sufficient to assume that professionals have a conscience or an innate sense of what's right and wrong?

Just as is the case with any strategic goals, making ethics and values tangible and measurable within an organization is necessary, in particular in multinational and multicultural environments. The explicit codification clearly communicates them, and makes them applicable to the members of the target audience. It creates a common foundation that all members can agree to. As stated in the beginning of this chapter, ethics is the formal study and encoding of moral standards!

The general term for such collections of formalized – and hopefully measurable – ethics and values is *Code of Conduct*, which often is defined as guidelines for responsible and desired behavior. Individuals usually do not use such a code, as they make their decision based on their individual value system that is often not explicitly coded or documented. However, individuals often refer to codes that are accepted by groups or organizations they belong to, which can be cultural, social, religious, or professional. This list is neither meant to be complete nor exclusive, but gives a reference to generally accepted high-level categories applicable in this discussion.

In their work, van de Poel and Royakkers (2011) distinguish between professional codes, formulated by professional associations, and corporate codes, formulated by companies. Among the important associations for the simulationists are the Society for Modeling and Simulation (SCS), the Association for Computing Machinery (ACM) Special Interest Group Simulation (SIGSIM), and the Simulation Society of the Institute for Operations Research and Management Sciences (INFORMS-Sim). These simulation societies not only collaborate on general simulation challenges, their technical committees focus also on association-specific solutions. The professional code of ethics for simulationists developed under the lead of SCS will be discussed in detail in the next section.

Another way to categorize codes of conducts is on the basis of objectives of the code. Frankel (1989) distinguishes them into three main categories:

- Codes that focus on the moral values of the target audience are *aspirational codes*. Not only do they state for the "outside world" what kind of values the profession stands for, but also appeal to the members of the profession to strive for these values as their guiding principles.
- The second group of codes focuses on applicable advice in concrete situations, helping members of the profession exercise good moral judgment. They are called *advisory codes*. They include general values and norms, but make them applicable.
- The *disciplinary codes* establish rules and boundaries for the members of the audience that are not meant to be crossed without repercussions or other negative impact on the violators.

In the domain of professional codes, the advisory codes are most often used, as they establish a moral foundation in the concrete boundaries of the profession. Engineering codes usually address several core domains:

- *Integrity and honesty* ensuring that services of the profession are provided faithfully and truthfully. This domain entails avoiding conflicts of interest, which in advisory codes are often spelled out in more detail.
- *Competency* ensures that a member of the profession continues to learn and stays up-to-date with his ability to provide the best services for the common good. This also requires professionals to be aware of the scientific philosophical foundations of their profession, as they build the ultimate foundation for competent solutions. Best practices and heuristics are necessary, but not sufficient for professionals.
- *Social obligations* address the relations to employers, colleagues, and customers or sponsors.
- *Responsibility towards the public* addresses all questions regarding the "public good," such as social challenges, environmental concerns, and safety, health, and welfare.

Codes of conduct are not without criticism. Among those challenges is the observation that organizations and associations may misuse the publically displayed code merely to create a favorable impression that is not based on actual facts. While the code displays a deep moral obligation to the public good, it is actually rarely applied to support organizational decisions. In practice, this kind of window-dressing creates only a short-term advantage, as the public criticism, if such practices are discovered and exposed, can be very harmful to the organizations and associations involved.

Another point of criticism is that codes often are too vague and are not helpful in formally prescribing desired behavior. The vagueness may even lead to conflicting guidelines, for example, if there is a discrepancy between objectives of the employer, the colleagues, and the customer. Finding the right balance between general applicability and specific guidelines is a general challenge. That is why moral sensibility is such an important characteristic for a professional.

A third point of discussion, as exemplified by Ladd (1991), is the observation that ethics requires individual moral judgment, not just blindly following a code. While this observation is clearly worthwhile to be discussed, the value of explicit codes as guidelines to learn the behavior and at least to serve as a general moral code governing accepted principles remains valuable in themselves.

In the author's opinion, the advantages of professional, aspirational, and advisory codes outperform the disadvantages. Assuming a common moral foundation in the era of multinational collaboration, which brings together professionals with a diverse cultural, social, and religious background, is in general not justified. Moral relativism implies that moral judgments are guided common historical, cultural, social, and religious constraints. In other words, there is no absolute moral global law. This philosophy, however, makes the need for a professional code of ethics even more important, as this is the only way to agree on a common set of ethics and values as they are not formalized in the new context of the professional organization or association.

3.4 A Code of Professional Ethics for Simulationists

3.4.1 Context for Developing the Code

The discussion on the need for a code of professional ethics for simulation experts in particular started in parallel with the rise of modeling and simulation as a tool supporting decision makers and later on as a tool for training and education.

One of the first articles on this subject was presented during the Winter Simulation Conference 1983 by John McLeod. He identified in McLeod (1983) the need to better address the extent to which information derived from simulation-based studies is valid and applicable to the decision of interest. The derived problem categories involve the selection of data, the modeler's interpretation of the dynamics of the system modeled, and the analyst's interpretation of the results.

3.4.2 Published Text of the Code

Enriched by many following discussions, the Society for Modeling and Simulation (SCS) sponsored a task group to recommend a Code of Professional Ethics for Simulationists focusing in particular on the needs for modeling and simulation experts (Table 3.2). Dr. Ören was tasked by the president of the society to spearhead the effort. This section comprises a verbatim copy of the resulting code of professional ethics for simulationists as published by Ören *et al.* (2002).

Table 3.2 Code of Professional Ethics for simulationists (Ören *et al.*, 2002).

PREAMBLE:
Simulationists are professionals involved in modeling and simulation activities and/or with providing modeling and simulation products and/or services. A detailed definition of the term "simulationist" is given by Ören (2000a,b). Rationale for this code is given in Ören (2002).

1. PERSONAL DEVELOPMENT AND THE PROFESSION
   As a simulationist I will:
   1.1 Acquire and maintain professional competence and attitude.
   1.2 Treat fairly employees, clients, users, colleagues, and employers.
   1.3 Encourage and support new entrants to the profession.
   1.4 Support fellow practitioners and members of other professions who are engaged in modeling and simulation.
   1.5 Assist colleagues to achieve reliable results.
   1.6 Promote the reliable and credible use of modeling and simulation.
   1.7 Promote the modeling and simulation profession; for example, advance public knowledge and appreciation of modeling and simulation and clarify and counter false or misleading statements.

2. PROFESSIONAL COMPETENCE
   As a simulationist I will:
   2.1 Assure product and/or service quality by the use of proper methodologies and technologies.
   2.2 Seek, utilize, and provide critical professional review.
   2.3 Recommend and stipulate proper and achievable goals for any project.
   2.4 Document simulation studies and/or systems comprehensibly and accurately to authorized parties.
   2.5 Provide full disclosure of system design assumptions and known limitations and problems to authorized parties.
   2.6 Be explicit and unequivocal about the conditions of applicability of specific models and associated simulation results.
   2.7 Caution against acceptance of modeling and simulation results when there is insufficient evidence of thorough validation and verification.
   2.8 Assure thorough and unbiased interpretations and evaluations of the results of modeling and simulation studies.

3. TRUSTWORTHINESS

As a simulationist I will:

3.1 Be honest about any circumstances that might lead to conflict of interest.

3.2 Honor contracts, agreements, and assigned responsibilities and accountabilities.

3.3 Help develop an organizational environment that is supportive of ethical behavior.

3.4 Support studies that will not harm humans (current and future generations) as well as the environment.

4. PROPERTY RIGHTS AND DUE CREDIT

As a simulationist I will:

4.1. Give full acknowledgement to the contributions of others.

4.2. Give proper credit for intellectual property.

4.3. Honor property rights including copyrights and patents.

4.4. Honor privacy rights of individuals and organizations as well as confidentiality of the relevant data and knowledge.

5. COMPLIANCE WITH THE CODE

As a simulationist I will:

5.1 Adhere to this code and encourage other simulationists to adhere to it.

5.2 Treat violations of this code as inconsistent with being a simulationist.

5.3 Seek advice from professional colleagues when faced with an ethical dilemma in modeling and simulation activities.

5.4 Advise any professional society that supports this code of desirable updates.

3.4.3 Translations and Organizational Adoptions

The code of professional ethics for simulationists were presented in English, but at the time this chapter is written it has been translated into French, Turkish, Italian, and Chinese by native speakers who also are active members of the simulation community.

In chronological order, at the time this chapter is written, the code has been adopted by

- Society for Modeling and Simulation International (SCS)
- Mcleod Institute of Simulation Sciences (MISS)
- McLeod Modeling and Simulation Network (M&SNet)
- Simulation Interoperability Standards Organization (SISO)
- Alabama Modeling and Simulation Council (AMSC)
- NATO Modeling and Simulation Group (NMSG)
- Modeling and Simulation Professional Certification Commission (M&SPCC)
- Meteksan Defense Industry Inc., Turkey
- Simulation Team, Italy
- Bulgarian Modeling and Simulation Association (Bulsim)

The number of organizations and associations interested in the code and adapting it is still increasing. It has become a foundational part of the profession of modeling and simulation.

3.5 Application Examples

This section will provide several examples for applications of the code of ethics in the domain of the profession of modeling and simulation. Many examples are based on author's experiences within this community as well as derived from discussions with colleagues and students. Several additional ones have been used as examples by Ören (2002) when giving a rationale for the code of ethics.

This list of examples is neither complete nor exclusive. It is meant to show many day-by-day situations where the simulationists may want to reflect on the values that are driving their decisions, and if they are aligned with the professional code. As such, the moral sensibility of simulation professionals – as addressed in the beginning of this chapter – shall be raised.

3.5.1 Characteristics of a Scholar

In a survey published in Tolk (2012a), a group of international peers was asked to identify for 12 characteristics if they are essential, optional, or not needed to identify someone as a scholar. The definition of these terms was based on literature research on the characteristics of a scholar. Interestingly enough, the characteristics with the highest ranking was *Ethics*, with 94% identifying it as essential, and the remaining 6% as optional. The following definition is given to the participants: A scholar follows strong professional ethics, rooted in honesty about the own work, accepting constructive criticism, treating others with respect, and not gaining personal advantage out of serving positions. The ranking of all 12 characteristics is shown in Table 3.3 (emphasis added for this section).

The group of contributors was too small to allow for general applicability of these results. However, it shows that the definition of a scholar clearly requires to be rooted in ethics as captured in this chapter. Interestingly enough, productivity and competitiveness, often used by tenure and pro-motion committees as a driving factor for their decisions, did not play such a clear decisive role in the survey results.

3.5.2 Collaboration as M&S Professionals

The following examples are not limited to M&S as a profession, but often applicable in the wider domain of science and engineering as well. However,

Table 3.3 Ranking of the characteristics of a scholar (Tolk, 2012a).

	Essential (%)	Optional (%)	Not needed (%)	Undecided (%)
Ethics	**94**	**6**		
Immersion	92	6		2
Disposition	86	11	2	
Authority	86	11	2	
Persistence	83	11	6	
Passion	80	17	3	
Productivity	57	34	9	
Definition	50	41	3	6
Recognition	37	51	9	3
Loyalty	37	49	14	
Connection	34	63	3	
Competitiveness	15	67	18	

as M&S itself is a very young discipline, and in many circles the validity of calling M&S a discipline in itself is even still discussed, highlighting some of these issues may help avoid unprofessional behavior that in established disciplines may not be evaluated with such scrutiny. As simulationists often still struggle to be recognized as their own profession, we have to be particularly careful to observe the moral codes of our profession.

As scientist, we produce models to help answering questions and establish a hypothesis in the form of a simulation system that produces output data based on the input data and the codified causality in the form of computable function of the system. These quasi-empirical data can then be compared with the real-world observations. Too often, however, the expected results are not accomplished. The temptation to "adjust" the data just a little bit, or to "calibrate" the results by excluding "outsiders" that do not fit the narrative is often high, in particular when the underlying simulation system is too complicated to make such "adaption of results" obvious to the peers. Such a behavior represents a direct violation of the code of ethics.

Unfortunately, this is not a fabricated example. Purposefully falsifying data is a trend observed too often in the broader community as well, as documented by Fanelli (2009), reporting about a study in which approximately 2% of the scientists admitted to have fabricated, falsified, or modified data or results. In related surveys asking about the behavior of colleagues, approximately 14% of unjustified data modifications were reported. Other scientific misconduct was even estimated to be higher, namely, slightly more than 30% for self-observations, and a little bit more than 70% for peer evaluations. Considering that these surveys ask sensitive questions, it appears likely that this is a conservative estimate of the true prevalence of scientific misconduct. In recent journal papers, more related challenges are discussed, such as the legal reliability in case of such fraud (Guerra-Pujol, 2015).

Another challenge for simulationists following the code of ethics is conducting peer reviews for journals and conferences. As professionals, we have to ask ourselves a series of hard questions: *Do we really use the same metrics when we evaluate papers from renowned experts as we do when evaluating the contribution of a novice? Do our research partners and friends from recent conference get some "extra credit" when we review their papers? Are papers that reference our own work less likely to be rejected by us than papers that ignore our research?* The code gives clear guidelines on how to behave. Recent studies show that there is indeed a bias in peer review that violates the high value of impartiality of such evaluations (Lee *et al.*, 2013). Some of these biases are human nature, others are systemic to the organization and processes, and need to be addressed. In particular, once recognized in the field as an expert, it becomes easy to believe, accredit, and communicate research that support one's views while ignoring or discrediting those results that do not. This is simply bad practice, and bad for the scientific discipline as well. The diversity of minds requires a multitude of possible opinions.

For practitioners of simulation in industry, the requirements to provide full disclosure of system design assumptions and known limitations and problems to authorized parties (2.5), and to be explicit and unequivocal about the conditions of applicability of specific models and associated simulation results (2.6) can become quite challenging, as they may result in competing objectives. On the one hand, the simulation industry is interested in providing reliable solution to increase the overall trust in the underlying methods and tools, on the other hand, they are interested in customers who mainly use the products provided by the represented company or organization. Being completely transparent about the limitations and constraints can lead to being eliminated from lucrative projects, so where should the line be drawn? Another aspect playing a role is the protection of intellectual property: Giving details away can easily play into the hands of the competition.

3.5.3 The Nature of Modeling and Simulation

Promoting the reliable and credible use of modeling and simulation is a tenet in the code of ethics for simulationists. This requires to clearly communicate the assumptions and constraints, which starts with a proper understanding of what modeling and simulation actually is. A colleague once stated anecdotally that the difference between a used car salesperson and a simulation salesperson is that the used car salesperson knows when he is lying. This is actually connected to the first enumerated element of the code, namely, to acquire and maintain professional competence and attitude. Many simulation professionals are focusing on the computational aspects of simulation as a software discipline, but that is only a facet of the profession. The philosophical underpinning of epistemological constraints is as important as the effects of numerical approximations in the coupling of nonlinear functions.

Modeling is the task-driven, purposeful simplification and abstraction of a perception of reality that is shaped by physical, ethical and cognitive constraints. It results in a formal specification of a conceptualization and can be conducted in several phases by individuals as well as by groups. The resulting model specifies the ontological boundary of what the model can compute. It defines the entities with their properties, behaviors, and relations. The model becomes the "reality" of an implementing simulation. Diallo *et al.* (2014) provide a formalism to introduce more rigor. Simulationists must apply such rigor to follow the high standards of professionalism, as in particular the other engineering disciplines require.

Simulation is the execution of a model. We are in particular interested in the execution of the model on a digital computer over time to gain quasi-empirical numerical insight into the behavior of complex systems. As stated before, this insight is limited to the modeled entities and phenomena, we cannot get insight outside of the model, as computable functions can only transform knowledge from input to output parameters, they cannot create new knowledge, as proven by Chaitin (1977). In addition, many implementation choices have to be made, and applying heuristics and numerical approximation may introduce errors into the simulation. Oberkampf *et al.* (2002) provide examples and introduce a framework to communicate such known challenges. Again, these works define the high bar of professionalism to be achieved by simulationists.

Users of simulation-based experiments and solutions cannot be expected to be held accountable to the same standards of simulation professionals. That is why validation and verification is important to provide trust into the provided results. The state of the art has been recently summarized by Sargent (2013). In addition, explainability and reproducibility of simulation results should be accomplished, as discussed by Uhrmacher *et al.* (2016).

It cannot be overemphasized that simulations are not a reproduction of reality. Simulations are based on models that represent an interpretation that postulates certain causalities, which are implemented as computable functions that work on a selected subset of input data to produce a selected subset of output data. It is appropriate to interpret *valid* simulations as executable *theories*. While it is still helpful and necessary to apply simulation to evaluate hypotheses, that is, evaluate assumptions about what *could be* without having the scientific proof that these assumptions are justified, the resulting simulations must clearly be understood – and presented as such when being used – to represent a so far unproven hypothesis, not a theory.

Gelfert (2013) explores the danger of "epistemic double standards" for the use of simulations and their underlying models: Whether simulations are used to support a case or to argue against one, professionalism requires to use the same sets of rigorous epistemic standards for their interpretations. Otherwise, simulations can easily be used to implement one's biases, and then the simulation outputs can be used to prove one's own points, which just went into building the model. This is a known phenomenon, referred to as experimenter's regress (Collins, 2002). Experimenter's regress leads to pathological science, as the experimenter has such strong belief in the hypothesis that any biased evidence they obtain strengthens their belief. Accordingly, Gelfert (2013) shows that it is often human nature that the M&S professional must be aware of, and it is the ethical responsibility to avoid such misuse of modeling and simulation, in particular for topics in which moral and epistemological considerations are deeply intertwined (Tolk, 2017).

3.5.4 Giving Back to the Community

This last section is of a more personal nature and intended to provide some thoughts for colleagues who are role models for their societies. Many of the professional M&S societies bestow higher membership ranks on distinguished members and provide recognitions and awards. Receiving such a rank – such as becoming a fellow of a simulation society – or a professional award – such as lifetime achievement awards – is not only a recognition, it also comes with a new set of responsibilities.

Distinguish members serve the society. This service includes in particular supporting young academics and colleagues who are starting within the society, as it is the new generation that will carry the discipline forward. It requires active engagement and encouragement, for example, in supporting presentations and publications, endorsing project, and so on. Constructive criticism is part of the process, but way too often it can be witnessed that distinguished members become unapproachable celebrities who behave like they "deserve" special attention and benefits. In some cases, young minds with new ideas are even negatively treated, as they do not follow the lead of the distinguished members and pursue alternative paths. But it is in the nature of science that new ideas have to be born to replace no longer applicable approaches, and the innovation rate of simulation is very likely to lead to the observation that one generation ideas, approaches, or methods may be completely overcome by the new ones.

Distinguished members are representatives of the society. They are expected to be the ambassadors for the societies in their presentations and publications. This includes an active support of the professional organization, not just the display of a plaque in the office. Distinguished ranks and awards are bestowed on members for their contributions, but they include the expectation that the member will continue such activities.

Being a representative and serving the society do not mean that shortcomings shall not be addressed. On the contrary, constructive criticism and academic dispute are driving forces in scientific development, but a professional will do this in a supporting way, being a mentor and leader for the society and the broader community. As a colleague of mine observed it well many years ago: If you are resting on your laurels, you are wearing them in the wrong place!

3.6 Summary

The *profession of M&S* requires honest and careful professionals who continuously learn about new research results to ensure reliable and credible simulation applications. As simulation applications are increasingly used for training, testing, and procurement decisions, mistakes or even slight variations outside the validity domain of applied models can have deadly consequences.

The following examples are given in Tolk (2012b, pp. 25–26):

- In April 2000, during a Marine Corps training mission in Arizona, the pilot of an MV-22 airplane dropped his speed to about 40 knots and experienced "vortex ring state" (VRS), a rotor stall that results in a loss of lift. Attempts to recover worsened the situation and the aircraft crashed, killing everybody on board. The pilot had 100 h in the Osprey simulator and nearly 3800 h of total flight time. However, none of his training or experience involved coping with a vortex ring. In January 2001, the General Accounting Office, in a presentation to the V-22 Blue Ribbon Panel, attested that the flight simulator used to train the soldiers in handling this aircraft did not replicate the VRS loss of a controlled flight regime.
- In March 2003, during Operation Iraqi Freedom, Patriot missiles shot down two allied aircraft and targeted another. The pilot and co-pilot

aboard the British Tornado GR4 aircraft died when the aircraft was shot down by a US Patriot missile. Another Patriot missile may have downed a US Navy F/A-18 C Hornet that was flying a mission over Central Iraq. Both cases were investigated and one cause of these failures stemmed from using an invalid simulation to stimulate the Patriot's fire control system during its testing.

While both examples come from the defense domain, their implications for other application domains are obvious: We do not want commercial airplane pilots to get bad training, medical simulation needs to be accurate, financial models that give wrong advice can lead to critical losses, environmental model can lead to severe environmental damage, and so forth. Simulation is a powerful tool, as it can provide insight into the dynamic behavior of complex systems, can help creating immersive virtual environments for training and education, and deliver valuable decision support. Simulation contributed significantly to new scientific discoveries, such as the Higgs boson particle (Atlas Collaboration, 2012). Overall, simulation is an established tool in many application domains. Trust and credibility, however, will be driven by the professionalism of the simulationists.

The code of professional ethics, as discussed in the context of this chapter, is a pivotal contribution and a fundamental building block for the *Profession of Modeling and Simulation*. Students, scholars, and practitioners must not only be aware of it, but they must also fill it with life on a daily basis.

References

Atlas Collaboration (2012) Observation of a new particle in the search for the Standard Model Higgs boson with the ATLAS detector at the LHC. *Physics Letters B*, 716 (1), 1–29.

Chaitin, G.J. (1977) Algorithmic information theory. *IBM Journal of Research and Development*, 21, 350–359, 496.

Collins, H. (2002) The experimenter's regress as philosophical sociology. *Studies in History and Philosophy of Science*, 33, 149–156.

Diallo, S.Y., Padilla, J.J., Gore, R., Herencia-Zapana, H., and Tolk, A. (2014) Toward a formalism of modeling and simulation using model theory. *Complexity*, 19 (3), 56–63.

Fanelli, D. (2009) How many scientists fabricate and falsify research? a systematic review and meta-analysis of survey data. *PLoS One*, 4 (5), e5738.

Frankel, M.S. (1989) Professional codes: why, how, and with what impact? *Journal of Business Ethics*, 8 (2), 109–115.

Gelfert, A. (2013) Climate scepticism, epistemic dissonance, and the ethics of uncertainty. *Philosophy and Public Issues*, 3 (1), 167–208.

Gert, B. and Gert, J. (2016) The definition of morality, in *The Stanford Encyclopedia of Philosophy* (ed. E.N. Zalta), Spring 2016 edn, Metaphysics Research Lab. C Available at http://plato.stanford.edu/archives/spr2016/entries/morality-definition/ (accessed September 2016).

Guerra-Pujol, E. (2015) Legal liability for research fraud. *Statistical Journal of the IAOS.* 10.3233/SJI-160303 (http://papers.ssrn.com/sol3/papers.cfm?abstract_id=2669118).

Ladd, J. (1991) The quest for a Code of Professional Ethics: an intellectual and moral confusion, in *Ethical Issues in Engineering* (ed. D.G. Johnson), Prentice Hall, Englewood Cliffs, CA, pp. 130–136.

Lee, C.J., Sugimoto, C.R., Zhang, G., and Cronin, B. (2013) Bias in peer review. *Journal of the American Society for Information Science and Technology*, 64 (1), 2–17.

McLeod, J. (1983) Professional ethics and simulation. *Proceedings of the 15th Winter Simulation Conference*, Vol. 1, pp. 371–374.

Oberkampf, W.L., DeLand, S.M., Rutherford, B.M. Diegert, K.V., and Alvin, K.F. (2002) Error and uncertainty in modeling and simulation. *Reliability Engineering & System Safety*, 75 (3), 333–357.

Ören, T.I. (2000a) Responsibility, ethics, and simulation. *Transactions*, 17 (4), 165–170.

Ören, T.I. (2000b) Educating the simulationists: conception of curriculum for simulation education. *Proceedings of the 2000 Winter Simulation Conference*, pp. 1635–1644.

Ören, T.I. (2002) Rationale for a Code of Professional Ethics for simulationists. *Proceedings of the 2002 Summer Computer Simulation Conference*, pp. 428–433.

Ören, T.I., Elzas, M.S., Smit, I., and Birta, L.G. (2002) A Code of Professional Ethics for simulationists. *Proceedings of the 2002 Summer Computer Simulation Conference*, 434–435.

Sargent, R.G. (2013) Verification and validation of simulation models. *Journal of Simulation*, 7 (1), 12–24.

Tolk, A. (2012a) What are the characteristics of a scholar? *SCS Magazine*, 2, 54–58.

Tolk, A. (2012b) *Engineering Principles of Combat Modeling and Distributed Simulation*, John Wiley & Sons, Inc., Hoboken, NJ.

Tolk, A. (2017) Bias ex Silico – Observations on Simulationist's Regress. *Proceedings of the 2017 Spring Simulation Mutliconference, San Diego, CA: Society for Modeling and Simulation*, pp. 314–322.

Uhrmacher, A., Bailsford, S., Liu, J., Rabe, M., and Tolk, A. (2016) Panel – reproducible research in discrete event simulation – a must or rather a maybe? *Proceedings of the 2016 Winter Simulation Conference, Piscataway, NJ*, IEEE, pp. 1301–1315.

van de Poel, I. and Royakkers, L. (2011) *Ethics, Technology, and Engineering: An Introduction*, John Wiley & Sons, Inc., Hoboken, NJ.

Part II

Education

4

M&S as a Profession and an Academic Discipline: A Contemporary View

John A. Sokolowski and Roland R. Mielke

Old Dominion University, Norfolk, VA, USA

4.1 Introduction

The history of modeling and simulation spans hundreds of years and has been utilized in many disciplines such as engineering, healthcare, transportation, and defense. This chapter does not intend to cover the entire gamut of modeling and simulation development. Instead, it focuses on the evolution of this technology from the early 1980s to 2016. This period represents significant advances in modeling and simulation and is the time period when it emerged as both a profession and an academic discipline. Prior to this time frame it was mainly a tool to help solve problems or design systems. It was not formally studied and only a few jobs existed where a person's main responsibility was to develop models and simulation or new modeling techniques to support their organization.

This chapter provides background material on what factors enabled the growth of modeling and simulation during this period. It also points out the three developmental paths that represent the majority of the modeling and simulation space. Following these areas we describe the growth in two key areas: industry and academia and also indicate where we see the future of modeling and simulation is headed.

4.2 Background

4.2.1 Modeling and Simulation Enabling Factors

There are three factors that, in hindsight, were necessary for modeling and simulation to develop as it did over the three decades covered in this

The Profession of Modeling and Simulation: Discipline, Ethics, Education, Vocation, Societies, and Economics, First Edition. Andreas Tolk and Tuncer Ören.
© 2017 John Wiley & Sons, Inc. Published 2017 by John Wiley & Sons, Inc.

chapter. We say "in hindsight" because in the early 1980s when significant work in modeling and simulation technology development started, these factors were not really considered by the developers. However, without them it is not likely that modeling and simulation would have developed as it did. These three factors are (1) computer technology, (2) network technology, and (3) problem space complexity. We provide background on each of these areas.

To support using modeling and simulation to simulate the real world, physical environment required the ability to represent that environment in computer code and then to execute that code, often in real time, to render a realistic representation of it. This representation included not only the physical behavior of the entity but also the rendering of that entity in some type of virtual representation to observe its behavior. The computational complexity of these calculations required improved computer technology to carry out the necessary program execution. While mainframe computer technology existed in the 1980s, it only provided printouts of data from the simulation. There was no virtual rendering of the physical environment to observe the simulation from a human perspective. This paradigm changed as the personal computer (PC) began to evolve in this decade. It was during this time that color graphics were introduced to aid in realistic visualization. PC chip speed also continued to increase providing for the ability to represent the physical world in real time or near real time. These two technology developments paved the way for simulation's use in virtual environment representation, which carried modeling and simulation beyond its scientific and engineering uses to many application areas outside these domains.

The second factor enabling the growth of modeling and simulation was networking technology. To allow a simulation to be shared across multiple computers required an efficient way to transmit data among them. The United States Department of Defense's Advanced Research Projects Agency (DARPA) had developed ARPANET that allowed for this transfer of information. ARPANET spurred the development of the TCP/IP network protocol for information transmission, which was put in place in 1983. So the second factor was now in place for the growth of modeling and simulation to take place.

The third factor that spurred development was problem complexity. Analytical solutions to complex mathematical equations did not exist for many mathematical models. The only way to solve these equations was through numerical approximations, which consisted of multiple executions of the mathematical model using small time steps to represent the continuous behavior of the system. So engineers and scientists began to rely on the simulation of these systems by this method to represent their behavior over time.

4.2.2   Industry Growth

The growth of modeling and simulation from an industry perspective began in the late 1970s as a result of the U. S. Department of Defense's (DoD) investment in this area. They initially contracted to build various combat simulators that were essentially one of a kind. These simulators cost millions of dollars and had limited interaction with one another. They were mainly used for skills training for individuals or small crews. Their graphics were cartoonish and did not immerse the user in a virtual environment that resembled reality. In 1978 Captain Jack Thorpe, an Air Force officer with a Ph.D. in psychology, envisioned having multiple simulators linked together emulating the experience of being in the actual vehicle in simulated combat conditions. However, it was not until 1983 that he was able to garner the support to begin designing and building such a system. This came about after he was assigned to DARPA. DARPA initiated the SIMNET program, which stood for SIMulator NETworking. The idea was to link together multiple training simulators to allow for more cost effective, large-scale training exercises not possible at this time.

The ability to develop SIMNET hinged on two technologies, networking and graphics display. At this point in time, ARPANET had matured to a point that its findings could form the basis for solving the networking portion of SIMNET. However, existing simulators and graphics engines cost more than the real systems they were meant to portray, so new technology had to be developed to make SIMNET affordable since the goal was to network several hundreds of these systems together at one time. Three companies were contracted to build this simulation system: BBN, Perceptronics, and Delta Graphics. BBN was involved with ARPANET so it was responsible for the networking design. Perceptronics was to design and build the physical simulator. And Delta Graphics, as its name implies, worked on the graphics engine design. The first two simulators were delivered to Ft. Knox in 1986 and SIMNET supported its first large-scale exercise in 1990.

BBN developed a networking protocol called Distributed Interactive Simulation (DIS) and its basic data unit protocol data unit (PDU). It became the main protocol to link simulators together throughout the 1990s. This protocol is still in use today along with high-level architecture (HLA) protocol, developed in the early 2000s.

SIMNET is known as a virtual simulation, that is, real humans operating simulated equipment. However, SIMNET also stimulated the development of constructive simulations, which encompass simulated systems and simulated humans. The simulated humans were in the form of semi-automated forces. These simulated forces had some artificial intelligence embedded so they could operate, in part, in an autonomous manner and did not need to be controlled by a human operator for part of their time. The funding and development of multiple constructive simulations to replicate mid to large war games spurred the development of several other companies. Some large defense contractors like General Dynamics, Northrop Grumman, and Lockheed Martin that were previously engaged in physical system development began to branch into simulation system development. Their role was mainly to support large-scale simulated exercise support and to integrate many standalone simulation systems. Smaller companies grew up around supporting both the virtual and constructive simulation efforts in DoD. The smaller companies specialized in a few aspects of simulation development such as automated forces, terrain generation, and protocol development. While too numerous to list all of them, a small sampling includes BMH, Sonalysts, MYMIC, and Aegis Technologies.

The need for modeling and simulation outside DoD also became apparent. One such area is medical and healthcare. In these areas it has been realized that the training capability can be enhanced through the use of simulation for many of the same reasons that DoD saw fit to use simulation. The medical community saw cost savings and greater student throughput via simulation. They did not have to pay for human subjects, known as standardized patients, and they could have students simultaneously examine their own simulated patient instead of having to take turns with a live patient. They could also practice different procedures that were either too risky to perform on live patients or had no live patients available with the requisite ailment. The same basic simulation technologies benefited this community. Networking is used to control multiple simulated patients at one time. High fidelity graphics are able to capture detailed anatomy to realistically represent all aspects of human disease to the point that the simulated system is almost indistinguishable from the real system. Many companies have been started in the medical and healthcare areas because the demand for these simulators is increasing and the amount of money available in the healthcare arena throughout the world exceeds that of defense budgets.

As an indicator of commercial growth in modeling and simulation, one has to look only at the number of job openings in this field. As of September 2016, there are about four thousand advertised jobs for modeling and simulation engineers. These jobs are in many different industries in addition to the ones already mentioned. This statistic reflects the growth of the modeling and simulation profession over the three decades covered by this chapter. One can see the impact that this technology has had on all aspects of life.

4.3 Academic Program Development and Evolution

4.3.1 Introduction

The development of contemporary, university level academic programs in modeling and simulation has evolved over the past thirty years along three relatively separate paths and has resulted in three different program types: (1) Modeling and Simulation (M&S) programs; (2) Computational Science and Engineering (CSE) programs; and (3) Computational Media (CM) programs. While each of the programs has a unique focus, all three programs address content generally considered to be modeling and simulation related.

M&S programs were developed primarily in response to the needs of the U.S. Department of Defense and the Manufacturing Industry. The Department of Defense uses modeling and simulation to develop and evaluate doctrine, evaluate the effectiveness of proposed new systems, and to more efficiently conduct training. The manufacturing industry uses modeling and simulation to design and develop new manufacturing processes and systems, to monitor and control the manufacturing supply chain and production facilities, and to evaluate the business case for investing in new production methods and technologies. Both of these sectors require modeling and simulation professionals having the capabilities to conduct extensive modeling and simulation studies and then make recommendations for decisions based upon the results of these studies.

CSE programs were developed primarily in response to the needs of the scientific research community and the engineering design community. Computation now is regarded as an equal and indispensable partner, along with theory and experimentation, in the advancement of scientific knowledge and engineering practice. Numerical simulation enables the study of complex systems and natural phenomena that are impracticable, or even impossible, to study by direct experimentation. The quest for ever higher levels of resolution and fidelity in such simulations requires enormous computational capacity and has provided the impetus for dramatic breakthroughs in computer algorithms and architectures. Thus, computational scientists and engineers must be capable of developing problem-solving methodologies and robust tools for the solution of very large and very complex scientific and engineering problems.

CM programs were developed primarily in response to the needs of video game studios, interactive media firms, and computer software development companies. Computational media includes the design and development of computer games for entertainment and education, the production of videos and movies requiring computer-generated special effects, and the design and implementation of user interfaces for software applications and computer-controlled products. Desired skill sets often include development of computer algorithms and programs, knowledge of human factors engineering, proficiency in the creation and use of digital media and art, and the capability to design and develop interactive, immersive virtual environments.

The purpose of this section is to explore the objectives and content of these three academic program types and then to compare and contrast the program curricula. This is done by investigating an academic program representative of each program type. We have selected Old Dominion University (ODU) as the sample M&S program, Harvard University as the representative CSE program, and Georgia Tech as the representative CM program. We conclude this section by offering some thoughts and observations about the potential future evolution of these programs.

4.3.2 Modeling and Simulation Programs

M&S refers to the use of models and simulations to develop data as a basis for managerial or technical decision-making (US Department of Defense, 2007, 2009). A model is a physical, mathematical, or otherwise logical representation of a system, entity, phenomenon, or process; the target of a simulation study is often called the simuland. A simulation is a method for executing a model. Historically, M&S meant using physical models operated in a real or modeled environment. Today, M&S most often refers to computer simulation. Mathematical or logical models are transformed to computer code and then executed on a digital computer to produce data. If it can be shown that the computer code accurately represents the model (the process of verification) and that the simulation output data are "close enough" to the simuland output data (the process of validation), then the simulation output data can be used in place of the actual simuland output data to reach conclusions concerning the simuland. Thus, the M&S process is useful because it facilitates evaluation of simuland performance without having to actually experiment with the simuland. This is especially important when the simuland exists only in concept or when experimenting with the simuland would be too difficult, too dangerous, or too costly.

It is often argued that the Department of Defense definition of M&S is insufficiently broad because, while it seems to address the application of M&S for experimentation, it appears to ignore the use of M&S for training and education. However, M&S is extremely important in training and education, especially where the objective is the training of skills or where the objective is training to conduct an assessment of alternatives followed by decision-making. In such cases, M&S can be used to immerse the student in a virtual representation of the real context in which the desired skills or assessment/decision-making must be exercised. The use of this approach, called virtual simulation, facilitates hands-on training and learning without having to utilize the real-world simuland, and doing so would be too expensive, too dangerous, or perhaps even impossible. Thus, it is important to recognize that the application of M&S for providing an experience is as important as using M&S for conducting experimentation.

The term M&S professional refers to individuals who have the knowledge and skills needed to develop models and simulations utilized to conduct experiments or create experiences with a simuland. Development of models requires knowledge of mathematics and subject matter expertise concerning the simuland. Thus, the ability to work cooperatively with subject matter experts is also very important. Transformation of a conceptual model to simulation code, and then the execution of that code on a computer platform, requires knowledge of computer science and skills in computer programming. The use of simulation output data to produce simuland performance information requires knowledge of system analysis techniques, while the use of graphs and images, perhaps animated, to aid in understanding and interpreting the output information requires knowledge from computer graphics and visualization. Since many simulands contain uncertainty and variability, knowledge of probability and statistics for input data modeling and output data analysis is required. Since many simulands contain humans as essential system components, the knowledge to computationally model human behavior and human decision-making is often needed. In addition, since most simulations are made to be used by humans, knowledge of human factors and computer interface design are required. Thus, the M&S professional must be educated in an extremely broad set of knowledge components from a number of disciplines, including mathematics, science, computer science, and engineering. So it is clear that the area of M&S is multidisciplinary. However, it is also clear that the M&S professional must have the capability to integrate this multidisciplinary knowledge to produce solutions to modeling and simulation problems. This is increasingly important as the complexity of simulation problems continues to expand, the requirements for problem solution resolution and fidelity increase, and the sophistication of available computer hardware and software grows. This is the reason why many universities now treat M&S as a separate and unique academic discipline. The capability to integrate M&S knowledge components from the several disciplines into M&S problem solutions for diverse problem domains is essential to the M&S discipline.

Historically, M&S has been viewed as an important research tool in numerous disciplines or application domains. Research in most domains often proceeds through a sequence of phases that include understanding, prediction, and control. In the initial phase, we are interested in understanding how events or objects are related. An understanding of relationships among objects or events then allows us to begin making predictions and ultimately to identify causal mechanisms. Finally, knowledge of causality enables us to exert control over events and objects. Research moves from basic to more applied levels as we progress through these phases. M&S is closely linked to all of these phases. At the basic levels, research is guided heavily by theory. Models are often used to represent specific instances of theories, discriminate among competing theories, or evaluate underlying assumptions. Likewise, simulations are used to test predictions under a variety of conditions or to validate theories against actual conditions. At the applied levels, simulations are also used to control events and objects. One of the primary uses for simulation is training where the goal is to control performance variability by improving operator reliability. Simulations in the form of mock-ups or prototypes are used in the creation of products and systems to validate predictions regarding operational requirements, specifications, and user/customer satisfaction.

Beginning in the mid 1990s, a second type of M&S professional began to emerge. Motivated by the rapidly growing use of simulation for training, analysis, and decision support by industry and government, these individuals are more interested in learning about M&S rather than just using M&S to study something else. Coming from backgrounds in mathematics, computer science, and engineering, these individuals are interested in the fundamental principles and theoretical foundations of M&S. They are anxious to investigate some of the major challenges of M&S: multiscale and multiresolution M&S, interoperability of simulations, composability of models, verification and validation, distributed and real-time simulation, and representation of increasingly complex and data-intensive system problems. In short, this group views M&S as a discipline. Their objective is to obtain a formal education in the M&S discipline and then to find employment opportunities as M&S scientists and engineers.

The growth of the view of M&S as a discipline is well documented in the literature. Since the late 1990s, a number of papers have been written stating the importance and urgency for developing educational programs in the discipline of modeling and simulation. These papers identify desirable program outcomes (Rogers, 1997), present suggestions for course and curriculum content (Oren, 2005; Petty, 2006), and describe potential approaches for, and challenges in, implementing a modeling and simulation program (Sarhoughian and Zeigler, 2000; Nance, 2000; Nance and Balci, 2001; Roberts and Ghosh, 2004). More recently, curricula (Szczerbicka *et al.*, 2000; Sarhoughian *et al.*, 2004) and models (Mielke *et al.*, 2009) for graduate modeling and simulation programs have been described. Graduate modeling and simulation programs have been started at several universities including the University of Alabama – Huntsville (2011), Arizona State University (2011), California State University – Chico (2011), Georgia Institute of Technology (2011), Old Dominion University (2011), and the University of Central Florida (2011). At the undergraduate level, several universities have developed tracks or concentrations focusing on narrow subareas of modeling and simulation as part of other degree programs. However, to date, only Old Dominion University has established an ABET accredited engineering program in modeling and simulation at the undergraduate level.

M&S Academic Programs at Old Dominion University

Old Dominion University has been one of the early leaders in developing M&S academic programs. Master's degree programs were started in 1998 and a doctoral degree program was established in 2000. Then in 2010, an undergraduate degree program was initiated. This program awards the Bachelor of Science Degree in Modeling and Simulation Engineering (BS-M&SE). Simultaneously, the Department of Modeling, Simulation and Visualization Engineering (MSVE) was established within the College of Engineering and Technology to administer all M&S academic programs. The first BS-M&SE degrees were awarded in May 2013 and the initial ABET accreditation visit for the program occurred in fall 2014. The establishment of an M&S academic department and the development of an undergraduate M&S program have been described in the literature. The department organization and its contribution to the overall mission of the university are described in Mielke *et al.* (2011). The initial planning for the undergraduate program is described in Leathrum and Mielke (2011), curriculum development is presented in Leathrum and Mielke (2012), and preparation for the initial ABET accreditation review is presented in McKenzie (2015).

Program Organization In designing M&S programs, we acknowledge that there are two primary student constituencies that must be served, the user constituency and the developer constituency. The user constituency consists of students who wish to utilize M&S as a tool to investigate another discipline. They need to know enough about M&S to select the best methodologies for their specific problem and then to apply these methodologies in an appropriate way. The developer constituency consists of students who wish to study M&S as a discipline. Their focus is to learn about the technical details of M&S and then to develop new M&S methodologies and to enhance existing M&S technologies. Upon graduation, these students are likely to seek employment in the M&S professional community as scientists, engineers, technical managers, and teachers.

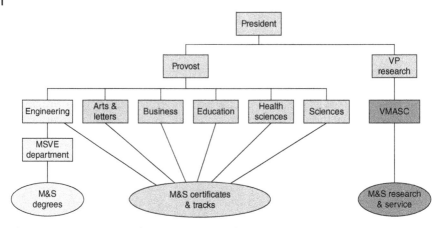

Figure 4.1 Organization of M&S programs and activities.

The view of M&S as a discipline leads naturally to the traditional department model for the developer constituency. These students first develop background in mathematics, computer science, and selected engineering topics, and then focus their core technical studies on the M&S Body of Knowledge (Petty, 2009). They emerge as technical specialists skilled in the design, development, and use of simulation technologies and methods. However, this educational path may not be attractive to the user constituency because of the extensive technical background requirements and the absence of an opportunity to focus deeply on a particular domain area. Thus, it is difficult to serve the needs of all potential M&S students with a "one-size-fits-all" approach. A multifaceted approach that addresses the needs and requirements of both student types is highly desirable.

The organizational structure of the M&S academic and research programs is shown in Figure 4.1. There are three important components to this structure. An academic department, the Department of Modeling, Simulation and Visualization Engineering, has been established within the College of Engineering and Technology. The MSVE Department administers academic degree programs in M&S at the bachelor's, master's, and doctoral levels, designed to support the needs of the M&S developer student constituency. In addition, the MSVE Department offers an undergraduate minor in M&S, and provides core courses for the graduate certificate programs, both designed to support the needs of the M&S user student constituency. All academic colleges offer graduate certificate programs. The graduate certificate programs consist of clusters of graduate courses designed to enhance the capability to utilize M&S as a tool in other disciplines and result in the award of a certificate of completion. Finally, the university has established a research center focused on M&S research activities. This center, called the Virginia Modeling, Analysis and Simulation Center (VMASC), is administered at the university level through the Office of Research to encourage participation by all academic colleges. It is a place where faculty and students from all disciplines can interact and work on cross-disciplinary projects.

Two university committees, the M&S Steering Committee and the M&S Executive Committee, have been established by the Provost's Office. The M&S Steering Committee consists of M&S faculty representing all six academic colleges. This committee is responsible for recommending policy and procedure and for operational issues spanning all M&S programs. The M&S Executive Committee consists of the dean or associate dean from each academic college. This committee is responsible for approving policy and procedure spanning all M&S academic programs. Together, these committees oversee and coordinate the cross-disciplinary activities for the M&S academic programs.

Undergraduate Modeling and Simulation Engineering (M&SE) Program The undergraduate M&SE program is designed to address four sources of program content requirements and goals: the ABET criteria for accrediting engineering programs (ABET Engineering Accreditation Commission, 2013), the literature defining an M&S body of knowledge (Petty, 2009), a set of discipline-specific student outcomes identified by program faculty, and university general education requirements. Discussion concerning the ABET criteria and the university general education requirements can be found in Leathrum and Mielke (2011), where the design of the curriculum is presented. The M&S Body of Knowledge and the discipline-specific student outcomes are briefly discussed here to help set the curriculum in context, though a more complete discussion is found in Leathrum and Mielke (2011).

The M&SE program prepares engineering graduates who can utilize modeling and simulation in various domains and for different applications, and who possess the foundation upon which to expand the current M&S Body of Knowledge. The M&SE program faculty has defined a set of essential knowledge and skills that they believe form the technical foundation for the discipline of modeling and simulation engineering. These are the concepts, principles, and methods that anchor the M&SE curriculum; they represent the fundamentals that every M&SE graduate must know and be able to use. The M&S essential knowledge and skills are stated as a set of student outcomes that are focused on the technical components of the M&S curriculum.

M&SE students who qualify for graduation will have the following:

1) An ability to communicate designs across technical and nontechnical boundaries.

2) An ability to model a variety of systems from different domains.
3) An ability to develop an input data model based on observed data.
4) An ability to select and apply appropriate simulation techniques and tools.
5) An ability to develop simulations in software.
6) An ability to apply the experimental process to acquire desired simulation results.
7) An ability to apply visualization techniques to support the simulation process.
8) An ability to use appropriate techniques to verify and validate models and simulations.
9) An ability to analyze simulation results to reach an appropriate conclusion.

The M&SE curriculum is first and foremost an engineering program having a focus on problem solving, design, and experimentation. The curriculum is designed with 127 credits; 32 credits of mathematics and basic science courses, 57 credits of engineering science and design courses, 32 credits of general education courses, and 6 credits of approved electives. The credit distribution is selected to satisfy ABET Criterion 5 (ABET Engineering Accreditation Commission, 2013) and the university's general education requirements. The curriculum is displayed in "showcase" format in Figure 4.2. In this display, the courses are distributed over eight semesters and the courses are sequenced to satisfy all prerequisite and corequisite requirements. The M&SE core technical courses are shown in Figure 4.3. In this display, the primary content components for each core course are shown.

Graduate Modeling and Simulation Programs The M&S graduate programs are administered through the MSVE Department. The MSVE Department offers programs of study leading to the degrees: Master of Engineering (ME) in M&S, Master of Science (MS) in M&S, Doctor of Engineering (DEng) in M&S, and Doctor of Philosophy (Ph.D.) in M&S. The ME and DEng programs are directed primarily at part-time students employed full-time in M&S industry who are seeking a more solid foundation in the discipline and/or preparing for technical leadership positions. The MS and Ph.D. programs are directed primarily at full-time students who are preparing for a career in advanced M&S research and/or academic positions.

The ME program is available only as a non-thesis option and is designed around a strong set of core courses addressing the foundation of the M&S Body of Knowledge. This program is available live and asynchronously

Course Number	Course title	Credits
Freshman year – First semester		
MATH 211	Calculus I	4
ENGL 110C	English composition I	3
CHEM 121N	Chemistry I & lab	4
ENGN 110	Engineering & technology Intro.	2
Gen Ed	Oral communication	3
Freshman year – Second semester		
MATH 212	Calculus II	4
CHEM 123N	Chemistry II	3
CS 150	Programming I	4
PHYS 231N	University physics I	4
MSIM 111	Information literacy for M&SE	2
Sophomore year – First semester		
STAT 330	Probability and statistics	3
PHYS 232N	University physics II	4
CS 250	Programming II	4
CS 252	Introduction to UNIX	1
MSIM 201	Introduction to M&S	3
Sophomore year – Second semester		
MATH 307	Differential equations	3
ENGL 231C	Technical writing	3
Gen Ed	Human creativity	3
Gen Ed	Literature	3
MSIM 205	Discrete event simulation	3
MSIM 281	M&S laboratory 1	1
Junior year – First semester		
CS 330	Object-oriented prog. & design	3
CS 381	Discrete structures	3
MSIM 320	Continuous simulation	3
MSIM 382	M&S laboratory 2	1
Gen Ed	Human behavior	3
Gen Ed	Option D course 1	3
Junior year – Second semester		
MSIM 331	Simulation software design	3
MSIM 383	M&S laboratory 3	1
MSIM 451	Analysis for M&S	3
MSIM 410	System modeling	3
Gen Ed	Interpreting the past	3
Elect	Approved elective 1	3
Senior year – First semester		
MSIM 441	Computer graphics & visualization	3
MSIM 487	Capstone design I	4
Gen Ed	Option D course 2	3
ENMA 401	Project management	3
MSIM 4xx	Approved MSIM elective 1	3
Senior year – Second semester		
ENMA 480	Engineering ethics	3
MSIM 488	Capstone design II	3
Elect	Approved elective 2	3
MSIM 4yy	Approved MSIM elective 2	3
Gen Ed	Impact of technology	3

Figure 4.2 Showcase M&SE curriculum.

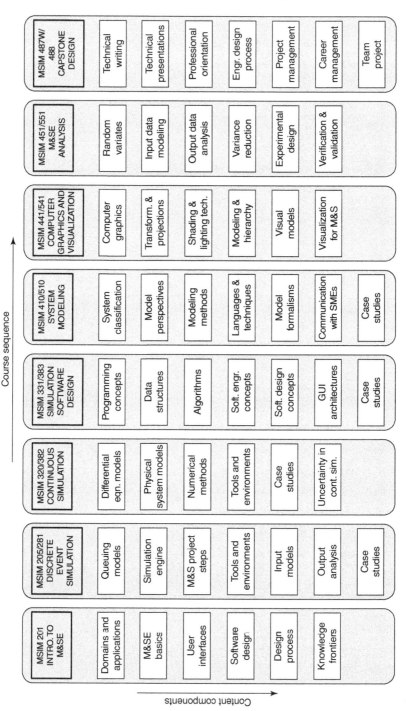

Figure 4.3 M&SE core courses with main content components.

through the Internet. The MS program is available only as a thesis option. The thesis research is designed to provide a research apprenticeship in which the candidate conducts guided research in an area of M&S. While some of the course work is available via asynchronous web delivery, students are expected to work with their supervising faculty during the completion of their thesis research. The DEng program is available only to M&S industry practitioners having at least 2 years of engineering experience. Candidates must have the cooperation and support of their employer. The program consists of a core of engineering management and leadership courses and advanced technical M&S courses. An applied project that demonstrates the candidate's ability to apply technical and managerial skills to the solution of a significant engineering problem must also be completed. Project activity requires periodic live interactions between the candidate and participating faculty and industry supervisors. The Ph.D. program focuses on developing the necessary skills and advanced knowledge to evaluate and conduct independent original research in an area of M&S. The goal of the program is to prepare students for careers in teaching and research in academic institutions, as well as the conduct or leadership of research and development in public and private organizations. The program requires the completion of four core courses and four elective courses selected to aid the dissertation research. The program also requires the successful completion of a progressive sequence of program examinations, including the Diagnostic Exam, the Qualifying Exam, the Dissertation Proposal, and the Dissertation Defense. Once again, some of the course work is available via asynchronous web delivery, but students are expected to be present on campus to work with their supervising faculty during phases of the dissertation research.

At present, the majority of students seeking to enter an M&S master's program have undergraduate degrees in something other than M&S. Entering students generally have academic backgrounds from mathematics, computer science, one of the natural sciences, or an engineering discipline. Thus, they have a good foundation in math and usually in computer science, but have a limited or perhaps no background in M&S. As a result, virtually all M&S master's programs develop master's curricula that include appropriate introductory M&S courses. At Old Dominion University, there is also a student cohort that has already completed an undergraduate degree in M&SE. Thus, there is the additional challenge to design a master's curriculum that accommodates students without previous M&S experience yet challenges all students.

The graduate MS/ME curricula at ODU are shown in flowchart form in Figure 4.4. Students who enter without an M&S background are required to begin their studies with a set of four introductory courses: model

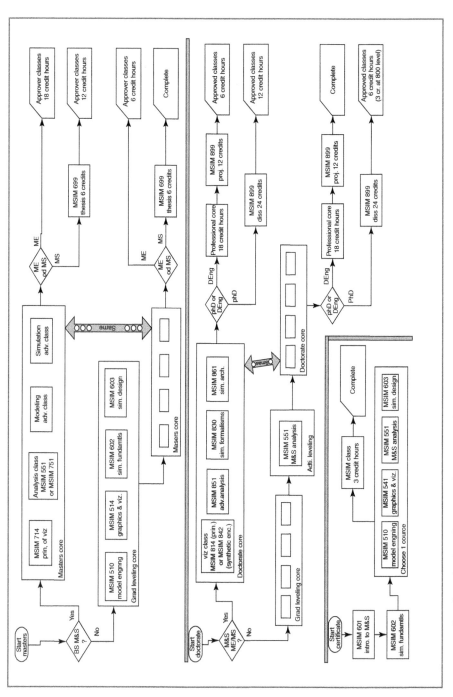

Figure 4.4 Flowchart display of ODU graduate curricula.

engineering; computer graphics and visualization; simulation fundamentals; and simulation software design. These courses cover modeling and simulation fundamentals sufficiently so that these students can then assimilate with the master's students who enter the program with an undergraduate M&SE degree. Both groups of students then complete the set of four master's core courses: principles of visualization; M&S analysis; an advanced modeling class; and an advanced simulation course. The remainder of the program consists of approved M&S elective courses and, for MS students, the master's thesis.

The graduate doctoral curricula are also shown in flowchart form in Figure 4.4. Entering students must have either completed an MS/ME degree in M&S or must complete a set of nine leveling courses, the four introductory courses, a course in M&S analysis, and the four master's core courses. All students then complete the set of four doctoral core courses: synthetic environments; advanced M&S analysis; simulation formalisms; and simulation architectures. The remainder of the program for DEng students consists of 18 credits of professional engineering core courses and 12 credits of DEng project. The remainder of the program for Ph.D. students consists of approved M&S elective courses and 24 credits of Ph.D. dissertation.

4.3.3 Computational Science and Engineering Programs

Computational Science and Engineering (CSE) refers to a multidisciplinary area, with connections to the sciences, engineering, mathematics, and computer science, that focuses on the development of problem-solving methodologies and robust tools for the solution of scientific and engineering problems (SIAM Working Group on CSE Education, 2016). It differs from its constituent disciplines because CSE focuses on the integration of knowledge elements from all of these constituent disciplines and as a result is distinct from any of them. Utilizing CSE knowledge to move from application area to computational results requires domain expertise, mathematical modeling, simulation methodologies, numerical methods, algorithm development, software implementation, program execution, verification and validation, data analysis, and visualization. Thus, it is clear that CSE is far more involved than simply using a commercial simulation tool to generate output data.

Another slightly different definition of computational science and engineering states that CSE is the field of study that exploits the power of computation as an approach to major challenges on the frontiers of natural and social sciences and all engineering fields (Harvard John A. Paulson School of Engineering and Applied Sciences, 2016). This definition reinforces the notion that CSE primarily is focused on the very large and complex problems that challenge our current capabilities to provide adequate computational power and to manage massive amounts of data. Many of the "Grand Challenge" problems enumerated by the National Academy of Engineering (National Academy of Engineering (NAE, 2016)) that focus on some of the most pressing concerns from the natural, physical, and social sciences, provide examples of these problems. The problems include weather and climate prediction, the engineering of new materials, combustion of fossil fuels and carbon sequestration, engineering the human brain, and the development of new medicines.

A limited number of CSE academic programs began to emerge during the late 1980s and the 1990s. These early programs were initiated mainly at universities that possessed strong science units and that were early to recognize the significance of emerging personal computer and supercomputer technologies to their research. However, it was not until the 2000s and 2010s that significantly more universities started CSE programs. Motivation for initiating these programs was the result of several national level studies and subsequent increases in government research funding for CSE research. In 2005, the President's Information Technology Advisory Committee (PITAC) Report (Belytschko, 2005) concluded that "Universities must implement new multidisciplinary programs and organizations that provide rigorous multifaceted education for the growing ranks of computational scientists the nation will need to remain at the forefront of scientific discovery." This document quickly was followed by another report in 2006 from the NSF Blue Ribbon Panel on Simulation-Based Engineering Science (SBES) (Simulation-Based Engineering Science, 2006). This report states that "seldom have so many independent studies by experts from diverse perspectives been in such agreement: computer simulation has and will continue to have an enormous impact on all areas of engineering, scientific discovery, and endeavors to solve major societal problems." Regarding education in computer simulation, the report states: "The old silo structure of educational institutions has become an antiquated liability. It discourages innovation, limits the critically important exchange of knowledge between core disciplines, and discourages the interdisciplinary research, study, and interaction critical to advances in SBES." Today, SIAM (SIAM Working Group on CSE Education, 2016) identifies at least 36 US universities that have started CSE graduate programs or tracks, and at least six US universities that have initiated CSE undergraduate programs. Internationally, many additional universities offer CSE programs.

While there are many similarities between M&S academic programs and CSE academic programs, there also are some significant differences. The focus of both programs is to use mathematical models and computer simulations to analyze the performance of systems or simulands. However, in CSE, the simulands of greatest interest are physics-based systems that usually are modeled as continuous-state, time-driven systems described by sets of differential equations. In M&S, the focus is on a much broader collection of simulands that include static stochastic systems, discrete-state event-driven systems, as well as continuous-state, time-driven systems. Thus, most CSE programs focus on the continuous simulation paradigm while most M&S programs focus on all three simulation paradigms, continuous simulation, discrete event simulation, and Monte Carlo simulation. Another significant difference is that CSE programs generally focus on experimental applications of modeling and simulation while M&S focuses on both the experimental and the experiential applications of modeling and simulation. In this sense, M&S has a broader focus than CSE.

CSE programs generally focus on problems that challenge our capabilities to provide and utilize computing capacity and our capabilities to utilize large supporting data sets. The objective is to find ways to solve larger and more complex problems or to solve faster problems that require more resolution and fidelity. As a result, CSE academic programs typically require significant coursework in topics such as computer architectures, parallel processing, high performance computing, and strategies for working with massive data sets. The goal is to expand the boundaries of what can be simulated. M&S programs address these same issues but, in the authors' opinions, to a lesser degree. In M&S, there is greater interest in finding a modeling and simulation solution that meets requirements and that is as simple and as cost effective as possible. Said another way, CSE is more focused on expanding the boundaries of science while M&S is more focused on finding good engineering solutions to an expanding collection of problems and issues having societal importance and value.

CSE Programs at Harvard University

CSE programs at Harvard University are offered through the John A. Paulson School of Engineering and Applied Science (SEAS) (John A. Paulson School of Engineering and Applied Sciences, Harvard University, 2016). To foster interdisciplinary research, SEAS does not have a traditional academic department organizational structure and does not award degrees by specific research area. Instead, graduate students work toward a degree in one of six subject areas: Applied Mathematics; Applied Physics; Computer Science; Computational Science and Engineering; Design Engineering; and Engineering Sciences. Two CSE degrees are offered, the Master of Science (SM) and the Master of Engineering (ME). In addition, a Secondary Field in Computational Science and Engineering is offered to be used as part of other SEAS doctoral degree programs. The initial version of SEAS, the Lawrence

Scientific School, was established in 1847. The current name and organization for SEAS was adopted in 2007 and the master's programs in CSE were started in 2013.

Program Organization The Graduate Programs in Computational Science and Engineering at Harvard University emphasize the following three focus areas:

- Mathematical techniques for modeling and simulation of complex systems
- Parallel programming and collaborative software development
- Efficient methods for organizing, exploring, visualizing, processing, and analyzing very large data sets

A single set of student learning outcomes is common across the CSE degrees. These student learning outcomes are presented in the following. Students who graduate from the program will have the ability to

- frame a real-world problem such that it can be addressed computationally;
- evaluate multiple computational approaches and choose the most appropriate one;
- produce a computational solution that can be used by others;
- communicate across disciplines;
- collaborate within teams;
- model systems appropriately with consideration of efficiency, cost, and available data;
- use computation for reproducible data analysis;
- leverage parallel and distributed computing;
- build software and computational artifacts that are robust, reliable, and maintainable; and
- enable a breakthrough in a domain of inquiry.

The CSE graduate programs are organized around four course groupings: CSE Core Courses; Applied Math Electives; Computer Science Electives; and Domain Electives. Degree requirements for each CSE program are stated by specifying the number of courses that must be selected from each course grouping. The course groupings are described in the following.

The CSE Core Course group consists of two courses from Applied Math and two courses from Computer Science:

- Applied Math
  - AM 205 – Advanced Scientific Computing: Numerical Methods
  - AM 207 – Stochastic Methods for Data Analysis, Inference, and Optimization

- Computer Science
  - CS 205 – Computing Foundations for Computational Science
  - CS 207 – Systems Development for Computational Science

The Applied Math Electives group consists of courses from Applied Math (three courses), Applied Computation (two courses), and Statistics (nine courses). These courses focus on mathematical modeling, advanced optimization, advanced statistical techniques, and statistical machine learning.

The Computer Science Electives group consists of twelve courses that address computational aspects of computer science. The list includes courses on data structures and algorithms, data systems, visualization, computer architectures, distributed computing, and machine learning.

The Domain Electives group consists of applied computation courses offered outside of Applied Math and Computer Science. Courses on this list are courses that apply the methods of computational science and engineering to a specific domain or class of problems. The courses on this list include

- AC 209 – Data Science
- AC 274 – Computational Fluid Dynamics
- AC 290 – Project-Based High Performance Distributed and Parallel Systems
- AC 297 – CSE Capstone Project
- AC 298 – CSE Interdisciplinary Seminar

Program Requirements In this section, the requirements for the three CSE degrees offered by SEAS are presented. The requirements are stated in terms of the number of courses, maximum and minimum, that must be selected from each of the four CSE course groupings.

The Master of Science Degree in CSE (SM-CSE) is designed to be a one-year program and requires the completion of eight courses. The program distribution is as follows.

- A minimum of three and a maximum of four courses from the CSE core course group
- A minimum of one and a maximum of four courses from the Applied Math Electives group
- A minimum of one and a maximum of four courses from the Computer Science Electives group
- A minimum of zero and a maximum of five courses from the Domain Electives group.

The final selection of courses requires advisor approval.

The Master of Engineering Degree in CSE (ME-CSE) is designed to be a two-year program and requires the completion of eight courses and a thesis.

The thesis requires most of the second year and is equivalent to eight courses. The course requirements are identical to the SM-CSE degree requirements. The thesis is expected to have a significant computational component in a domain that fits with the student's background, course selection, and interest.

The Secondary Field is available to any student enrolled in a Ph.D. program in the Graduate School of Arts and Sciences. The student's plan of study must be approved by the CSE Program Committee and by the student's home department. The Secondary Field requires the completion of four courses; the required course distribution is selected as follows.

- A minimum of two and a maximum of four courses from the CSE core course group
- A minimum of zero and a maximum of two courses from the Applied Math Electives group
- A minimum of zero and a maximum of two courses from the Computer Science Electives group
- A minimum of zero and a maximum of one course from the Domain Electives group.

4.3.4 Computational Media Programs

Computational Media (CM) degree programs at US universities are offered under a variety of titles: computational media; computational media design; interactive computing; digital + media; media arts and science; art, culture, and technology; arts and technology; and others. In addition to having different titles, these programs also have significant differences in the program content and objectives.

During the 1980s and 1990s, a number of computational media programs were initiated by computer science departments, computer engineering departments, or computer gaming departments in response to the needs of the emerging video game industry, interactive media firms, and computer software development companies. These academic programs are often structured as multidisciplinary programs and include partners from art departments having a digital art component or psychology departments having a human factors component. The focus of these programs includes design and development of computer games for entertainment and education, the production of videos and movies requiring computer-generated special effects, and the design and implementation of user interfaces for software applications and computer-controlled products. Required knowledge and skills include programming, computer architectures, computational algorithms, computer graphics and visualization, human factors engineering, digital media and art, and virtual environments. Since the 2000s, additional focus areas, including artificial intelligence, machine learning, information assurance, cyber security, and the analysis and utilization of massively large data sets have been added to these programs at many universities.

During the late 2000s and early 2010s, another type of computational media program began to emerge. These programs, often administered from arts and humanities colleges, are interdisciplinary programs that include partners from the sciences, the social sciences, engineering, and especially computer science. For these programs, computational media refers to an inter-disciplinary field that is concerned with computation as a medium for creative expression and communication (University of California – Santa Cruz News-center, 2014). Computational media combines the theories and research approaches of the arts and humanities with the processes and technologies of computer science to analyze, explore, and enable the computer as a medium for creative expression. Examples of instructional activities and research efforts in these emerging programs include the following.

- Participatory culture studies that explore the role of information and communication technologies in the shift from a top-down culture to a culture of participation and social engagement. An important component of this work concerns the emergence of social media and the role of social media in promoting greater participation in culture, social activism, and politics.
- Performative technology studies that explore new methods for combining media and technology to create new visual, aural, and connective material of performance. A component of this work concerns the fusing of digital media, communication networks, and interactive systems with lighting and sound design, performer movement, and stage and set design to create real-time shared multimedia experiences.
- Playable media studies explore the use of computer games and the creation of new media forms that invite and structure play. A component of this work combines game design and artificial intelligence with writing, art, and media to engage audiences, tell stories, and shape social space.

A recent report, based on a workshop supported in part by the National Science Foundation, the National Endowment for the Humanities, and the National Endowment for the Arts, presents in more detail the development of a vision for the future of the computational media discipline (Wardrip-Fruin and Mateas, 2014).

CM Programs at Georgia Tech

Georgia Tech established the College of Computing as one of its five (now six) academic colleges in 1988 (Georgia Institute of Technology, 2016a).

This college evolved from the School of Information and Computer Science and was the first college of computing in the United States. In 2007, the college reorganized with two main units, the School of Computer Science and the School of Interactive Computing. In 2010, a third unit, the School of Computational Science and Engineering, was added. The Bachelor's Degree in Computational Media, and several closely related graduate programs, are administered through the College of Computing.

Undergraduate Program Organization   The undergraduate degree in Computational Media is a joint offering between the College of Computing and the School of Literature, Media, and Communication (LMC) within the Ivan Allen College of Liberal Arts. The program was developed in 2005 in recognition of computing's significant role in communication and expression. It now is one of the fastest growing undergraduate programs at Georgia Tech. The program objectives for the Computational Media program are stated in the following (Georgia Institute of Technology, 2016b).

Computational Media graduates, within 5 years of graduation, will impact the local, national, and global communities by

- designing and implementing compelling digital artifacts for business, education, the public sector and entertainment;
- contributing to the development of new genres and forms of media based on a knowledge of the cultural significance as well as the computational affordances of digital media;
- having the flexibility to adapt to ongoing changes in the field of digital media over their careers; and
- communicating complex ideas and concepts through computing mediums in a multitude of diverse environments.

The Computational Media program follows the ABET accreditation guidelines for Computer Science programs. These guidelines state that CM students demonstrate proficiency in nine required ABET student learning outcomes; the CM program has added four additional program-specific student learning outcomes. These student learning outcomes are stated in the following.

Students who successfully complete the CM program will demonstrate the following:

a) An ability to apply knowledge of computing and mathematics appropriate to the discipline
b) An ability to analyze a problem, and identify and define the computing requirements appropriate to its solution

c) An ability to design, implement, and evaluate a computer-based system, process, component, or program to meet desired needs

d) An ability to function effectively on teams to accomplish a common goal

e) An understanding of professional, ethical, legal, security and social issues and responsibilities

f) An ability to communicate effectively with a range of audiences

g) An ability to analyze the local and global impact of computing on individuals, organizations, and society

h) Recognition of the need for and an ability to engage in continuing professional development

i) An ability to use current techniques, skills, and tools necessary for computing practice

(CM 1) Understand and apply the principles and affordances of computing for creative expression

(CM 2) Understand the historical and cultural forces that have led to the rise of digital media

(CM 3) Create digital artifacts with an awareness of the history, audience, and context

(CM 4) Appreciate and evaluate future trends in the development of digital media

Undergraduate Program Requirements Computational Media students have considerable flexibility in selecting the focus of their undergraduate curriculum within the "thread" system utilized at Georgia Tech. Each CM student must complete the course work for one of the three possible College of Computing threads: Intelligence; Media; or People. In addition, each student must complete the course work for one of the four possible LMC threads: Film, Performance and Media; Game Studies; Interaction Design and Experimental Media; or Narrative Studies. All choices of CS and LMC threads lead to a 122 credit program having the following high-level course distribution:

- Core Required Courses   46 credits
- CS Required Courses   31–40 credits
- LMC Required Courses   30 credits
- Required Capstone Courses   6 credits
- Free Elective Courses   0–9 credits

An overview of the courses contained in each course distribution area is presented in the following. CS required courses vary slightly depending upon the CS thread that is selected; the listing presented here is for the Media Thread. Similarly, the LMC required courses vary slightly depending upon the LMC thread selected; the listing presented here is for the Narrative Studies Thread.

Core Required Courses – Four mathematics courses (linear algebra, differential calculus, integral calculus, and multivariable calculus), two science courses with laboratories, two English writing courses, and seven courses to satisfy the university general education program requirements (three social science courses, two humanities courses, one government course, one physical education course).

CS Required Courses – Introduction to Computing, Object-Oriented Programming, Data Structures and Algorithms, Discrete Math for CS, Media Device Architectures, Objects and Design, Computing and Society, Computer Graphics, and two courses from a set of courses on media technology (Video Game Design, Introduction to Visualization, Computational Journalism, Computational Photography, Digital Video Special Effects, Computer Animation, Advanced Algorithms).

LMC Required Courses – Introduction to Computational Media, Studies in Fiction, Interactive Narrative, two courses from a set of courses on design or communication (Constructing the Moving Image, Communication and Culture, Principles of Interactive Design), and five courses from a set of courses on computational media or literature.

Students who complete this CM program option will have knowledge of the following:

- The principles and technologies for software that acts as an interface between humans and reactive virtual environments that may include information, shapes, animations, simulations, sounds, and shared spaces
- Performance acceleration techniques for the acquisition, processing, transfer, and rendering of the various media
- Representation schemes for various media and the corresponding data structures and algorithms

In addition, students will be able to

- design, implement, and test environments where humans interact with 3D models and/or audio;
- design and implement data structures for these media environments;
- design and implement architectures controlling the interface between hardware and software in media devices;
- build discrete element simulations; and
- describe the impact of presentation and user interaction on exploration using rich media

Master's Program in Human–Computer Interaction Human–computer interaction (HCI) is the study of how humans interact with computers.

It focuses on activities involved in the design, development, and evaluation of computer systems, with a goal of understanding how computers and technology affect people and society. At Georgia Tech, the Master of Science degree program in Human–Computer Interaction (HCI) is a multi-disciplinary program offered collaboratively by four schools: Industrial Design; Interactive Computing; Literature, Media and Communication; and Psychology (Georgia Institute of Technology, 2016c). Students may enter the HCI program through any of the four participating units. The choice of unit usually reflects the student's academic background and the intended area of degree specialization. The degree requirements include completion of 30 credits of course work, six credits of a master's project, and a corporate summer internship following their first year of study.

The required course work includes nine credits of common core courses, 9–12 credits of courses in the specialization, and 9–12 credits in a broader set of elective courses. The common core courses consist of CS 6753 – HCI Professional Preparation and Practice (seminar), PSYC 6023 – Psychology Research Methods for HCI, and CS 8803 – Human Computer Interactions. The specialization courses vary by area of specialization; the courses for the Interactive Computing Specialization consist of three credits from a list of software development courses and six credits from a list of design, evaluation, and cognitive modeling courses. The elective courses may be selected from a wide variety of areas, including Architecture, Music Technology, Industrial and Systems Engineering, Human–Robot Interaction, Human Factors, Management of Technology, and Cognitive Science.

Doctoral Program in Human-Centered Computing Human-centered computing (HCC) refers to the interdisciplinary science of designing computational artifacts that support human endeavors (Georgia Institute of Technology, 2016d). The doctoral program in HCC at Georgia Tech brings together studies in three related areas: human–computer interaction; social computing; and cognition, learning, and creativity. The degree requirements include the completion of required coursework, a qualifying examination, a dissertation that starts with a dissertation proposal and ends with a dissertation defense, and service as a teaching assistant twice during the program of study.

The required coursework includes three core courses, two courses from an area of HCC specialization, and one course from a second area of specialization. In addition, three courses must be selected from a Ph.D. minor. The HCC core courses consist of three CS courses: CS 6451 – Introduction to Human-Centered Computing; CS 6452 – Prototyping Interactive Systems; and CS 7455 – Issues in Human-Centered Computing. The specialization courses are selected from sets of courses representing five subareas: Artificial Intelligence; Cognitive Science; Human–Computer Interaction; Learning Science and Technology; and Social Computing. The minor courses are selected from sets of courses, outside of the HCC program, representing at least six areas: History, Technology and Society; Industrial Design; Industrial and Systems Engineering; Literature, Media and Communication; Psychology; and Public Policy.

The qualifying exam is required of all HCC doctoral students. The purpose of the exam is for the student to demonstrate competency in: basic computing concepts and methods; written research communication; oral research communication; core HCC knowledge; knowledge in the selected HCC specializations; and design and evaluation of human-centered systems.

4.4   Future Program Evolution

The importance of modeling and simulation has grown dramatically over the past thirty years. Facilitated by advances in computer technology and computer networks, and motivated by the requirement to support the development of ever more complex problem solutions, modeling and simulation has emerged as one of the most exciting disciplines of the twenty-first century. Today, modeling and simulation is used by government and industry to test and evaluate new concepts, to design new systems and processes, and to manage the acquisition and distribution of materials and products. Modeling and simulation is used by scientists to investigate natural phenomena and man-made systems, to understand and control complex social, political and economic systems, and to expand our knowledge of the world around us. Modeling and simulation is even used to entertain and educate people of all ages. It is not surprising that now modeling and simulation is viewed as a professional discipline and academic programs have been established to teach and disseminate the knowledge and skills required to participate in that profession.

What is surprising is that the modeling and simulation profession, and the educational programs that were established to serve that profession, evolved along three somewhat disparate paths and seem to have retained that separation even as the profession has grown. Graduates from M&S programs, CSE programs, and CM programs each seem to be in high demand by potential employers. In fact, the demand seems to be far greater than the number of graduates being produced. Obtaining reliable quantitative data on demand is difficult because the North American Industry Classification System (NAICS) has no classification designation for modeling and simulation. Over 100 US universities have established one or more of the modeling and simulation academic programs described here, but these programs often are undersubscribed; there is capacity to enroll additional students utilizing existing faculties and facilities.

Our experience has shown that when middle school and high school students are introduced to modeling and simulation, they often become interested enough to consider entering the profession. This happens for two reasons. First, virtually all students utilize computers for many facets of daily life – schoolwork, entertainment, and social networking. Playing with computation today is like tinkering with physical things in years past. Students become very excited and motivated when presented with the opportunity to expand their computational worlds. Second, modeling and simulation seems to them to be a natural approach to investigate the formulation of solutions to many of the "grand challenge" problems. They are enthusiastic about working to improve the human situation. The problem is that too few students actually have the opportunity to be properly introduced to modeling and simulation. The modeling and simulation academic programs are relatively new programs; most were initiated within the past dozen years. The public often is not aware of the existence of these programs. Students do not understand the meaning of the words modeling and simulation or computational science and engineering. Compounding the problem is that there is no professional organization or society that represents collectively the modeling and simulation academic programs or the industries and organizations that utilize the graduates of these programs.

Future growth of the modeling and simulation enterprise would benefit from utilizing a team approach. Any single academic program type, M&S, CSE, or CM, presently lacks the critical mass and industrial following necessary to properly promote the modeling and simulation profession. However, if all of these programs act together, and are joined by their collective industrial supporters, we have the opportunity to send our message to a much larger portion of the population. Our message is that the modeling and simulation profession is an essential component of our country's leadership position in science and engineering, and a very exciting and rewarding career path for students having an interest in science, technology, engineering, and mathematics (STEM).

The importance of modeling and simulation is borne out by the increasing number of jobs advertised across the country. Companies are creating positions for modeling and simulation engineers because they realize that this capability can aid in addressing complex problems faced by their companies. These problems range from internal management efficiencies to product development and optimization of all types of business practices. While we have not seen a trend of dedicated modeling and simulation businesses spring up across the country, it is clear that companies and governments are hiring individuals with this expertise directly to leverage this capability in solving the problems already described. It is clear that the educational system will need to meet the demand for these professionals in the foreseeable future.

References

ABET Engineering Accreditation Commission (2013) *Criteria for Accrediting Engineering Programs, 2014–2015 Accreditation Cycle,* Available at http://www. abet.org (accessed May 23, 2013).

Arizona State University (2011) http://www.asu.engineeringonling.com (accessed May 25, 2011).

Belytschko, T. and Lazowska, E. (2005) Computational Science: Ensuring America's Competitiveness. *President's Information Technology Advisory Committee (PITAC) Report,* Available at http://vis.cs.brown.edu/docs/pdf/ Pitac-2005-CSE.pdf (accessed August 19, 2016).

California State University – Chico (2011) http://www.ecst.csuchico.edu/ ~mcleod/degree/html (accessed May 25, 2011).

Georgia Institute of Technology (2011) http://www.msrec.gatech.edu (accessed May 25, 2011).

Georgia Institute of Technology College of Computing (2016a) http://www.cc. gatech.edu (accessed September 4, 2016).

Georgia Institute of Technology College of Computing (2016b) http://www.cc. gatech.edu/academics/degree-programs/bachelors/computational-media/ objectives (accessed September 4, 2016).

Georgia Institute of Technology College of Computing (2016c) MS-HCI Website, http://www.cc.gatech.edu/master-science-human-computer-interaction (accessed September 7, 2016).

Georgia Institute of Technology College of Computing (2016d) PhD-HCC Website http://www.cc.gatech.edu/human-centered-computing-phd-program (accessed September 7, 2016).

Harvard John A. Paulson School of Engineering and Applied Sciences (2016) Graduate Programs https://www.seas.harvard.edu/academics/graduate (accessed July 7, 2016).

John A. Paulson School of Engineering and Applied Sciences, Harvard University (2016) Graduate Programs https://www.seas.harvard.edu/academics/graduate (accessed July 7, 2016).

Leathrum, J.F. Jr. and Mielke, R.R. (2011) A Bachelor of Science in Modeling and Simulation Engineering. *International Symposium on Engineering Education and Education Technologies, Orlando, FL,* July 19–22, 2011.

Leathrum, J.F. and Mielke, R.R. (2012) Outcome-based curriculum development for an undergraduate M&S program. *AutumnSim 2012, Conference on*

Education and Training Modeling and Simulation (ETMS'12), San Diego, CA, October 28–31, 2012.

McKenzie, F. (2015) Preparing for the EAC-ABET visit for a novel undergraduate program in Modeling and Simulation Engineering. *Proceedings of the 13th Latin American and Caribbean Engineering Institutions (LACCEI) Conference, Santo Domingo, Dominican Republic,* July 29–31, 2015.

Mielke, R., Scerbo, M., Gaubatz, K., and Watson, G. (2009) A model for multidisciplinary graduate education in modeling and simulation. *International Journal of Simulation and Process Modeling,* 5 (1), 3–13.

Mielke, R., Leathrum, J., and McKenzie, F. (2011) A model for university-level education in modeling and simulation. *M&S Journal,* 6 (3), 14–23.

Nance, R.E. (2000) Simulation education: past reflections and future directions. *Proceedings of the 2000 Winter Simulation Conference,* pp. 1595–1601.

Nance, R.E. and Balci, O. (2001) Thoughts and musings on simulation education. *Proceedings of the Winter Simulation Conference,* pp. 1567–1570.

National Academy of Engineering (NAE) (2016) Grand Challenges for Engineering Available at http://www.engineeringchallenges.org (accessed August 19, 2016).

Old Dominion University (2011) http://eng.odu.edu/msve (accessed May 25, 2011).

Oren, T.I. (2005) Toward the Body of Knowledge of modeling and simulation. *Proceedings of the Interservice/Industry Training, Simulation and Education Conference (I/ITSEC), Paper 2025, Orlando, FL,* December 2005.

Petty, M.D. (2006) Graduate Modeling and Simulation Overview Course. Final Technical Report, Contract N00014-031-0948, Defense Modeling and Simulation Office, March 2006.

Petty, M. (2009) Modeling and simulation Body of Knowledge topics. Certified Modeling & Simulation Professional Program – Management Plan, Modeling and Simulation Professional Certification Commission, Version 1.0, June 12, 2009, Available at http://www.simprofessional.org.

Roberts, C.A. and Ghosh, S. (2004) A proposed model for an undergraduate engineering program in modeling and simulation. *Proceedings of the International Conference on Simulation Education, Western Simulation Multi-conference,* pp. 38–44.

Rogers, R. (1997) What makes a modeling and simulation professional: the consensus view from one workshop. *Proceedings of the Winter Simulation Conference, Piscataway, NJ,* 1997.

Sarhoughian, H.S. and Zeigler, B.P. (2000) Towards Making Modeling and Simulation into a Discipline, Available at www.acims.arizona.edu.

Sarhoughian, H., Cochran, J., Collofello, J., Goss, J., and Zeigler, B. (2004) Graduate education in Modeling & Simulation: rationale and organization of an online masters program. *Proceedings of the Summer Computer Simulation Conference.*

SIAM Working Group on CSE Education (2016) Graduate Education for Computational Science and Engineering, Available at https://www.siam.org/students/resources/report.php (accessed July 26, 2016).

Simulation-Based Engineering Science (2006) Report of the National Science Foundation Blue Ribbon Panel on Simulation-Based Engineering Science, National Science Foundation, May 2006, Available at https://www.nsf.gov/attachments/106803/public/TO_SBES_Debrief_050306.pdf (accessed August 19, 2016).

Szczerbicka, H., Banks, J., Rogers, R., Oren, T., and Sarjoughian, H.Z. (2000) Conceptions of curriculum for simulation education. *2000 Winter Simulation Conference*, pp. 1635–1644.

University of Alabama – Huntsville (2011) http://catalog.uah.edu (accessed May 25, 2011).

University of California – Santa Cruz Newscenter (2014) UC Santa Cruz creates new Department of Computational Media, October 13, 2014, Available at http://news.ucsc.edu/2014/10/computational-media.html (accessed August 24, 2016).

University of Central Florida (2011) http://www.ist.ucf.edu/phd (accessed May 25, 2011).

US Department of Defense (2007) Department of Defense Directive Number 5000.59: DoD Modeling and Simulation (M&S) Management.

US Department of Defense (2009) Department of Defense Instruction Number 5000.61: Modeling and Simulation (M&S) Verification, Validation, and Accreditation.

Wardrip-Fruin, N. and Mateas, M. (2014) Envisioning the Future of Computational Media. *Center for Games and Playable Media, University of California – Santa Cruz*, https://games.soe.ucsc.edu/envisioning-future-computational-media-systems-project (accessed September 1, 2016).

5

Academic Education Supporting the Professional Landscape

Margaret L. Loper and Charles D. Turnitsa

Georgia Tech Research Institute, Atlanta, GA, USA

5.1 Overview

Continuous education of modeling and simulation (M&S) professionals and short courses to improve the knowledge of the workforce have become as important as the initial academic education, because the turnaround time for new technologies in M&S is as fast as in other high tech professions. Overall, the changing professional landscape requires people going back for education more often, as well as linking M&S to K-12 education for developing the pipeline early. To support continuous education, M&S curriculums have been developed at a variety of academic institutions. In some cases, agreements exist that allow an easy transfer from one institution to another, but this is more the exception than the rule. For better support of M&S as a profession, the idea for a distributed professional education catalog would not only make the various curriculums comparable but also actually help professionals take required courses at their university of choice. This chapter looks at recent and current developments in professional education and makes recommendations on necessary next steps.

5.2 The Science of Modeling and Simulation

Modeling and simulation can be a confusing term, as decision makers sometimes struggle with differentiating "simulation science" from "science that uses simulation" (Weisel, 2011). In the world of the Department of Defense (DoD) Acquisitions, M&S spans the continuum from concept exploration to technology design to engineering and manufacturing to

The Profession of Modeling and Simulation: Discipline, Ethics, Education, Vocation, Societies, and Economics, First Edition. Andreas Tolk and Tuncer Ören.

production and deployment and finally to sustainment and support (Department of Defense, 2015). Most organizations are interested in M&S technologies that can be used to support analysis for operational requirements, acquisition, test and evaluation, performance prediction, training, and so on. In essence, M&S supports the systems engineering life cycle, as well as every aspect of scientific investigation engineers and scientists pursue. With rare exception, the increasing capability of this technology, the broad reach of its application, and the growing expectations of new systems have challenged not just individuals, but many organizations in their attempt to understand, benefit from, and manage this thing called M&S.

Based on the increasingly rapid development and continuing maturation of the methods, resources, and theories associated with the creation and use of computational models and computer-based simulations, it is easy to appreciate that M&S has become a methodological activity, that is, a science, like chemistry, biology, or physics. In this regard, it makes sense to speak in terms of the discipline of M&S. Consider that what is now called computer science was the synthesis and specialization of mathematics and engineering (as well as some other disciplines). Similarly, the synthesis of computer science, communications, engineering, mathematics (especially statistics), and other areas are emerging into a methodological area of study that we might call M&S science. This maturation of M&S science is evidenced by the increase in academic and professional education now offered that leads to advanced degrees and professional certifications in M&S.

However, a science has various disciplines; just as there is a great deal of difference between biologists who study plants (botanists) and those who study animals (zoologists), there is a great deal of difference between M&S scientists who, for example, work in the domain of training and those who work in the domain of systems engineering. Until recently, many organizations have treated M&S as a domain, where in actuality it is better described as a science. Perhaps this is why it has been so difficult for many organizations to achieve the progress in their efforts that M&S should provide. Any technical organization with the mission of developing system solutions to challenging problems must have expertise in a variety of areas. We need experts in the science of M&S, especially when dealing with highly complex systems and systems-of-systems (Holland and Bachman, 2014).

5.3 Modeling and Simulation Education

As M&S becomes increasingly important, there is a significant and growing need to educate, train, and certify M&S practitioners, researchers, and faculty. Efforts to meet that need have taken a number of forms: academic degree programs, non-degree professional education, and professional certifications. In this context, an academic degree refers to a bachelors, masters, or doctorate program; non-degree professional education refers to skills and knowledge attained for career advancement through facilitated learning; and professional certification refers to a designation earned to ensure qualification to perform a job or task.

A few universities offer M&S as an academic discipline with a degree program. Graduate-level programs are currently offered or have been offered by the University of Central Florida, Old Dominion University (ODU), the University of Alabama in Huntsville (UAH), and the Naval Postgraduate School (NPS). These programs follow a traditional academic approach where students enroll in a graduate program and take a series of courses that lead to a degree. It is common for the curriculum to be based on the theory and science of the underlying subject area, and for the degree to culminate with a formal project, thesis, or dissertation.

An alternative form of M&S training is professional (or continuing) education. These courses differ from academic degrees in that they focus on applied learning rather than theory. They may also focus on a narrower area of study. Professional education courses are taught by universities and industry, and come with differing levels of accreditation. In this context, accreditation refers to the standards against which the student's course performance was assessed, which is often measured by continuing education units. There are more choices in this space than for academic M&S degrees. In addition to ODU, UAH, and NPS, courses are offered by the Georgia Tech Research Institute, George Mason University, and the Army Simulation and Modeling School, to name but a few. These professional education programs serve an important role in M&S education. Many practitioners and researchers do not have the time nor desire to pursue a traditional academic degree program; however, they need to learn new skills and knowledge to be effective in their jobs.

5.3.1 Professional Versus Academic Education

The World Economic Forum's Future of Jobs Report predicts widespread disruption to business models and labor markets due to the "Fourth Industrial Revolution" unleashed in part, by the advances in fields such as artificial intelligence, machine learning, and robotics (World Economic Forum, 2016). How can individuals equip themselves with the knowledge and skills throughout their life that are critical for success in a world where the jobs available as well as the skills needed both change at a rapid pace? One approach is through continuing and professional education.

Continuing professional education refers to "the education of professional practitioners, regardless of their practice setting, that follows their preparatory curriculum and extends their learning . . . throughout their careers. Ideally this education enables practitioners to keep abreast of new knowledge, maintain and enhance their competence, progress from beginning to mature practitioners, advance their careers through promotion and other job changes, and even move into different fields" (Queeney, 1996). This is not a new concept; ongoing education for professional practitioners at one time was provided through apprenticeships and guild systems of the middle ages, and it was an informal adjunct of professional practice into modern times (Queeney, 2000).

To be effective, continuing and professional education must contribute to the support of ongoing professional competence. Competence is the whole of knowledge (knowing) and skills (performing) that people have at their disposal and that they can use efficiently and effectively to reach certain goals in a wide variety of contexts or situations (Kirschner *et al.*, 1997).

Academic and professional education contribute a certain level of achievement toward competence. An academic education involves the study of a subject within a discipline and often includes a research focus. The purpose is a contribution to the learner's specialized knowledge of a subject and not necessarily the application thereof. The purpose of professional education is to impart knowledge, understanding, and practical experience to the learner to enable the learner to apply the knowledge in a practical manner, in a professional practice. This leads to a different set of skills, each with different purposes and contexts for the world of work (McIntyre, 2014).

In order to prove competence in an occupationally directed professional qualification, it must be proven that the learner has knowledge and understanding of the theory (foundational competence), that the learner has the ability to apply that knowledge and understanding practically (practical competence), and that the learner has the ability to apply that knowledge, understanding, and practical skill in an ever-changing environment (reflexive competence). This idea is part of the understanding of the competencies gained in different types of education, and it goes to the heart of the ascending levels of learning, in the cognitive domain, as seen through Bloom's Taxonomy (Bloom, 1956).

- Foundational competence is gained through learning knowledge about a subject (recognition and recall of specific facts and specifics) and through comprehension (the ability to interpret knowledge and paraphrase the given information about a task or problem).
- Practical competence is gained through learning application (where the learner can process information, and apply it, in a situation different from

the context it was learned within), and through learning analysis (being able to separate complex ideas into parts; systemic thinking applied to problem solving; understanding relationships among parts of a complex problem).

- Reflexive competence is gained when the learner is able to synthesize what they have learned (combining elements of what they have learned so far together in ways to form new knowledge), and finally is able to apply all they have learned to evaluation (being able to make decisions, judging problems and solutions, and making selections from the field based on applicable criteria, and rationale gained from understanding the principles involved).

Adults that pursue professional education often have different motivations for continuing their education, different from those pursuing academic degrees. These motivators include the following (ICSD, 2016):

- *Relevancy:* Content that directly relates to their current job responsibilities and/or personal interests.
- *Practicality:* Information that is practical rather than factual or intellectual, for example, training versus education.
- *Critical thinking:* Applying knowledge to open-ended questions and new situations.
- *Active learning:* Engaged and participating in the learning process, rather than watching and taking notes.
- *Respect:* Treated as professionals; their contributions are valued and they are recognized for their workplace expertise.
- *Desire for mastery:* Desire to become proficient, not just "pass the test." Mastery is promoted by incorporating all levels of Bloom's Taxonomy: know, understand, apply, analyze, synthesize, and evaluate.

5.3.2 DoD Modeling and Simulation Workforce Development

The DoD has a growing need for an educated M&S workforce. This need includes users, developers, managers, and executive-level personnel, who can effectively apply M&S to DoD requirements. In a strategic planning session in 2011, the DoD M&S Steering Committee (M&S SC) reaffirmed the vision and supporting goals originally established for the DoD M&S Enterprise in 2007[1]:

M&S Vision: Empower DoD with M&S capabilities that effectively and efficiently support the full spectrum of the Department's activities and

---

1 http://www.msco.mil/strategicVision.html

operations. End State: A robust M&S capability enables the Department to more effectively meet its operational and support objectives across the diverse activities of the services, combatant commands, and agencies. A defense-wide M&S management process encourages collaboration and facilitates the sharing of data across DoD components, while promoting interactions between DoD and other government agencies, international partners, industry, and academia.

In doing this, the M&S SC once more established that a trained workforce – military, government civilian, contractor – with relevant skills and fully in tune with the department's current and emerging objectives are essential to achieving (i) an effective and efficient modeling and simulation capability and (ii) the skills to manage and apply those capabilities. The annual reaffirmation of those vision and goals has, unfortunately, only generated a minimal investment at the DoD corporate level to establish a workforce development capability – and in the development of a strategy to put those capabilities and skills in place.

During the development of a recent DoD M&S Enterprise strategy, many of the M&S SC members identified workforce development as a persistent gap. The strategic priority is the need for a competent, seasoned, career government, and industry workforce. They identified a number of obstacles in the path of achieving resolution of this gap and thus solving the long-term professional recruitment and M&S career development issues:

- Some will say the problem begins in elementary or middle school with the lack of solid math, science, and engineering education that carries through to professional degree levels.
- A lack of standard M&S curriculum content at all levels.
- Designation of the workforce with a unique career code and the designation of core competencies in M&S.

As a result, DoD finds itself in much the same position – at least concerning workforce development – as it was in 2007, or even earlier:

- No DoD-wide workforce development strategy.
- No DoD enterprise-level investment in M&S-specific workforce development.
- A very limited DoD enterprise-level education foundation – primarily two very outdated Defense Acquisition University (DAU) continuous learning modules, a M&S staff officers course with limited offerings, an outdated online Essentials of Modeling and Simulation Course, a handful of best practices guides online that can be used for self-training, and occasional tutorials at semiannual and annual professional gatherings.

• Only one service – the Army – with a fully resourced and managed plan to develop its military and civilian workforce; even there, however, many commands have not yet built this training into its human resource management processes.

Given budget cuts – and even more anticipated "efficiencies" in the very near future – it is doubtful that a significant change to the situation described immediately above will take place. The highest priority effort would be coordinating with the DAU to update M&S-related continuous learning modules to reflect recent, major, changes to acquisition policies and methods. The M&S SC could also reestablish an active and interested stakeholder process for both the M&S foundation and community-specific modules. This approach would allow a broader set of users across DoD to leverage the mature infrastructure of DAU continuous learning capabilities for course enrollment and content maintenance – and establish a means by which future modules (e.g., M&S for logistics, training M&S, etc.) could be crafted without having to redevelop the foundation material each time.

These are small steps with potentially huge payoff – especially in the context of broader M&S education efforts ongoing in current or evolving academic departments across the country or even within DoD STEM (science technology engineering and math) efforts. Doing nothing should not be an acceptable option – failure to educate the government M&S workforce (customers, managers, users and developers) further impedes making significant progress in removing the many obstacles preventing interoperability and reuse of current and future M&S capabilities. These fledgling efforts must be constructed such that DoD policies, management process, standards and standardization support efforts, and core infrastructure for asset visibility and reuse are made visible through these workforce development efforts – and kept current and relevant to the demand of the workforce and the DoD.

5.3.3 M&S Science and the Body of Knowledge

To support the range of professionals that need M&S education, the DoD published the M&S Workforce Body of Knowledge (BoK) (Department of Defense, 2008). The BoK provides standardized language and associated knowledge base for users, developers, managers, and executive-level personnel to effectively apply M&S to DoD requirements. The BoK is considered a starting point for defining the core knowledge and skills that a member of the DoD M&S workforce performs.

The goal of the M&S BoK is to form a comprehensive and concise representation of concepts, terms, and activities needed to make up a professional M&S domain. As such, the M&S BoK represents the common understanding of relevant professionals and professional associations. Although the discussion on the necessity to create the M&S BoK was initiated more than a decade ago (Szczerbicka *et al.*, 2000), we still have not reached an agreement that can be used to drive common curricula, aligned professional education, or generally accepted professional certificates. Without this foundation, we do not have a set of core competencies that define what every M&S professional needs to know. This danger was addressed in a 2008 Winter Simulation Conference panel on Sustaining the Growth and Vitality of the M&S Discipline (Yilmaz *et al.*, 2008): We have no common definition for M&S professionals or M&S experts; therefore, everybody can claim to be one without an option to be held responsible to this claim. However, the Certified Modeling and Simulation Professional (CMSP) certification, described elsewhere in this volume, is an initial attempt to remedy this omission.

One of the main reasons may be the failure of academia to recognize M&S as its own discipline and drive the development of the M&S BoK more actively. The reason for this reluctance may be that every traditional discipline already applies M&S as a tool to drive their own discipline forward. This viewpoint is also reflected in the academic journal environment: The majority of M&S journals focus on applications. For example, biologists, military training specialists, transportation engineers, and aerospace designers all have some M&S application knowledge. While discipline-driven viewpoints of M&S as an engineering approach is without question justified – M&S applications methodologically helped drive solutions in many disciplines – they fall short when it comes building a basis for professional education.

The M&S BoK should serve as a foundation for defining the core knowledge needed by the M&S workforce. However, most M&S courses (academic and professional) were created prior to the development of the BoK. This raises interesting questions about how it might be used in education. Since academic courses are governed by institutional procedures, the BoK might be used for reference but could not be required. However, the DoD could require professional education courses be assessed against the BoK to ensure a common level of education across the different options available to students.

5.4 Current State of M&S Professional Education

A working professional in a position, or career that is focused on modeling and simulation has all of the typical pathways to further their education. This can take the shape of relying on foundational information from a bachelor's level college degree. For M&S that would typically be in a field such as computer science, computer engineering, applied mathematics, systems engineering, or something related. It is possible that professionals might know early enough that they are interested in M&S and that they would pursue a bachelor's degree specifically related to M&S, such as the Modeling, Simulation, Visualization and Engineering program at Old Dominion University (one of the first bachelor's degrees specifically aimed at M&S). Additionally, a professional might choose to pursue a graduate-level degree, such as master's degree (science, arts, or engineering) in a more specific topic, perhaps even Computational Science or M&S specifically. However, it is more likely that working professionals, whose career comes to M&S, will decide to enhance their knowledge, and their functionality within the career path, through professional education. What follows is a description of some of the options available, currently, for professional education in M&S topics.

Professional education can take several different forms. Professional education for M&S can take the form of instruction in a number of different specific tools and software packages. It can take the form of short courses, or a certificate program that focuses on specific techniques, or even a smaller slice of the information that would be studied as part of a graduate degree. It can also take the form of preparation of knowledge for the specific needs of an organization, or a professional certification.

5.4.1 Software Training

One of the most direct ways in which a professional can take advantage of M&S education is through becoming educated in specific tools and software packages. Such education is valuable for specific job-related skills, and allows the professional to become more proficient at using the tools to complete their job tasks. It also imparts knowledge on how to develop more meaningful models, or simulations that can produce more reliable, or more useful data – through this the professional becomes aware of new ways to apply the software, and how to get better results by using the application. A wide variety of different software tools come under this category, but a few names that stand out in the M&S community might be Simio, AnyLogic, Arena from Rockwell Automation, or MatLab and Simulink from Math-Works. These tools, and other like them, have training sessions and education sessions available from the software manufacturing company, and many of the more popular software products have classes available from third party sources. The rewards to the professional, and their organization, are immediate and can exhibit benefits for an immediate project, and subsequent projects. This sort of education is very tactical and applied, in that it can be targeted directly at problems that might arise in ongoing projects, or to fill skill gaps within teams. A limitation of this sort of training is that it does not teach more generalizable principles about modeling and simulation, but it is not intended to. The use and benefits of specific M&S tool training are as described above – they provide immediate benefits to the trainee and the organization by giving the ability to apply powerful M&S tools to projects in a more efficient way, and to gain stronger results from the use of the trained software.

5.4.2 Short Courses

If a professional is exploring options for education beyond instruction on a specific tool, one of the options in the current environment is short courses. These are instructional courses, typically taking a few days (sometimes up to a week, or very rarely, longer), that give a broader more general topic of instruction to a student, beyond just instruction in a software package. A week is a short time, even if it is full-time instruction on one topic, so it is not the same level of breadth that a semester-long graduate course would offer, but it is enough time so that a very strong introduction into a topic can be given, along with history, consideration of the value of the topic, some techniques, and insight into when and where the topic is best employed.

Many M&S topics are offered by universities that teach modeling and simulation topics as graduate course, and it may be the same faculty members that are also teaching the short courses. But, as mentioned, the information in a short course is more focused, and more likely to get to the point of applicability of knowledge faster than in a graduate course that might dwell more on background and theory. Exceptions to these trends exist, however, and most institutes that offer short courses make descriptive information about the contents of the course available to a would-be customer.

A list of short courses is below. This is not exhaustive in terms of the sources of such courses, nor is it exhaustive in the topics that the listed sources themselves offer. But it is an example of the variety of topics in such courses, as they apply to M&S. Interested individuals should contact the organizations identified to find out more information, if they want to see a current description of offerings.

- Averill M. Law & Associates – seminars in multiple topics, including discrete event simulation, and others
- DiSTI – courses on distributed simulation technologies, graphics programming, and others

- George Mason University – modeling throughout the life cycle, M&S basics, and others
- Georgia Institute of Technology – fundamentals and topics courses, across engineering fields
- Massachusetts Institute of Technology – topical modeling courses on traffic and many other areas
- Old Dominion University – fundamentals, topics courses, and others through the VMASC center
- University of Southern California – biomedical modeling topics through their BMSR program
- University of Alabama in Huntsville – fundamentals and topics courses, across engineering fields

Some of these organizations are universities, or research centers affiliated with universities, while others (such as Averill M. Law & Associates or DiSTI) are corporations that have an interest and effective capability to present M&S training courses.

5.4.3 Certificate Programs

An option for a professional who wants to gain more information into the theory and basis for knowledge that makes up the various aspects of M&S is the certificate program. A certificate program is a structure of education offered by a university, or research center, that is composed of education units that are either semester-long courses or the equivalent (so, typically more than a short course, and of a deeper topic than a software instructional program). Rather than requiring the full number of courses, however, that make up a full graduate degree, a certificate program is much more focused, and only requires typically four to six classes. It is only the course work, there are no preliminary courses (if the professional meets the requirements to engage the program), and there is usually no requirement for a thesis or capstone project.

The benefits for the professional are that they are receiving university-level education in the topics that are central, or core, to the aspect of M&S the certificate program deals with, and the certificate does not require the same level of commitment (either time or intensity) that a graduate degree requires. The only drawbacks are that while the professional becomes educated to a high level, they do not gain whatever job or career benefits might come from having a graduate degree.

A number of universities, including some of the same that offer short courses, also offer certificate programs in M&S. There are differences, of course, in the approach that universities take. Georgia Tech, for instance, has an M&S certificate program that has several required courses in the foundations of modeling and simulation. Then there are a wide variety of different topical courses that can be taken, the majority of which focus on particular application areas that a working engineer or scientist would want to focus on (Loper *et al.*, 2011). Columbus State University offers a similar certificate program, requiring six academic units, two of which are core M&S foundational courses, but the additional four courses are based on applicable computer science topics that an M&S software professional would need to be aware of, at the graduate level (TSYS, 2016). One of the differences between the two programs is that the academic units in the Georgia Tech program are offered as (roughly) week-long short courses, with a large amount of academic knowledge delivered in a short but intense period of study to the students – but in a traditional face-to-face instructional environment. A number of other certificate programs follow this same formula. The Columbus State University program is based on traditional semester-long courses, but they are delivered entirely online (asynchronously), so students can attend the courses from anywhere, and working professionals can access the information at their own schedule. As with the Georgia Tech approach, other educational organizations exist that follow the Columbus State University model.

One final topic that should be addressed under the heading of short courses is the concept of a preparation course for a professional certificate program. These exist for a broad variety of different professional certification programs, especially in the various professions, including engineering. In the M&S community, the Certified Modeling and Simulation Professional is the professional certification program, and exists for both career professionals involved in M&S technical services (such as engineers, computer scientists, analysts, and others) and for career decision makers involved in M&S project management. The CMSP is defined elsewhere in this text, so would not be described at length here, but just as with other professional certification programs, there are some short courses and refresher courses available to provide the professional thinking about taking the certification testing with a refresher of knowledge on the topics that she or he has learned during their career in M&S. University of Alabama in Huntsville has provided such a course of study, as an example.

5.4.4 Organization-Specific Education

In addition to the previous options, one of the most direct approaches to continuing education and development of M&S professionals is by the organizations that require the services of those individuals. This can take many forms, but in each case it is tailored to cover the aspects of M&S that are of interest to the organization. Some of the methods in which an organization can focus the continuing education on a professional in M&S is through workplace education programs.

This organization-specific education can be in-house tutorial sessions, which are often semiformal or nonformal in their structure, the most common case being the instruction of junior professionals in a topic that a more senior professional is expert in. It can also be a more formal field of study, or can be specifically applied education programs, including some of those previously mentioned above (software education, short courses, etc.), but tailored or specifically selected, to focus on the educational needs of the organization. There is some ability of an organization to specifically tailor more traditional education, from the general M&S education (whether in core principles or in applied areas) to what the organization has specific need of – by offering internships. In this way, the student is getting the benefits of traditional academic education, but is also simultaneously gaining experience through internship with the organization, and the ability to see first-hand the types of work and projects that exist – and to learn specifics about M&S from the more senior professionals and mentors that the intern is working with.

Finally, it is possible that an organization will have its own specific needs for professional education, that it creates its own educational unit. An exemplar instance of this is what the Newport News Shipbuilding company has created, in its Apprentice School. The Apprentice School is a technical school that teaches advanced trade education to students, who can then apply that knowledge to professional and career jobs at Newport News Shipbuilding. While this education takes the form of the usual trades you would expect at a shipyard, such as machinist, foundry, pipefitting, and welding – it also covers a variety of advanced education topics, including modeling and simulation. Students who excel in the world class Shipbuilder Curriculum and their own specific shipbuilding education track are eligible to apply for an advanced discipline training. This represents further instruction beyond the level of the usual curriculum. One of the advanced training options is in the "Modeling and Simulation Program Analyst" track, and includes information on a variety of different M&S topics, including 135 h of study in the areas of core M&S topics, discrete event simulation, and other areas (Apprentice School, 2016). The description of the specific field is as follows:

Modeling and Simulation (M&S) is an advanced process that allows analysis of data to assess designs, improve processes, and make critical business decisions. M&S apprentices work and learn in every stage of

M&S study, including conceptual modeling, software development, technical artistry and animation, data collection, simulation construction and output analysis. M&S apprentices design and build computer simulations, identify problems and improve operations in a real-world environment. M&S apprentices participate in a 3-year rotation plan where they learn to apply innovative simulation practices in the area of ship construction and Research & Development projects.

Not every organization is able to engage specific, in-house professional education by providing a specific program of study. However, organizations that employ several M&S professionals would find it beneficial to support the exchange of technical knowledge, and even tutorial sessions among their bullpen of M&S workers. Organizations that can engage interns would benefit in all the usual ways from engaging young M&S professionals who are just starting their career to reaping the added benefits of specific M&S knowledge that is pertinent to the organization, through the internship process. Finally, an organization like Newport News Shipbuilding that has the enterprise breadth and strength to support its own educational facility is certainly in a good place to provide, with other technical education, a focus on modeling and simulation. Again, this provided great benefit to the organization. Reliance on the various M&S education resources (such as the body of knowledge, or attempts to standardize curriculum, or the topics of study required for the CMSP) is a great way for an organization to organize its nascent efforts, as modified by the specific needs of the organization itself.

5.4.5 U.S. Army M&S Professional Support

As an excellent example of many different approaches to the nurturing and education of M&S professionals, an examination of the approach that the U.S. Army takes with its two different M&S career tracks is presented here. The U.S. Army, of course, is an enterprise of nearly unique dimensionality and size, but it is presented as an aspirational example, of which enterprises and organizations providing guidance and education of M&S professionals can benefit from.

Simulation operations is the art and science of applying live, virtual, and constructive simulation technologies in support of military operations, training, and acquisition activities that include testing, experimentation, and analysis (Department of Defense, 2008). The U.S. Army has developed two M&S career titles to develop and train its workforce: the Functional Area (FA) 57 Simulation Operations designation for military officers and the Career Program (CP) 36 Analysis, Modeling & Simulation designation for civilians.

The FA57 is an officer with operational experience who understands military operations and training. They develop, plan, coordinate, and execute exercises at all levels of command: battalion, brigade, division, combatant command, interagency, and multinational. FA57s are experts in modeling, simulation, and Army Battle Command Systems (ABCS) and facilitate the training and operational environment for commanders to conduct first-class mission planning and mission rehearsal exercises. FA57 currently has 217 authorizations in the Active Army and 209 in the Reserve Component with ranks ranging from Captain to Colonel.

FA57 training and education are among the best available in the Army:

- The Simulation Operations Course trains students on what simulations are and how best to utilize them in support of any organization.
- The Battle Command Officer Integration Course provides students with hands-on instruction on the ABCS. Students learn the inputs, outputs, and architecture of this "system of systems" and how best to employ it during both training and operations in support of the Commander.
- The Army Knowledge Management Qualification Course introduces students to knowledge management basic tools and techniques that they may apply in their units.
- The School of Advanced Military Studies (SAMS) is where an officer's planning skills are honed to perfection. SAMS graduates are given a special skill identifier and required to commit to at least one payback tour as a planner within a military headquarters upon graduation.
- The Advanced Simulation Course introduces senior FA57 officers to cutting-edge topics in the M&S career field. This course engages students on how to think and manage current and future challenges that await them as senior leaders of the FA57 functional area.
- The Intermediate Level Education and the Advanced Operations Warfighting Course prepares field grade officers with a warrior ethos and warfighting focus for leadership in Army and Joint, Interagency, Intergovernmental and Multinational organizations executing full-spectrum operations.
- All FA57s are eligible to work and train with an industry partner in the M&S field for up to 1 year.
- All FA57s are eligible to compete for fully funded Advanced Civil Schooling, which provides officers the opportunity to pursue a master's degree or Ph.D. from one of several universities.

The CP36 is the Department of Army's civilian Analysis, Modeling and Simulation career program, for training, educating, and developing civilian human capital in a systematic fashion. The CP36 Army Civilian Training, Education, and Development Systems (ACTEDS) was approved on April 15,

2006. Analysis, modeling, and simulation is pervasive throughout the Army, and is found in the acquisition, analysis, operations, testing, training, experimentation, and intelligence communities.

CP36 civilians work in a wide variety of organizations, including program offices; research labs; technology, development, and engineering facilities; analysis centers; test ranges; logistics centers; headquarters; and training centers and ranges. CP36 careerists support M&S activities throughout the acquisition life cycle and in the analysis, experimentation, intelligence, operations and plans, testing, and training communities. Dedicated educational programs and training enable M&S professionals to apply current and emerging technology with credibility and success.

CP36 training and education include the following options:

- Army Simulation and Modeling School, which includes the Simulation Operations Courses, Advanced Simulation Course, Battle Command Officer Integration Course, and the National Training Center Right-Seat-Ride Program.
- A CP36 may be approved to pursue a bachelor's or master's degree program funded by the career program. A careerist must be assigned to a permanent civilian position for 3 years before requesting this opportunity.
- Developmental Assignments can vary in length from 14 to 90 days, for the purpose of learning about a new organization, broadening understanding of Army analysis, M&S, or developing a specific skill or ability in a different analysis, modeling, and simulation community.
- Army Greening Assignments are intended to provide civilians opportunities to learn more about the operational Army.

5.5 A Look Forward

Computer-based M&S has emerged as a mature discipline with a wide variety of deep theoretical, methodological, and applied subject areas. The field unquestionably has widespread impact on society today, but as the practice of M&S has evolved, no clear consensus has emerged concerning guiding principles in M&S education and curriculum development. While we are not advocating a single, standardized approach to M&S education, consensus within the community concerning goals, objectives, and content in M&S programs is highly desirable, and will aid in the enhancement and development of new M&S education programs. We believe it is time to establish a consensus in the community concerning standards for curricula development in the M&S field.

5.5.1 A Combined Approach to M&S Competence

A new and perhaps a better approach to academic and professional qualifications would be for the two to work together to produce a skilled workforce that has both an academic knowledge of the job and the real-world skills to perform the tasks required.

The most striking difference between these forms of qualification is perhaps that a professional qualification, due to the nature of the training and the fact that it is built on practice analysis, offers a merit of competence and expertise. It therefore certifies that, having completed the course or training, the graduate has the essential knowledge and skills to perform the duties required of his/her profession (McIntyre, 2014). In contrast, an academic qualification does not certify competence and is not based on a systematic or formal practice analysis; it certifies that the student has successfully learnt the theory behind the practice. For this reason, should human error lead to damages, there is no recourse for an academic institution.

Simulation was perceived a success story in so many application domains so rapidly that academia did not find time to deal with the scientific basis before solving problems. In order to continue to grow, we need this common M&S theory that is application independent. While the focus of M&S has been simulation in the past, what makes M&S really special and distinguishes it significantly from other software engineering disciplines is the modeling part. Solving problems on the conceptualization level in a common theory has the potential to drive new methods and results in new solution in several traditionally independent application disciplines. University-based M&S continuing education programs must help to educate professionals in such M&S-specific thinking resulting in new solutions, not in application discipline-specific point solutions. A common M&S theory developed, maintained, and extended by a community of M&S science will allow the academically meaningful transfer of solutions between different application disciplines. Trial and error, which is the only approach when engineering solutions are not rooted in a common theory, will be replaced by academic rigor based on scientific insight (Loper *et al.*, 2011).

Combining these two thoughts, we can see that there is a defined, and useful, difference between professional education and academic education (McIntyre, 2014). And we can see that the special quality of M&S from other information science disciplines is the reliance on models and modeling (Loper *et al.*, 2011). From this we can derive that model-based thinking can, and perhaps should, be applied to both professional education, as well as the theory behind academic education. Furthermore, this may be the link between the two different types of education, when looked at through a hybrid lens. One case for this is the thinking behind model-based systems engineering, where the principles of using models and model-based approaches to systems engineering have proven to be a very effective way to address many of the questions that the systems engineering process raises – professional training in the techniques and tools used for model-based systems engineering, but it might be an even better approach to give a student in the discipline (or a professional, getting some quantity of professional training) an appreciation for the theory behind how models represent referent systems, and the link between a model and the questions that is being used to answer. In this hybrid view, both approaches benefit. The academic perspective benefits from having actual applications to prove its theories, and to use as case studies. The professional education perspective benefits from having the theoretical basis of modeling, and its unique contribution, being part of what the professional learns.

An example of the hybrid approach can be seen at the Clemson University International Center for Automotive Research (CU-ICAR). There, researchers and students of all levels are participating in research in many different aspects of engineering in the life cycle of automobiles and automobile manufacturing. There is a very close tie with industry, with industry partners from virtually all of the large-scale manufacturers of automobiles, and also with providers of engineering and M&S tools – such as Siemens. However, while this is an excellent source of professional education, it is also tied to an academic program – the Department of Automotive Engineering at Clemson University. In their program catalog, the department has listed 38 different courses at the graduate level (master's and doctorate). Of those, the catalog specifically mentions modeling, simulation, computer-aided engineering, or a very similar term in nearly half of the courses (14 out of 38, and it is hinted at in another half dozen courses). So, model-based engineering and model-based thinking are part of the program here – and yet, it is strongly influenced by state-of-the-art industrial partnerships, and what a professional engineer would have to know. A very good example of the hybrid approach – bringing the profession into the academic, and allowing academia to assist with professional education.

5.5.2 Principles of M&S Education

M&S educational programs should provide broad coverage of the discipline and all of its facets; they must prepare graduates for taking on work and responsibilities as M&S professionals. Education programs must cover a broad range of M&S users, ranging from technology developers to persons responsible for creating models and representing them in computer software to those with a stronger application-specific focus (who are heavier users of M&S technology).

Below we outline a set of guiding principles for M&S programs that were proposed in 2000 by the Georgia Tech Modeling and Simulation Research and Education Center (MSREC) leaders and MSREC's advisory board (Fujimoto, 2000). These principles form a basis in formulating effective M&S education programs moving forward.

Principle 1: A solid grounding in fundamentals is essential.
A successful educational program in M&S must be solidly grounded in fundamentals of mathematics, statistics, and the physical sciences. The physical sciences provide the fundamental laws underlying many simulation models that are essential to accurately represent real-world phenomena, such as the trajectory of objects through space. Discrete and continuous mathematics provide the underlying framework, language, and tools for describing and manipulating such phenomena. Probability and statistics provide the essential tools for the creation of valid models as well as the analysis of the results produced by the simulations.

Principle 2: Basic knowledge and skills in computing fundamentals are important.
A practical and theoretical grounding in computing is necessary to create, analyze, and utilize software realizations of models. A solid background in data structures (e.g., list processing), numerical methods, and algorithms forms the core computing requirements, and must be supplemented with knowledge and practical skills in software architectures, operating systems, and multiple programming paradigms (e.g., object-oriented programming). An appreciation of advanced computing topics such as distributed systems, human–computer interaction, visualization, and data storage as they apply to simulation is also valuable.

Principle 3: Tight coupling with application domains must be maintained.
M&S that is not tied to a specific problem domain has limited, if any, value. As such, students only versed in M&S techniques without domain experience will continually be limited in what they can accomplish in attacking real-world problems. There is no substitute for knowledge of the vocabulary, abstractions, processes, and tools commonly used in a specific domain. While students clearly cannot be expected to have deep insight into many different domains, it is reasonable to require students to be well versed and knowledgeable in at least one.

Principle 4: Exposure of students to a broad range of core M&S topics is essential.
A solid education in a broad range of core M&S topics is essential to the effective application of sound techniques to specific problems, as well

as to the development of the ability to span multiple domains. Specific topics include model creation and evaluation (including multiresolution modeling), model execution (including distributed and real-time execution), experimental design, output analysis, random number generation, system architecture, and verification and validation.

Principle 5: Fluency in multiple modeling paradigms is a key to intellectual development.

Modeling is the very basis for defining and communicating the purpose and function of a simulation. A successful and effective simulation execution depends on an appropriately modeled abstraction of the problem. It is thus critical that students be exposed to many different modeling paradigms. A large number of modeling paradigms are currently in use. These paradigms cover a wide spectrum, including discrete and continuous models, process and data models, abstraction, classification, aggregation, and generalization. A student with limited exposure to modeling techniques may be unable to apply the most natural construct for a given problem.

Principle 6: Students should understand the full M&S life cycle.

Students should be immersed in the entire life cycle of a simulation study several times during their education. Steps in the life cycle of a simulation study typically include problem formulation and planning, data collection and model definition, model verification, implementation and validation, design of experiments, data analysis, and documentation and presentation of final results. Other aspects of the life cycle that are becoming more important include integration and interoperability issues.

Principle 7: Effective communication skills are a prerequisite for success.

While not specific to M&S programs per se, the need for M&S practitioners to have excellent communication and presentation skills, both written and oral, merits mention. The inherent, multidisciplinary nature of modeling and simulation makes the need for excellent communication skills that span multiple domains all the more important.

5.5.3 Professional Education Course Catalog

There is a growing need for a trained workforce – military, government, industry – with relevant skills and fully in tune with the current and emerging objectives. This is essential to achieving an effective and efficient modeling and simulation capability. While several universities offer academic M&S degree programs, the time and expense of earning these degrees often limit the number of people that go through these programs. Professional education is a viable alternative for gaining M&S skills and knowledge. William F. Waite was an earlier supporter of developing professional education courses to support the M&S community. In the 1990s, numerous courses emerged, offered by a wide range of university and commercial groups (including his company). As a result, Bill recognized there was a need to have a foundation against which courses could be assessed. The DoD M&S BoK provides one such measure. As more courses emerged, a discussion about which of these courses could be used as preparation for the CMSP began. In a discussion in 2013, Bill suggested that instead of certifying professional education courses for the CMSP, the community should instead create an M&S professional education catalog to indicate which courses meet the criteria for gaining competence in a specific M&S knowledge or skill area. This distributed education approach would enable learners to select courses that meet their competency needs based on the delivery approach and location of courses. This professional education catalog has not yet emerged, but remains a goal for the M&S community to embrace.

Acknowledgments

We would like to acknowledge Dr. Thomas Holland and Mrs. Jane Bachman, authors of the 2014 Marine Corps Systems Command M&S Education Summit final report. Section 5.1 of this chapter is paraphrased from Section 2.1 of their report. We would also like to thank Dr. Mikel Petty and Greg Reed who reviewed a preliminary draft of this chapter. Their support is gratefully acknowledged.

References

Apprentice School (2016) Modeling & Simulation Program Analyst. Available at http://www.as.edu/programs/adv/modelingsim/index.html (accessed October 2016).

Bloom, B.S. (1956) *Taxonomy of Educational Objectives: Book 1 – Cognitive Domain*, Longman, New York.

Department of Defense (2008) Modeling and Simulation Body of Knowledge (BOK). Final report. Available at http://www.msco.mil/msLibrary.html.

Department of Defense (2015) Operations of the Defense Acquisition System 5000.02 Interim Instruction. Available at http://www.navysbir.com/docs/500002p.pdf (accessed November 2016).

Fujimoto, R. (2000) Principles for M&S education. *Simulation and Technology Magazine*, SISO.

Holland, T. and Bachman, J. (2014) MCSC Modeling & Simulation (M&S) Education Summit Report. Final Report for the Marine Corps Systems Command (MCSC) Systems Engineering, Interoperability, Architectures & Technology (SIAT) workshop.

ICSD (2016) *The GTRI Information & Cyber Sciences Directorate Short Course Newsletter*, Vol. 2, No. 2.

Kirschner, P., Van Vilsteren, P., Hummel, H., and Wigman, M. (1997) The design of a study environment for acquiring academic and professional competence. *Studies in Higher Education*, 22 (2), 151–171.

Loper, M.L., Henninger, A., Diem, J.W., Petty, M.D., and Tolk, A. (2011) Educating the workforce: M&S professional education. *Proceedings of the Winter Simulation Conference*, pp. 3968–3978.

McIntyre, G. (2014) *Academic vs professional qualifications*. The Skills Portal. Available at http://www.skillsportal.co.za/content/academic-vs-professional-qualifications (accessed October 5, 2016).

Queeney, D.S. (1996) Continuing professional education, in *The ASTD Training and Development Handbook*, 4th edn (ed R.L. Craig), McGraw-Hill, New York.

Queeney, D.S. (2000) Continuing professional education, in *Handbook of Adult and Continuing Education* (eds A.L. Wilson and E. Hayes), PN John Wiley & Sons, Inc., New York, pp. 375–391.

Szczerbicka, H., Banks, J., Rogers, R.V., Ören, T.I., Sarjoughian, H.S., and Zeigler, B.P. (2000) Conceptions of curriculum for simulation education. *Proceedings of the Winter Simulation Conference, Orlando, FL*, Volume 2, (eds J.A. Joines, R. R. Barton, K. Kang, and P.A. Fishwick, pp. 1635–1644.

TSYS School of Computer Science (2016) *Certificate Program in Modeling and Simulation*. Available at https://cs.columbusstate.edu/ (accessed, October 2016).

Weisel, E.W. (2011) Towards a foundational theory for validation of models and simulations. *Proceedings of the Spring 2011 Simulation Interoperability Workshop*, Orlando FL.

World Economic Forum (2016) *The Future of Jobs*. Available at http://www3.weforum.org/docs/WEF_Future_of_Jobs.pdf (accessed November, 2016).

Yilmaz, L., Davis, P. Fishwick, P.A. Hu, X. Miller, J.A. Hybinette, M. Ören, T.I. Reynolds, P. Sarjoughian, H. and Tolk, A. (2008) Sustaining the growth and vitality of the M&S discipline. *Proceedings of the 40th Conference on Winter Simulation*, pp. 677–688.

6

The Certified Modeling and Simulation Professional Certification and Examination

Mikel D. Petty,[1] Gregory S. Reed,[2] and William V. Tucker[3]

[1]*University of Alabama in Huntsville, Huntsville, AL, USA*
[2]*Torch Technologies, Huntsville AL, USA*
[3]*Simulationist.US, Huntsville, AL, USA*

6.1 Overview and Introduction

The Certified Modeling and Simulation Professional (CMSP) certification is a professional certification organized and administered by the National Training and Simulation Association (NTSA). The CMSP certification identifies individuals who have attained a significant degree of knowledge and experience in modeling and simulation (M&S). To earn the CMSP designation, a candidate must show evidence of education and experience in M&S, provide references supporting his or her certification, sign a statement of ethics, and demonstrate substantial expertise in M&S by passing a comprehensive certification examination. The current CMSP examination's scope is defined by a consensus-based M&S "body of knowledge" topic index intended to cover the essential knowledge of the M&S professional discipline. The examination is based on a "question bank" of more than 2000 questions drawn from every topic and subtopic in the body of knowledge. Every question is explicitly traceable to a published, publicly available, and peer-reviewed source in the M&S literature. Each candidate is challenged with an automatically generated examination instance consisting of questions selected from the question bank that is customized to the type of CMSP certification (user/manager or developer/technical) that the candidate is seeking. An online examination system allows candidates to attempt the examination conveniently and intuitively.

This chapter describes the overall CMSP certification and then explains in some detail the examination that is part of the certification requirements.

The Profession of Modeling and Simulation: Discipline, Ethics, Education, Vocation, Societies, and Economics, First Edition. Andreas Tolk and Tuncer Ören.

This chapter is organized into three main sections. Section 6.2 provides an overview of the CMSP certification, describing its intent, requirements, governance, and history. Section 6.3 details the CMSP examination's topical content, question characteristics, and examination instance structure. Section 6.4 describes the online system that has been developed to administer the examination.

6.2 Introduction to the CMSP Certification

The CMSP certification is a professional certification organized and administered by the NTSA. Award of the CMSP designation is intended to recognize individuals who have attained a significant degree of knowledge and experience in M&S.

6.2.1 Overview of the Certification

The maturation of a professional discipline often brings the institution of a professional certification for practitioners of the discipline and the acceptance by the larger community of that certification as an indicator of expertise and competence by those who have attained it. Professional certifications help propagate the use of best practices, they explicitly or implicitly establish a "body of knowledge" that defines what practitioners in the discipline are expected to know, and they identify a pool of certified individuals who have been assessed as qualified in the discipline (Lewis and Rowe, 2010). Accepted professional certifications already exist for many professional disciplines, including aviation, law, medicine, project management, and finance.

The CMSP certification is intended to identify individuals who have attained a significant degree of knowledge and experience in M&S. To achieve CMSP certification, a candidate must show evidence of sufficient education and experience in M&S, provide letters of reference supporting their certification, sign a statement of ethics relating to the practice of M&S, and demonstrate substantial expertise in M&S by passing a comprehensive certification examination (Lewis and Rowe, 2010). That examination is the focus of Sections 3 and 4 of this chapter.

As M&S matures as a discipline, it is anticipated that employers and sponsors seeking M&S professionals will recognize the value of and preferentially consider those who have attained professional certification in M&S via the CMSP designation, and that the designation will become an industry standard (Lewis and Rowe, 2010).

6.2.2 History and Governance

In the 1960s, John McLeod, the founder of the Society for Computer Simulation (now the Society for Modeling and Simulation International), established a working group to review the state of the M&S profession. At that time, he considered the idea of a professional certification for M&S professionals, but set it aside due to concerns that the required program to enforce a code of ethics would be too difficult to implement.

In the 1990s continuing interest in the emergence of M&S as a profession led to the organization of a panel session of M&S educators at the 2000 Winter Simulation Conference. That group recommended that a professional certification program for M&S professionals be established. Partially as a result of that recommendation, a small group of industry leaders, including William F. Waite, established what is now known as the Modeling and Simulation Professional Certification Commission (M&SPCC). Working under the sponsorship and oversight of the NTSA, the CMSP certification is governed by the M&SPCC, which consists of nine members (three each from industry, academia, and government) (Lewis and Rowe, 2010). The M&SPCC's mission is to develop and promote a professional certification program for M&S professionals that recognizes a significant level of knowledge and competency (Gross, 2014). The M&SPCC is, therefore, responsible for the development, provision, promulgation, and sustainment of the CMSP certification in general and the CMSP examination in particular. The initial M&SPCC consisted of members representing professional societies with an interest in M&S (the National Training and Simulation Association, the Society for Modeling and Simulation International, and the Simulation Interoperability Standards Organization), universities with education programs or offerings relating to M&S (Old Dominion University, the University of Central Florida, and the University of Arizona), corporations offering M&S products and services, and government agencies that use M&S (Lewis and Rowe, 2010). The M&SPCC's organizing principles are transparency (an open process), quality (examination traceable to a body of knowledge), confidence (certification is reliable evidence of competence), and ethics (ethical standards for M&S professionals are asserted) (Gross, 2014).

The CMSP certification was first developed and offered to the community in 2001 (Gross, 2014). The initial examination questions were developed by a group of acknowledged M&S experts, largely based on their personal expertise and experience. The examination covered many topical areas within the M&S body of knowledge by virtue of the range of backgrounds in the experts chosen to develop the questions. Since that examination was offered to the community over 200 individuals have been designated as Certified Modeling and Simulation Professionals. Due to strong support from William F. Waite, many of the certificates were earned by employees of AEgis Technologies.

Over time, the initial examination was criticized for a number of shortcomings, for example, insufficient testing of systems engineering skills. In addition, because there was only one examination, the questions were becoming fairly widely known in the M&S community. It became clear that the examination and the certification process needed to be updated.

Recognizing the need for a renewal of the examination's content, the NTSA and M&SPCC commissioned a collaborative community effort beginning in late 2009 to develop a new version of the examination. The new version improved on the initial version in the following ways:

1) It intentionally attempted comprehensive coverage of the emerging M&S body of knowledge.
2) The questions in the new version were updated to reflect advances in the field since 2001.
3) All questions in the new version were based on authoritative published sources.
4) A distinction was made between two different types of M&S professionals, defined earlier, with related but different examinations for each type.
5) There were sufficient examination questions available that no two candidates were likely to get the same, or even substantially similar, examinations.

The overall certification program was renewed as well, in accordance with accepted guidelines for professional certification programs, including the American Society for Testing and Material standard ASTM E2659-09, which provides a standard set of requirements for a certification program, and International Organization for Standardization/International Electrotechnical Commission standard ISO/IEC 17011, which is an accreditation standard for certificate issuers. The renewal process was completed and the new examination was first used for certification in 2012 (Gross, 2014).

6.3 Examination Content and Questions

This section details the CMSP examination's structure and intent, topical content, question characteristics, and examination instance structure.

6.3.1 Structure and Intent

The CMSP examination is a professional certification examination for M&S professionals. However, not all M&S professionals work at the same level or on the same types of activities. To reflect that, in the current version of the CMSP examination, two types of CMSP professionals are acknowledged and two types of certification are offered: User/Manager and Developer/ Technical.

The CMSP User/Manager examination is intended and designed to identify persons with the knowledge required for the following:

1) To employ and explain key terms, definitions, and concepts in M&S.
2) To apply important principles of M&S practice, including simulation ethics, business considerations, and related communities of practice.
3) To understand and work effectively within typical and important uses of M&S, including application areas and domains of use.
4) To identify, assess, and select relevant simulation technologies, including modeling paradigms and implementation architectures, for a specific application.
5) To determine whether the use of simulation is, or is not, appropriate for a specific application.
6) To plan, initialize, and execute simulation runs or trials to satisfy project requirements.
7) To analyze, interpret, and apply the results of simulation runs in the context of an application.
8) To manage aspects of projects involving the use or development of simulation models and systems.

The CMSP Developer/Technical examination is intended and designed to identify persons with the knowledge required for the following:

1) To employ and explain key terms, definitions, and concepts in M&S.
2) To apply important principles of M&S practice, including simulation ethics, business considerations, and related communities of practice.
3) To understand and work effectively within typical and important uses of M&S, including application areas and domains of use.
4) To design and develop simulation models of various types, including mathematical, logical, structural, and conceptual.
5) To identify the underlying mathematical issues associated with many simulation models, including numerical evaluation algorithms, digital discretization, and numerical precision.
6) To implement simulation models as executable software and verify those implementations.

7) To validate simulation models using suitable methods and assess the suitability of a model for a specific application.

8) To design and implement technical infrastructures needed to support simulation systems.

Note that the first three items in the two lists are common to the two certification types, the remaining five items in each list are distinct.

The overall goal of the examination is to ensure that successful candidates have a representative understanding of the full spectrum of M&S, that is, that they have knowledge across the body of knowledge. It has been suggested that because of its breadth, the CMSP examination is more challenging in some ways than other, more focused, professional certifications, but it is not true that a formal education in M&S is required to pass the examination. The overall goal of the examination renewal process is to better serve the M&S community, by helping to identify and differentiate those who are genuinely professional M&S practitioners.

6.3.2 Topics and Subtopics

The CMSP examination is based on a consensus-based topic index intended to cover the essential parts of the M&S body of knowledge. The topic index used for the examination was adapted and extended from the SimSummit M&S Body of Knowledge Index (Lacy *et al.*, 2010). SimSummit was an ongoing effort organized by William F. Waite to promote consistency and collaboration in the M&S community (Waite, 2010). By basing the revised examination's content on a community consensus body of knowledge index, the reliability and credibility of the examination were enhanced. Some of the changes made to the SimSummit topic index for the CMSP examination were motivated by the availability of suitable published sources and the testability of the content for each topic. The CMSP topic index is organized into eight topics, each of which is subdivided into several subtopics; there are a total of 54 subtopics. The topics and subtopics of the CMSP topic index are shown in Table 6.1.

6.3.3 Question Counts, Sources, Formats, and Attributes

CMSP examination questions were developed for each of the subtopics. In total, the examination question bank contains 2007 questions. On average, there are approximately 40 questions for each subtopic in the topic index. Every topic has at least 30 questions and some topics have more than 100 questions. Table 6.1 details the number of questions for each topic and subtopic in the question bank.

Table 6.1 M&S topic index topics and subtopics, and number of questions in the examination question bank for each. Note that the number of questions for each topic is the total of the questions for the topic's subtopics.

Topics and subtopics	Questions
1. Concepts and context	226
1.1 Fundamental terms and concepts	147
1.2 Categories and paradigms	43
1.3 History of M&S	36
2. Applications of M&S	185
2.1 Training	24
2.2 Analysis	30
2.3 Experimentation	22
2.4 Acquisition	55
2.5 Engineering	13
2.6 Test and evaluation	41
3. Domains of use of M&S	321
3.1 Combat and military	39
3.2 Aerospace	20
3.3 Medicine and health care	60
3.4 Manufacturing and material handling	20
3.5 Logistics and supply chain	19
3.6 Transportation	42
3.7 Computer and communications systems	27
3.8 Environment and ecology	23
3.9 Business	18
3.10 Social science	20
3.11 Energy	3
3.12 Other domains of use	30
4. Modeling methods	503
4.1 Stochastic modeling	52
4.2 Physics-based modeling	36
4.3 Structural modeling	1
4.4 Finite element modeling and computational fluid dynamics	50
4.5 Monte Carlo simulation	81
4.6 Discrete event simulation	69

(*continued*)

Table 6.1 (Continued)

Topics and subtopics	Questions
4.7 Continuous simulation	22
4.8 Human behavior modeling	84
4.9 Multiresolution simulation	72
4.10 Other modeling methods	36
5. Simulation implementation	448
5.1 Modeling and simulation life-cycle	21
5.2 Modeling and simulation standards	30
5.3 Development processes	26
5.4 Conceptual modeling	1
5.5 Specialized modeling and simulation languages	54
5.6 Verification, validation, and accreditation	99
5.7 Distributed simulation and interoperability	28
5.8 Virtual environments and virtual reality	23
5.9 Human–computer interaction	111
5.10 Semi-automated forces/computer generated forces	34
5.11 Stimulation	21
6. Supporting tools, techniques, and resources	37
6.1 Major simulation infrastructures	29
6.2 M&S resource repositories	3
6.3 M&S organizations	5
7. Business and management of M&S	133
7.1 Ethics and principles for M&S practitioners	21
7.2 Management of M&S projects and processes	68
7.3 M&S workforce development	5
7.4 M&S business practice and economics	36
7.5 M&S industrial development	3
8. Related communities of practice and disciplines	154
8.1 Statistics and probability	37
8.2 Mathematics	42
8.3 Software engineering and development	20
8.4 Systems science and engineering	55

Importantly, every question is drawn from and explicitly traced to a published, peer-reviewed, and publicly available source in the M&S literature. Over 175 different sources were used for the questions. Those sources include journal articles, conference papers, books, and book chapters. Although no single source was used for a majority, or even a significant minority, of the questions, the four sources from which the largest number of questions were drawn were (Banks, 1998; Greasley, 2008; Sokolowski and Banks, 2010; Tolk, 2012). Despite their value in many situations, inherently transient sources such as web pages, Wikipedia articles, presentations and technical reports posted online, and other similar sources were intentionally not used as sources for the examination questions.

Approximately, 75% of the questions are multiple choice, with four answers (one correct and three incorrect). The remaining 25% are True–False (which is a special case of multiple choice, with only two answers, one correct and one incorrect). No other types of questions are used. These types were used to support objectivity in scoring the examination and to facilitate automatic generation and online delivery of the examination. Some of the questions include diagrams, images, and mathematical formulas.

In addition to the question itself and its answers (correct and incorrect), each of the questions has several additional attributes: a unique identifying number, the source for the question, the specific page number within the source from where the question was drawn, and the question's author. Moreover, each of the questions is categorized in the following three ways:

1) Subtopic in the topic index (and thus implicitly also topic)
2) Certification type (User/Manager, Developer/Technical, and Core)
3) Difficulty (very easy = 1, easy = 2, moderate = 3, difficult = 4, very difficult = 5)

Table 6.2 shows counts of questions in each certification type and Table 6.3 shows counts of questions in each difficulty level in the question bank. Table 6.4 shows several actual questions from the question back with all of the questions' attributes exposed. A candidate taking the examination

Table 6.2 Counts of questions in each certification type. Core questions may appear on both types of examinations.

Certification type	Questions
Core	751
User/Manager	440
Developer/Technical	816

Table 6.3 Counts of questions at each level of difficulty.

Difficulty	Questions
Very easy	145
Easy	492
Moderate	923
Difficult	391
Very difficult	56

Table 6.4 Sample questions from the CMSP examination.

Attribute	Value
Question number	8.545
Question	Which of the following is *not* a use of simulation?
Correct answer	Justify decisions already made based other criteria
Incorrect answer	Describe and analyze the behavior of a system
Incorrect answer	Ask and answer "what if" questions about a system
Incorrect answer	Help in designing new systems
Type	Core
Difficulty	2 (easy)
Topic	1.1 Fundamental terms and concepts
Source	Banks, J. (1998) Principles of simulation, in *Handbook of Simulation: Principles, Methodology, Advances, Applications, and Practice*, John Wiley & Sons, Inc., New York, NY, pp. 3–30.
Page number	3
Question author	M. Petty
Question number	6.20
Question	In simulating a physical system governed by partial differential equations, ___ can be used to facilitate estimation of derivatives.
Correct answer	Fourier analysis
Incorrect answer	The Graham–Schmidt process
Incorrect answer	The downhill-simplex method
Incorrect answer	Gauss–Jordan elimination
Type	Developer/Technical
Difficulty	5 (very difficult)
Topic	4.2 Physics-based modeling
Source	Kaplan, W. (1991), *Advanced Calculus*, 4th edn, Addison-Wesley, Redwood City, CA.

Table 6.4 *(Continued)*

Attribute	Value
Page number	530
Question author	W. Colley
Question number	9.78
Question	Which of the following terms best describes use of models and simulation by the military for the purposes of obtaining insight into the cost and performance of military equipment?
Correct answer	Requirements and acquisition
Incorrect answer	Exploration of advanced technologies and concepts
Incorrect answer	Training
Incorrect answer	Geonavigation
Type	User/Manager
Difficulty	3 (moderate)
Topic	3.1 Combat and military
Source	Smith, R. D. (2009) *Military Simulations & Serious Games*, Modelbenders Press, Orlando FL.
Page number	38
Question author	S. Barbosa
Question number	8.10
Question	Which of the following terms is best defined as "the process of determining whether an implemented model is consistent with its specification"?
Correct answer	Verification
Incorrect answer	Validation
Incorrect answer	Accreditation
Incorrect answer	Calibration
Type	Core
Difficulty	2 (easy)
Topic	5.6 Verification, validation, and accreditation
Source	Petty, M. D. (2010) Verification, Validation, and Accreditation, in *Modeling and Simulation Fundamentals: Theoretical Underpinnings and Practical Domains* (eds J. A. Sokolowski and C. A. Banks), John Wiley and Sons, Inc., Hoboken, NJ, pp. 325–372.
Page number	330
Question author	M. Petty

will not see all of the supplementary attributes, just the question itself and its possible answers.

6.3.4 Question Quality Control

A group of expert external reviewers, which did not include any of the question developers, reviewed each of the questions. All of the reviewers were knowledgeable in M&S and many of them had received the CMSP designation. The reviewers examined the questions for correctness, clarity, and relevance, and they determined if each question's categorizations for type, topic, and difficulty were appropriate. Based on their feedback, the questions needing revision were revised before the examination was opened to the public.

6.3.5 Examination Instances

A unique instance of the CMSP examination is generated for each candidate by randomly choosing questions from the question bank. As mentioned earlier, there are two different types of CMSP certification, User/Manager and Developer/Technical. The candidate selects the type of certification he or she is seeking. In addition, the candidate may select up to one-third of the subtopics within each of four specific topics (applications, domains, modeling methods, and implementation) for exclusion from his/her examination instance. This exclusion option is intended to compensate for the very broad nature of the M&S body of knowledge. The M&SPCC determined that candidates cannot reasonably be expected to have extensive knowledge in all aspects of M&S.

Based on the candidate's chosen certification type and topic exclusions, a generated examination instance will have the following structure:

1) A total of 100 questions.
2) All questions either from the candidate's selected certification type or from "Core."
3) No less than 10 questions from each of the eight topics.
4) No questions from any of the candidate's excluded subtopics.
5) Average difficulty of all questions in the instance no less than 2.5 and no greater than 3.5.

6.3.6 Degree of Difficulty

The question development team strove to establish a degree of difficulty in the questions such that it is reasonable to expect that a candidate who passes the examination has a broad understanding of M&S, with substantial depth in some specific topics. The examination is meant to be difficult enough that the CMSP designation represents a noteworthy degree of knowledge and accomplishment, but not so difficult that experienced and knowledgeable M&S practitioners are unable to pass it.

As mentioned, the examination is customized to each candidate in two ways: (1) candidates select the type of certification they are seeking, and (2) candidates may exclude a limited number of subtopics from their examination instance. Nevertheless, an examination instance will have a broad scope, and it is unlikely that even a well-educated and experienced candidate will be able to simply sit down and answer the test accurately over a period of a few hours based on their existing knowledge. To the contrary, it is expected that each candidate will search for some of the answers. For that reason, candidates are given 14 days to complete it. The possibility that a candidate could complete some parts of the examination by searching for or looking up the answers over the 14-day examination period is entirely intentional; the M&SPCC wanted the process of preparing for and taking the examination to be itself a beneficial learning experience (Gross, 2014). Nonetheless, if necessary candidates may retake the examination; a new instance is generated for each attempt.

6.4 Examination Delivery System

In parallel with the question development, a new online examination delivery system was designed and implemented to allow CMSP candidates to attempt the examination conveniently and intuitively. This section describes the website and supporting software for administering the examination. First, the method for selecting the implementation approach is explained. Next, the examination procedure from the candidate's perspective is described. Then, the Web site's development process and features are explored in more detail. Finally, the process of populating the question database is discussed.

6.4.1 Trade Study on Existing Examination Software Packages

At the outset of the implementation effort, a trade study of 13 existing software packages for online testing was conducted to determine whether any met the needs of the CMSP examination process. The requirements for the examination software are summarized in Table 6.5. The table additionally illustrates the rarity of some of the required features. Some features, such as Web-based delivery and third-party support and hosting, were fairly common among the packages. However, the examination had certain

Table 6.5 A summary of requirements for the examination software. Requirements are listed in descending order of the number of packages that met the requirement. A total of 13 packages were examined. The requirements are assigned numbers for easy reference.

#	Requirement	Packages that met this requirement
1	Web-based delivery and scoring of questions	10
2	Third-party customer support	10
3	Easy-to-moderate effort of creation/setup/ installation	9
4	Third-party hosting	8
5	Easy-to-moderate effort of question entry and maintenance	8
6	Use questions in CMSP format, including images and formatting	7
7	Select a number of random questions	7
8	Select questions by category	7
9	Low price (per test, per student, and/or base price)	5
10	Automatically select questions from multiple categories	2
11	Allow multiple-day time window; allow user to pause and resume	2
12	Support three different categorizations (type, difficulty, topic)	1
13	Select questions to meet an average difficulty level	1

unique and specific needs not provided by many packages. Most important of these were the ability to select questions to meet an average difficulty level as discussed earlier and support for three different categorizations of questions. Both of these features were supported by one package each; however, they were not supported by the same package. Some packages were different from the CMSP examination in scale or scope; for example, some supplied an infrastructure to offer entire courses online.

Another option would have been to modify an existing package. In general, the primary problem with modifying existing packages to suit the requirements is that it requires developers to study and understand the existing code, the structure of the databases it creates, and how its various software components interact. This approach was considered carefully for many packages where source code was available. For example, one particular package was unable to support requirements 10, 12, 13, and 11. Modifying this particular package to support three different kinds of question categorization would have required its mechanism for question categorization and storage to be revised, and a mechanism for user question session saving would need to be implemented to meet requirement 11. Moreover, the organization that developed its software does not allow it to be rebranded, meaning that CMSP would not have been able to use its own branding and identity. Another package was more module-based, meaning that some modifications could be made in a modular fashion. This module-based design allowed for somewhat easier additions and enhancements to the code. By default, it did not meet requirements 2, 4, 5, and 6. Its primary disadvantage was a complicated codebase. Owing to the lack of third-party software meeting all of the CMSP examination's needs, the team decided to develop a custom system.

Section 6.4.2 explains the Web site's features and procedure from the candidate's perspective before subsequent sections delve into development of the Web site in more detail.

6.4.2   Examination Process from a Candidate's Perspective

A candidate for CMSP certification initiates the process by paying the required registration fee using an external Web site controlled by the CMSP certification administrators (NTSA). After the candidate has paid the fee, an administrator creates an account for the user in the CMSP online system. The Web site automatically generates a random password for the user and sends him or her an email with login instructions.

Upon first logging into the CMSP Web site, the candidate is prompted to provide some basic credentials, including their name, contact information, and type of certification sought (User/Manager or Developer/Technical). The candidate is then presented with a complete list of question subtopics and prompted to choose subtopics that he or she wishes to exclude from his or her examination (referred to as "exclusions"). Once the candidate has provided this information, he or she is presented with the "home" page to which he or she will return upon subsequent logins. The home page contains information and navigation links pertaining to the candidate and to his or her examination instance; this includes a link to start or resume the examination, the number of questions that he or she has answered thus far, and the amount of time remaining in the examination period.

When the candidate starts the examination, the system automatically generates an examination instance for the candidate. The system selects the questions for the instance in a process that implements selection logic specified by the M&SPCC. The database of potential questions is filtered to select Core questions and either User/Manager or Developer/Technical questions as appropriate for the candidate, then to remove questions in the subtopics excluded by the candidate. From this pool, 10 questions are chosen from each of the eight topics in the topic index and 20 additional questions are selected "at large" from any nonexcluded subtopic. The questions are then sorted by major topic. The average difficulty of the questions selected for an instance is also considered. If an examination instance's average question difficulty is less than 2.5 or greater than 3.5, the examination instance generation process is repeated. (Note that this is average difficulty; individual questions on an instance can, and likely will, be less or more difficult.) Because the questions were written collectively with varying difficulty in mind and the question set is large, the average difficulty test rarely fails; even when it does, examination instance generation is not computationally expensive and the regeneration is very fast.

Once an examination instance has been generated for the candidate, he or she is then directed to the examination. This interface, shown in Figure 6.1,

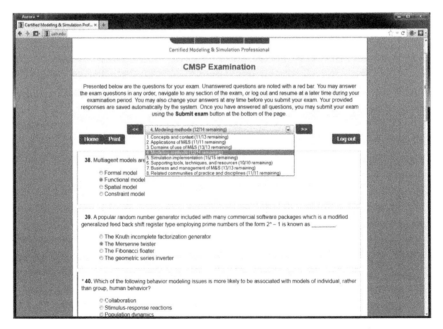

Figure 6.1 Screenshot of the examination Web site, showing navigation functions, features such as the print button and counts of unanswered questions remaining in all major topic areas, and three actual sample questions. Question 40 has not yet been answered by the candidate; the Web site indicates that fact with a red bar and asterisk.

is intended to be as intuitive and unrestrictive as possible while providing necessary functionality. The Web site assumes that each question is either true/false or has four potential answers (one correct, three incorrect). The order of the possible answers is randomized when presented to the user. The Web site also displays any images or equations associated with each question. Questions not yet answered by the candidate are noted with a red bar and asterisk, and the Web site also shows the number of questions left unanswered in each topic using a dropdown list. The candidate can navigate to any major topic using the arrow buttons or the dropdown list. The candidate is free to respond to questions in any order and can change answers at any time. Once he or she selects an answer for a question, whether or not it has been answered previously, the response is automatically recorded by the system.

The candidate must complete all 100 questions on his or her examination instance in order to proceed. He or she may log out and later resume the examination as often as desired during the examination period, which starts when the examination instance is created and lasts for 14 days. This period can be extended by the Web site administrators if approved by the M&SPCC on a case-by-case basis. Once all of the questions have been answered, the "Submit Exam" button located near the bottom of the page becomes active. When the candidate presses this button, the examination is submitted for grading, and he or she is directed to an optional comment form where he or she can submit feedback regarding the examination. Once the examination has been submitted, it is marked as such on his or her "home" screen, and the candidate may not return to the examination unless allowed access by a Web site administrator.

Upon submission, the Web site system automatically scores the examination and then emails the administrators, informing them how many questions the candidate answered correctly and whether he or she passed or failed the examination. At least 85 of the 100 questions must be answered correctly to pass the examination. The administrators then inform the candidate of his or her status manually, usually via email.

6.4.3 Examination Web site System Structure and Administrative Features

The Web site is constructed using the PHP and JavaScript scripting languages, the MooTools AJAX framework, and standard technologies such as HTML and CSS. The Subversion version control system and a set of custom scripts allow development versions of the Web site system, which may contain experimental and not yet thoroughly tested features, to coexist alongside the Web site system used by candidates and administrators.

Aside from the features already mentioned, the Web site also contains an administrative frontend that allows Web site administrators to add or modify users, examination instances, and questions in the question database. The frontend provides an instant search feature, which allows administrators to dynamically filter the list of questions to find questions of interest to them. The frontend also allows administrators to export the entire set of questions in the standard comma-separated value (CSV) format. It also provides aggregate reporting of question statistics, including the number of times each question has been included in an examination instance and the number of times each of the possible answers has been selected by a user. These features are intended to support question auditing and categorization.

6.4.4 Question Bank and Database Format

The question database used by the system is intended to be the official version of the question bank for the M&SPCC's purposes. Data items associated with each question, such as the question text, identification number, correct answer, incorrect answers, source, and author, are stored in mySQL format, while images and equations are stored in a directory. Question text is written and displayed using a sanitized HTML markup, which allows questions to reference images, italicize appropriate key words such as "not," and display subscripts and superscripts as necessary. Together, the mySQL database and images directory constitute the entirety of the question bank. The version control scheme for the question bank stores both questions and images, and scripts allow images and the database to be rolled back to previous versions if necessary.

To allow question authors to work in a familiar environment, questions are originally written and stored in Microsoft Word documents. The Web site administrative frontend provides the ability to import questions into the question bank using an import script. This script parses Word documents, extracting fields associated with each question in the process, and populates the bank with questions provided in the Word format.

6.5 Future Work

Future work related to the CMSP examination includes the following:

1) Additional questions will be developed for those subtopics with relatively few questions.

2) As candidates attempt the examination, correct/incorrect statistics will be accumulated for each question, and questions that prove to be unduly troublesome will be revised or replaced.

3) The topic index will be extended and additional questions will be developed as the M&S body of knowledge expands, and questions developed for the new topics and subtopics.

Acknowledgements

Financial support for the development of the CMSP examination questions and the online examination system was provided by the National Training and Simulation Association and The Boeing Company. Frederick L. Lewis and Patrick T. Rowe (NTSA) provided essential direction during the examination development process, with James A. Robb (NTSA) providing continuing guidance. Substantial contributions were made to the topic index by William F. Waite (AEgis Technologies) and David C. Gross (Lockheed Martin). A group of anonymous reviewers led by John A. Sokolowski (Old Dominion University) reviewed the questions for accuracy, consistency, and relevance. Glenn S. Nesbitt, Joshua M. Ciardelli, and Justin K. Watson (UAH) contributed to the development of the CMSP Web site system software. J. Mark McCall (Simulation Interoperability Standards Organization) gave permission to reuse portions of an earlier description of the CMSP examination (Petty et al., 2012). Margaret L. Loper and Charles D. Turnitsa reviewed a preliminary draft of this chapter. Their support is gratefully acknowledged.

References

Banks, J. (1998) *Handbook of Simulation: Principles, Methodology, Advances, Applications, and Practice*, John Wiley & Sons, Inc., New York, NY.

Greasley, A. (2008) *Enabling a Simulation Capability in the Organisation*, Springer-Verlag, London.

Gross, D.C. (2014) Preparation course for the certified modeling and simulation professional examination. *Presentation at the 2014 Interservice/Industry Training, Simulation, and Education Conference, Orlando FL*, December 1–5, 2014.

Lacy, L.W., Gross, D.C., Ören, T.I., and Waite, W.F. (2010) A realistic roadmap for developing a modeling and simulation body of knowledge index. *Proceedings of the Fall 2010 Simulation Interoperability Workshop, Orlando FL*, September 20–24, 2010, pp. 240–248.

Lewis, F. and Rowe, P. (2010) The certified modeling & simulation professional (CMSP) program: why it was created, where it stands now, and what you can do to support it. *SCS M&S Magazine*, 1 (1), 16–20.

Petty, M.D., Reed, G.S., and Tucker, W.V. (2012) Topics, structure, and delivery of the new certified modeling and simulation professional examination. *Proceedings of the Spring 2012 Simulation Interoperability Workshop, Orlando FL*, March 26–30, 2012, pp. 188–195.

Sokolowski, J.A. and Banks, C.A. (2010) *Modeling and Simulation Fundamentals: Theoretical Underpinnings and Practical Domains*, John Wiley & Sons, Inc., Hoboken.

Tolk, A. (ed) (2012) *Engineering Principles of Combat Modeling and Distributed Simulation*, John Wiley & Sons, Inc., Hoboken.

Waite, W.F. (2010) SimSummit – shaping the future of modeling and simulation, part 1 – establishing the SimSummit forum. *SCS M&S Magazine*, 1 (1), 21–32.

Part III

Society

7

Modeling and Simulation Societies Shaping the Profession

Robert K. Armstrong[1] and Simon J.E. Taylor[2]

[1]Eastern Virginia Medical School, Norfolk, VA, USA
[2]Brunel University London, London, UK

7.1 Overview and Introduction

Wendell Berry said, "A proper community, we should remember also, is a commonwealth: a place, a resource, an economy. It answers the needs, practical as well as social and spiritual, of its members – among them the need to need one another (Berry, 2003)." While Berry's quote is referencing the social communities of neighborhoods and churches, the value found in sharing the journey extends to borderless communities – societies and associations – as well. This chapter focuses on the various societies and associations that have and remain committed to shaping the modeling and simulation profession.

Societies and associations – the terms are often used interchangeably, and throughout this chapter referred to collectively as organizations – are member-supported groups formed to unite and inform people who work in the same occupation or that have similar professional or personal interests.[1] These organizations can provide a strong sense of community and a base and impetus for coordinated activity. Typically, people join societies and associations to work with people who share their interests. A sense of belonging is key to involvement; a sense of belonging to a greater community improves motivation, health, and happiness – while also fostering teamwork while the community works toward a common set of goals. In a sense, the societies and associations people join for professional and personal interests are microcosmic representations of the general

---

1 Retrieved from https://www.thebalance.com/professional-association-1736065[2] (accessed September 26, 2016).

The Profession of Modeling and Simulation: Discipline, Ethics, Education, Vocation, Societies, and Economics, First Edition. Andreas Tolk and Tuncer Ören.

societies that we inhabit as human beings – the collection of people, infrastructure, and government that promote the relatively ordered world where we work, play, and sleep.

7.2 The Importance of Societies and Associations

Societies and associations bring together people in agreement regarding the organization's common theme. Differing points of view, however, can provide opportunities for conflict. This conflict can be negative or positive. "Conflict is positive when it causes people to consider different ideas and alternatives; results in increased participation and more commitment to the decisions and goals of the group; results in issue clarification and/or reassessment; or helps build cohesiveness as people learn more about each other. Conflict is destructive when it leads to bullying, harassment, or discrimination; diverts energy from more important issues and tasks; polarizes groups so that cooperation is reduced, or destroys the morale of people, or reinforces poor self-concepts.[2]" The challenge for all societies and associations is to manage conflict while growing the organization and meeting member needs. Positive conflict is typically necessary to advance the goals of any organization.

Societies and associations tend to offer both tangible and intangible benefits to their members. Tangible benefits include services, products, information, and discounts. Traditional benefits offered by societies and associations are typically most important to the members, and include specialized learning, access to collective knowledge, a professional community, and advocacy. Intangible benefits include networking, a sense of belonging, a common purpose, and volunteering opportunities.[3] Depending upon the way societies and associations are structured, and upon their business model, nonmembers may be able to take advantage of organization-created benefits. The informational output of some organizations is too valuable to restrict to "paying customers," and so access is not restricted.

Societies and associations can have a significant positive impact upon economies. According to data gleaned from the Internal Revenue Service (IRS), in the United States – in 2013, the IRS recognized nearly 67,000 trade and professional associations, many of which are societies and associations. Also in 2013, membership organizations generated a payroll of nearly $51 billion. In the same year, trade and professional associations generated $142 billion in revenue

---

2 Retrieved from http://onpace.osu.edu/modules/transitioning-to-the-workplace/ addressing-conflict-in-the-workplace/positive-and-negative-conflict (accessed September 31, 2016).

3 Retrieved from http://www.asha.org/associates/Why-Do-Associations-Exist/ (accessed September 25, 2016).

and assets. Membership associations in the United States hold an estimated $306 billion in assets. This speaks to the economic benefit that societies and associations can have on a greater community, region, or country.

So, the assumption from this data is that societies and associations are useful entities that provide benefits to members, nonmembers, and the greater community. How do societies and associations impact an emerging technical discipline – for instance, a relatively new professional area like modeling and simulation?

7.3 The Organization of the Modeling and Simulation Profession

To better understand how societies and associations are shaping the modeling and simulation profession, it is important to understand how the profession is organized. To some, modeling and simulation is an area of study, much like computer science. To others, it represents the family of tools that can be applied to solve challenging problems. To others, still, it embodies the delivery of a substitute for an experience that is too resource intensive to actually provide as real-world events. For the purpose of this chapter, the profession of modeling and simulation encompasses all three of these interpretations – something to study (the academic endeavor), something to use (a tool), and something to do (the experience).

Modeling and simulation is a discipline with different application domains. The greater community of "simulationists" – modeling and simulation professionals – includes people who work to define the academic foundations of the discipline (science that contributes to modeling and simulation theory), people who look for general methods to solve general problems (engineering processes rooted in modeling and simulation theory), and people who solve real world problems (modeling and simulation tool development). Three commonly referenced modeling and simulation application domains are support for analyses, support to systems engineering activities, and providing or supporting training and education activities. There could be other descriptions of modeling and simulation domains, but these are typical.

While foundational concepts and theories are, by their very nature, critical to modeling and simulation, the need for solutions to real user problems continues to drive modeling and simulation. The science of modeling and simulation and the use of modeling and simulation are definitively intertwined. Organizing solutions requires a supportive environment to share ideas. Societies and associations serve as safe havens for technical and nontechnical discussion of concepts and theories as well as the development and application of modeling and simulation.

Members of modeling and simulation organizations are developers, supporters, vendors, users, and consumers. Organizations tend to be focused on either the discipline of modeling and simulation or the use of modeling and simulation engineering for application development specific to an industry area.

In the United States, creators, users, and consumers of modeling and simulation (and analysis) can be found in all the major industry sectors listed in the 2014 Industry Economic Output table below. Interestingly, every major industry sector in the United States has a direct or indirect need for modeling and simulation. Several of these industry sectors can lay claim to at least one modeling and simulation society or organization.

Industry sector	2014 Economic output	
	$ Billions	Percent
Total	31,295.9	100.0
Goods-producing, excluding agriculture	8,169.1	26.1
Mining	651.9	2.1
Construction	1,166.6	3.7
Manufacturing	6,350.5	20.3
Service-providing	21,184.2	67.7
Utilities	400.4	1.3
Wholesale trade	1,538.6	4.9
Retail trade	1,560.5	5.0
Transportation and warehousing	1,031.3	3.3
Information	1,553.4	5.0
Financial activities	3,987.9	12.7
Professional and business services	3,335.0	10.7
Educational services; private	327.7	1.0
Healthcare and social assistance	2,130.5	6.8
Leisure and hospitality	1,162.3	3.7
Other services	637.5	2.0
Federal government	1,133.1	3.6
State and local government	2,385.9	7.6
Agriculture, forestry, fishing, and hunting	511.9	1.6
Special Industries	1,430.8	4.6

Source: From the U.S. Department of Labor, Bureau of Labor Statistics. Available at http://www.bls.gov/emp/ep_table_202.htm.

7.4 Societies and Associations Focused on Advancing the Discipline of Modeling and Simulation

The discipline of modeling and simulation, arguably still emerging, is the focus of several international societies and associations. These organizations are fostering the collaborative development of much of the core science with which modeling and simulation tools are built. These organizations are not industry specific. An examination of their members will show a multitude of academics from a myriad of backgrounds, along with members of industry who are building tools and using modeling and simulation to solve complex problems outside the boundaries of a specific industry sector. These are the people who are expanding upon and sometimes laying the foundation for the more widespread and consistent use of modeling and simulation throughout both industry and society.

7.4.1 Society for Modeling and Simulation International

Most notable is the *Society for Modeling and Simulation International* (SCS, www.scs.org). "SCS is the premier technical Society dedicated to advancing the use of modeling & simulation to solve real-world problems; devoted to the advancement of simulation and allied computer arts in all fields; and committed to facilitating communication among professionals in the field of simulation. To this end, SCS organizes meetings, sponsors and co-sponsors national and international conferences, and publishes the *SIMULATION: Transactions of The Society for Modeling and Simulation International* and the *Journal of Defense Modeling and Simulation* magazines.[4]"

Management of SCS is performed by a Board of Directors, and Executive Director and a President's Council. The SCS President and President-Elect of the Board are member elected. Members pay dues and provide the majority of the workforce effort required to meet the goals of the Society. The Board of Directors, considered the SCS Officers, is typically made up of core leaders (President, President-Elect, Past President, Treasurer, Secretary, Executive Director, and Chairman), Vice Presidents (Conference, Membership, Publications, Education), and several Directors at Large. The Presidents Council consists of all past Presidents of the Society. Permanent staff includes the Executive Director, Accountant, Administrative Assistant, and Publications Manager/Editor.

---

4 Retrieved from http://scs.org/about/ (accessed September 31, 2016).

There are several standing SCS Committees, which are important in that they represent the critical areas of focus for the organization. The purpose of these committees is explained further.

- *Governance Committee:* Ensures the governance of SCS is correct and consistent by maintaining all organizational policies. This committee makes sure that organizational leadership remains bound to established, agreed upon policies and procedures.
- *Strategic Planning Committee*: Develops, maintains, and evaluates SCS's five-year-plus plan.
- *Audit Committee*: Ensures organizational financial compliance to SCS policies, best practices, and laws.
- *Personnel Committee*: Provides oversight to personnel policy for SCS employees.
- *Nominating Committee*: Manages the process to select and vet Board of Directors candidates.
- *Fellow Selection Committee*: Reviews, considers, and nominates SCS Fellow candidates.
- *Presidents Council Committee*: Provide support to the SCS President at his/her discretion.
- *Awards and Recognition Committee*: Manages process to recognize worthy professionals for their accomplishments.
- *Education and Workforce Development Committee*: Supports the effort to transform modeling and simulation into a recognized discipline and profession.
- *Conference Committee*: Overall responsibility for SCS conference activities.

SCS hosts multiple annual conferences. Conferences are important opportunities to meet many like-minded people in a very short period of time. They help to personalize topics by linking the human with the activity. Conferences grow an individual's professional network and help to build their knowledge base.

- POWERPLANTSIM: Focused on the special needs of nuclear and fossil power plant simulation.
- SPRINGSIM/SUMMERSIM: These two conferences bring together leading experts in various domains of modeling and simulation.
- AUTUMNSIM: A joint multiconference in conjunction with the Asia Simulation Conference.
- SIMAUD: Focused on simulation in the fields of architecture, urban design.

Education and workforce development is a critical component of SCS. Student chapters are encouraged and supported. Chapters can be found in the U.S. (University of Central Florida and Old Dominion University) and internationally (Latvia, China, and Turkey). Student chapters are critical to the development of future modeling and simulation scientists, educators, and practitioners.

Publications are a critical component offered by SCS. Their publications attract authors from all areas of modeling and simulation, including academia, government, and commercial sectors. SCS publishes two regular journals.

The Journal of Defense Modeling and Simulation: Applications, Methodology, Technology is published quarterly. It is a refereed, archived journal focused on advancing the practice, science, and art of modeling and simulation in defense/federal sectors.

SIMULATION: Transactions of the Society for Modeling and Simulation International is published monthly. It is a refereed, archived journal with distinct sections: theory and applications. The publication acceptance rate is 30%. Special issues are considered on a case-by-case basis.

Theory-focused articles reflect original work in areas such as

- model execution
- simulation interoperability
- modeling techniques, languages, and development systems
- analysis methodologies and techniques
- verification, validation, and accreditation
- randomness in simulations
- techniques for real-time simulation

Application-focused articles reflect original work depicting mature applied experiments and development. Application areas of interest include the following:

- Computer science
- Engineering
- Natural and life sciences
- Defense
- Social and cognitive sciences

Special issues of the journal *SIMULATION* concentrate on areas of theory and application that are certainly relevant but typically not on areas that would support a monthly publication cycle. There are also issues that focus the articles on a slightly more defined topic. Such areas include medical simulation, driving simulations, optimization in industry and engineering, sustainability modeling, complex social system modeling, software tools, virtual environments, rare event simulation methodologies and applications, supply chain modeling, and distributed computing. Special issues help to provide a concentration of thought and ideas for SCS members, and provide insight into areas to which some members are otherwise never exposed.

SCS also publishes a monthly *M&S Newsletter*. This is a general interest publication and is accessible by the general public. Topics are broad and planned to be of interest to the entire modeling and simulation community. This newsletter replaced the SCS M&S Magazine in 2014.

7.4.2 Simulation Interoperability Standards Organization

Another organization with cross-domain focus is the *Simulation Interoperability Standards Organization* (SISO, http://www.sisostds.org//). SISO is an international organization based in Orlando, Florida. Orlando is the historical home of a number of U.S. Department of Defense (DoD) program and acquisition offices specializing in the training and education of military personnel.

SISO formed because of the challenges and successes that came from early DoD efforts to make disparate simulations interoperable. Given the costliness of live training exercises – especially large-scale exercises – DoD leaders needed to artificially represent battlefield activities and the impacts of operational decision making in as realistic a fashion as possible. Modeling friendly and enemy weapon systems and the operational environment, and simulating model activities over time, appeared to be a smart approach. Early on, both hardware and software were minimally adequate to meet the realism needs of DoD. A glaring issue, however, was the fact that multiple different models and simulations needed to be used to represent even the simplest training scenario and environment, and, few, if any, of these models and simulations were built to be interoperable. This was not due to developer laziness; it was due to the fact that no one had ever asked for interoperable simulations. They simply did not exist – yet.

Standards are usually appreciated, but the scope of their benefits often is not. Standards are documented agreements, developed through consensus, that contain precise criteria that serve as rules or guidelines for products, processes, services, and materials (like standard types of electrical plugs, cable television connectors, and video formats). They facilitate uniformity that enables, ideally, worldwide use of a product. Standards remove technical barriers. For an emerging discipline such as modeling and simulation, standards were and are critical. Standards help developers by giving them shortcuts, so to speak. A software developer is no longer required to create a specific, proprietary solution, but can, instead, rely upon the standard to guide relevant features and capabilities. Since standards are created in consensus, developer needs are, therefore, vetted when standards are approved.

Efforts to create DoD simulation interoperability standards, supported by governments and used by developers, led to a series of Distributed Interactive Simulation (DIS) Workshops from 1989 through 1996. DIS was developed as an Institute of Electrical and Electronics Engineers (IEEE) standard. As DIS became more mature, so did support for the standard. SISO was born, holding its first sponsored DIS workshop in 1997 in Orlando, Florida. In 2003, IEEE granted SISO status as a recognized IEEE Sponsored Committee; SISO now maintains two major IEEE standards: IEEE 1278 (Distributed Interactive Simulation) and IEEE 1516 (High Level Architecture for Modeling and Simulation).[5] These two standards are critical to not only DoD simulations, but also to any set of simulations existing on a distributed network and required to be interoperable. SISO also manages several other International Organization for Standardization (ISO) and the International Electrotechnical Commission (IEC) standards, as well as their own standards. Both of these standards are recognized and followed by the simulation development community.

Like SCS, SISO is a volunteer-based organization, with fiscal oversite provided by a Board of Directors. The Executive Committee (Chair, Vice Chair, Secretary, several members) is the policy body providing overall guidance and leadership. Workshops, which are critical to the collaborative nature of the Society, are managed via the Conference Committee (Chair, Vice Chair, Secretary, and several other members). The Standards Activity Committee (Chair, Vice Chair, Secretary, and several other members) oversees the core SISO functions of consensus-based standards management. SISO is a membership-supported organization. Their business model provides access to standards information regardless of membership status, which is different from many societies and associations, but which makes sense given their standards development mission. Membership affords voting rights, free downloads, and the right to run for office, and is required to attend SISO Conference and Workshops. SISO is also supported by sponsorships, typically from commercial partners, with public/international sponsors providing support as their policies and laws permit.

SISO maintains an active Web site (www.sisostds.org) that includes a publicly accessible document library, as well as information on standards activities and workshops (past and upcoming). The SISO Standards Activity Committee creates and shares all their standards, guidance, references, and administrative products via their Web site.

---

5 The history of SISO, retrieved from https://www.sisostds.org/AboutSISO/Overview/TheHistoryofSISO.aspx (accessed October 3, 2016).

Standards development has an impact across multiple simulation domains, each of which participates in or benefits in some way from SISO efforts. SISO has many valued affiliates, including industry associations (the National Training and Simulation Association (NTSA), Simulation Australasia, and the aforementioned SCS), industry conferences (Interservice/Industry Training, Simulation and Education Conference (I/ITSEC), and the International Training and Education Conference (ITEC)), and professional organizations (the Military Operations Research Society (MORS) and the International Test and Evaluation Association (ITEA)). SISO and its membership align closely with these affiliated organizations. Memorandums of Agreement are in place with select affiliates, streamlining the process of collaboration and sharing.

7.4.3 Association of Computer Machinery Special Interest Group on Simulation and Modeling

The Association of Computer Machinery (ACM, www.acm.org) hosts a Special Interest Group (SIG) on Simulation and Modeling (ACM SIGSIM, www.acm-sigsim-mskr.org/) that has been important to the growth of modeling and simulation over the years. SIGSIM started as a Special Interest Committee (SIC) in 1967 and was chartered as a SIG in 1969 after fantastic growth. SIGSIM is one of 37 active SIGS in the ACM. The organization is managed by elected officers, and utilizes an advisory board for strategic direction.

SIGSIM is focused on the following areas of modeling and simulation: discrete, continuous, Monte Carlo, system dynamics, game-based, agent-based, artificial intelligence-based, and virtual reality-based model representation; distributed, parallel, and web-based model execution; models composed to support live exercises, experimentation, demonstration, and trials; and hardware, human, and software in-the-loop conditions. SIGSIM hosts a modeling and simulation knowledge repository at their Web site that serves the comprehensive modeling and simulation community. SIGSIM sponsors several conferences including the ACM SIGSIM Conference on Principles of Advanced Discrete Simulation (ACM SIGSIM PADS) and ACM Conference on Modeling, Analysis and Simulation of Wireless and Mobile Systems (MSWiM). SIGSIM also cosponsors the Winter Simulation Conference and the IEEE/ACM International Symposium on Distributed Simulation and Real Time Applications. The organization also offers courses on introductory simulation principles via their Web site.

One particularly noteworthy conference should be mentioned because it is cosponsored by several important modeling and simulation societies and associations: The Winter Simulation Conference (WSC). WSC is the premier international forum for disseminating recent advances in the field of system simulation. In addition to a technical program of unsurpassed scope and quality, WSC provides the central meeting place for simulation practitioners, researchers, and vendors working in all disciplines in industry, service, government, military, and academic sectors. The conference dates back to 1967. Cosponsors include

- American Statistical Association (ASA), Technical Cosponsor
- Arbeitsgemeinschaft Simulation (ASIM), Technical Cosponsor
- Association for Computing Machinery: Special Interest Group on Simulation (ACM/SIGSIM)
- Institute of Electrical and Electronics Engineers: Systems, Man, and Cybernetics Society (IEEE/SMC), Technical Cosponsor
- Institute for Operations Research and the Management Sciences: Simulation Society (INFORMS-SIM)
- Institute of Industrial Engineers (IIE)
- National Institute of Standards and Technology (NIST), Technical Cosponsor
- The Society for Modeling and Simulation International (SCS)

The Winter Simulation Conference is also supported by the WSC Foundation. The purpose of the WSC Foundation (WSCF) is to develop and manage a fund to help ensure the continuance and high quality of the conference. Management of the fund allows for (a) providing loans or grants from the fund to the WSC Board of Directors when the WSC Board makes such requests and the WSCF deems it prudent and wise to do so; (b) allocating assets of the fund to balance return, risk, and liquidity; and (c) accepting donations to the fund and raising money for the fund as and when it is deemed appropriate. The WSCF operates independently of the WSC Board of Directors, and is also independent of the annual WSC Committees. The conference also supports free access to all papers ever published in the conference.

7.5 Societies and Associations Focused in Specific Areas or Industries

The number of societies and associations focused on advancing the discipline of modeling and simulation may be considered relatively small. There are societies and associations with a slightly more specific industry focus, and that support a community of users who seek to solve specific problems by leveraging modeling and simulation; these users typically do not need to validate scientific theory or develop relevant standards. The worldwide list of societies and associations that fit into this category – organizations supporting the use of modeling and simulation to solve industry-specific problems – is large. The following is not a comprehensive list, but serves to highlight those societies and associations that are helping to shape the modeling and simulation profession.

7.5.1 Operations Research Societies and Associations

Operations research as a modern discipline grew out of military planning activities in the First and Second World Wars, when leaders sought quantitative analysis to support complex decision-making. Operations research organizations have a significant positive impact on the modeling and simulation profession. Operations research is defined as "a discipline that deals with the application of advanced analytical methods to help make better decisions. The terms management science and analytics are sometimes used as synonyms for operations research."[6] While its own formal discipline, operations research overlaps other disciplines. Operations research relies heavily on computational and statistical methods, and benefits tremendously from the computer science field. Relevant to this chapter, the subdisciplines of operations research include modeling and simulation, which has its own roots in the field of computer science. The tools and mechanisms used by operations research professionals are not only for operations research, but can also be leveraged across a multitude of industries and sectors. This makes societies and associations that focus on operations research important to the field of modeling and simulation; they are shaping the future by their aggressive and comprehensive development of models and use of simulations to support their efforts.

Three organizations of note include the Institute for Operations Research and the Management Sciences (INFORMS, www.informs.org), the Military Operations Research Society (MORS, www.mors.org), and The Operational Research Society (www.theorsociety.com).

7.5.1.1 Institute for Operations Research and the Management Sciences

INFORMS, headquartered at the University of Maryland in Catonsville, Maryland, is the largest organization in the world focused on this area. Their mission is to grow the science behind, development, and use of OR, management science, and analytics. INFORMS publishes 14 scholarly journals, organize national (the United States) and international

---

6 INFORMS: What is Operations Research? Retrieved from https://www.informs.org/About-INFORMS/What-is-Operations-Research (accessed October 5, 2016).

conferences for academics and professionals, and provide analytics certification and continuing education to members. They provide a robust Web site with many resources available to members.

INFORMS hosts the INFORMS Simulation Society (I-SIM). The INFORMS simulation section became the INFORMS Simulation Society in November 2004. While they are the newest "Society" in INFORMS, the section was started as the College on Simulation and Gaming in The Institute for Management Science (TIMS) in 1963, long before the ORSA/TIMS merger to form INFORMS in 1995. The Societies page on the INFORMS web site describes their mission in brief:

"The INFORMS Simulation Society provides a focus within INFORMS for the field of simulation; encourages the development and dissemination of knowledge in simulation and related fields; recognizes outstanding contributions both in technical innovation and in service to the profession; promotes communication and interaction among individuals and organizations who share an interest in simulation, including the sponsoring of conferences on simulation and sessions at INFORMS meetings; encourages students to study topics in simulation and to pursue related professional careers; and supports the continuing education of all simulation professionals."[7] I-SIM cosponsors the Winter Simulation Conference and their own conference series.

7.5.1.2 Military Operations Research Society

MORS is a U.S.-based society that is tailored to meet the OR needs of the United States DoD. MORS is over 50-years-old, initially a symposium sponsored by the United States Navy's Office of Naval Research. MORS was formed in 1966. Today, MORS holds annual meetings as well as classified and unclassified workshops. They publish a quarterly magazine and a refereed journal, supporting original authors and republishing relevant historical documents.

7.5.1.3 Operational Research Society

The Operational Research Society is a U.K.-based organization. While nearly every country has some form of operations research society or association, this one is unique. The United Kingdom can rightfully claim to have originated modern operations research during the Second World War, making this society important today. A critical mission of this society is the promotion and investment in operations research throughout the United Kingdom. Members perform volunteer and pro bono work in

---

7 INFORMS Simulation Website, https://www.informs.org/Participate-in-a-Community/Societies-and-Sections/ISIM (accessed October 13, 2016).

schools and organizations, respectively, with the intent of helping others appreciate the importance of well executed operations research. This society sponsors numerous conferences and workshops and publishes a number of journals as well as a monthly magazine called "Inside O.R." – all promoting and informing readers across a number of academic and industry sectors.

In terms of modeling and simulation, the Operational Research Society's simulation operations research group runs special interest meetings during the year and a biannual Simulation Workshop. Founded in 2002 by Simon J. E. Taylor and Stewart Robinson, the workshop attracts around 100 academics and practitioners to discuss simulation-related issues. The Operational Research Society (with Springer) also supports the *Journal of Simulation*, also founded by Simon Taylor and Stewart Robinson in 2006, a high quality journal publishing a wide variety of papers related to discrete simulation.

7.5.2 Healthcare Societies and Associations

Healthcare modeling and simulation can be characterized as human-based (standardized or simulated patients), device-based (manikins, part-task trainers, virtual simulations), and multimodal (some combination of the two). A standardized or simulated patient (SP) is a person who has been coached to represent (simulate) a patient, including history, body language, physical findings, and emotional and personality characteristics (Barrows 1987). Manikins and part-task trainers can be physical and virtual patient models that provide individuals and small teams with relevant physiological context.

Modeling and simulation societies and associations in the international healthcare domain are important because they are all focused to a large degree on the education of healthcare students and training of healthcare practitioners. Their members use models and simulations to impart knowledge and skills; they know the value of healthcare modeling and simulation tools. These organizations are a critical part of the collaborative process required to establish practitioner standards of performance – standards that will be met in part through the use of simulation and simulators to educate, train, and eventually validate individual and small team performance. The value of simulations in this context is to impart training and education that meets a standard. While there is interest in the science behind the simulation, the simulation is only useful if it can be used to effectively support acquisition and validation of technical skills and practical knowledge.

The first medical specialty to require a simulation-based course to periodically reassess individual performance was anesthesiology. A course to meet that need was developed by the American Board of Anesthesiology and managed by the American Society of Anesthesiologists. The course was groundbreaking because it was the first time that simulation was mandated to assess practitioners. Interestingly, other specialty groups (like surgeons) are investigating how to best incorporate simulation to assess and validate performance.

Four established societies and associations lead the way in establishing best practices for healthcare simulation. All four are member supported and serve the international healthcare community.

The Society for Simulation in Healthcare (SSH, www.ssih.org) is a U.S.-based international society that accredits simulation centers, certifies healthcare simulation professionals, holds an annual conference and numerous regional events and workshops, and publishes the *Journal of the Society for Simulation in Healthcare* on a monthly basis.

The Society in Europe for Simulation Applied to Medicine (SESAM, www.sesam-web.org) is based in Copenhagen. Like SSH, SESAM accredits simulation centers and educational institutions and holds annual meetings. SESAM does not publish a journal, but does highlight compelling literature to their members.

The International Nursing Association for Clinical Simulation & Learning (INACSL, www.inacsl.org) serves the largest practitioner community. Their focus is on improving patient safety. The Association has developed Standards of Best Practice to provide their communities input to the science of simulation and development of solutions. INACSL hosts an annual conference and periodic workshops, and publishes a journal; they do not accredit programs or institutions.

The Association of Standardized Patient Educators (ASPE, www.aspeeducators.org) supported development of the SP Methodology, which is the standardized approach to case development, SP training, and SP feedback to the student. ASPE provides educational content at their annual conference, and hosts a virtual learning center for members. ASPE does not accredit programs or institutions.

7.5.3 Environmental Societies and Associations

The International Environmental Modelling and Software Society (iEMSs, www.iemss.org), based in Switzerland, promotes the development and use of environmental software tools. Like most of the organizations mentioned throughout this chapter, they organize conferences and workshops, and publish a newsletter and a journal. iEMSs focuses on environmental data mining, distributed environmental modeling, complexity and feedback, and ecosystem services.

7.5.4 Social Science Societies and Associations

The Computational Social Science Society of the Americas (CSSSA, www .computationalsocialscience.org) is based in Washington, D.C. They promote their namesake, computational social science, which is "the science that investigates social and behavioral dynamics through social simulation, social network analysis, and social media analysis."[8] Their counterpart in Europe is the European Social Simulation Association (ESSA, www.essa.eu. org), and in Asia, the Pan-Asian Association for Agent-based Approach in Social Systems Sciences (www.paaa.asia/). These organizations play an important role; social behavior and societies in total are tremendously difficult to model accurately. These models are important to better understand how to motivate citizens to organize toward accomplishing specific goals.

7.5.5 Defense Societies and Associations

The National Training and Simulation Association (NTSA, www .trainingsystems.org/) connects the international defense industry (makers of things) with the United States DoD and other military services (buyers of things) while creating opportunities for technical interchange. NTSA events such as the Interservice/Industry Training, Simulation and Education Conference are catalysts for learning, collaboration, and development by exposing attendees to both state-of-the-art needs and solutions.

While technically not a society or association, the North Atlantic Treaty Organization (NATO) Modelling and Simulation Centre of Excellence (MSCOE, www.mscoe.org) fits within this discussion. MSCOE seeks to provide a consistent modeling and simulation framework for NATO countries – a challenging task given the sovereign status of every NATO member. MSCOE also seeks to align NATO with the U.S. DoD modeling and simulation standards and activities, a desire driven by the continued deployment of NATO coalition forces and the need for compatible operational capabilities.

7.6 National Societies and Associations

Many countries have modeling and simulation societies and associations that, while not large enough to impact the shape of the industry or immediate direction of the field, do serve to bring together like-minded

---

8 Computational Social Science, retrieved from https://computationalsocialscience.org/ (accessed October 10, 2016).

academics and professionals seeking to collaborate and network on an area that they are typically passionate about. The passion for modeling and simulation is honest and borderless. It comes from knowing what can be simulated, relatively inexpensively, via computer, as compared to what tends to be unaffordable when conducted in the real world. Academic excellence is not restricted to large populations or countries with high gross domestic products. An academic or professional in a smaller country can still perform simulation experiments and learn about his or her world relatively inexpensively, making simulation very appealing.

National societies and associations shape the modeling and simulation profession by fostering collegiality. Colleagues have the opportunity to find partners and collaborators, and entrepreneurs and professionals have the opportunity to connect with academics working in basic and applied research. Researchers and professionals are no longer alone in their interests. Growth is encouraged, and new theories and technologies are developed and shared. The profession continues to mature. Modeling and simulation users benefit.

The following table includes a partial list of national modeling and simulation societies and associations not included previously.

Country/region	Name	Web site
Australia/New Zealand	MSSANZ: Modelling and Simulation Society of Australia and New Zealand Inc.	www.mssanz.org.au
Japan	JSST: Japan Society for Simulation Technology	www.jsst.jp/e/
The Netherlands	DSSH: Dutch Society for Simulation in Healthcare	www.dssh.nl/en
Singapore, China, India, Malaysia, Japan, Korea Philippines, Brunei	Pan Asian Simulation Society in Healthcare	passh.org/
Europe	ESSA: European Social Simulation Association	www.essa.eu.org/
Europe	Federation of European Simulation Societies (EUROSIM)	www.eurosim.info
Europe	EUROSIS: European Multidisciplinary Society for Modelling and Simulation Technology	www.eurosis.org

Country/region	Name	Web site
China	HKSSiH: Hong King Society for Simulation in Healthcare	hkssih.org.hk/
Australasia	Simulation Australasia	www .simulationaustralasia .com
North America	North American Simulation and Gaming Association (NASAGA)	www.nasaga.org
Latin America	ALASIC: Asociación Latinoamericana de Simulacíon Clínica	www.alasic.org
Switzerland	SAGSAGA: Swiss Austrian German Simulation & Gaming Association	www.sagsaga.org
Europe	EUROSIM: Federation of European Simulation Societies	www.eurosim.info
EUROSIM Member Societies		
Germany	ASIM - German Simulation Society	www.asim-gi.org
Spain	CEA SMSG: Spanish Modelling and Simulation Group	
Croatia	CROSSIM: Croatian Society for Simulation Modelling	
Czechoslovakia	CSSS: Czech and Slovak Simulation Society	www.fit.vutbr.cz/ CSSS/
Belgium, The Netherlands	DBSS: Dutch Benelux Simulation Society	www.dutchbss.org
Belgium, France	FRANCOSIM: Société Francophone de Simulation	www.sofrasims.fr/
Hungary	HSS: Hungarian Simulation Society	
Italy	ISOMERES: Italian SOciety for Modeling in Engineering REsearch and Simulation	
Italy	LIOPHANT SIMULATION - international modelling and simulation group (Italy)	www.liophant.org/
Latvia	LSS- Latvian Society for Simulation	www.itl.rtu.lv/imb/ index.php? lang=en&id=11
Kosova		www.ka-sim.com/

Country/region	Name	Web site
	KA-SIM - Kosova Society for Modeling and Simulation	
Poland	PSCS: Polish Society for Computer Simulation	www.ptsk.pl/
Denmark, Finland, Norway, Sweden	SIMS: Simulation Society of Scandinavia	www.scansims.org/
Slovenia	SLOSIM: Slovenian Society for Simulation and Modelling	
The United Kingdom, Ireland	UKSIM: United Kingdom Simulation Society (UK, Ireland)	uksim.info/ uksim2016/ uksim2016.htm

7.7 The Future

The future of modeling and simulation societies and associations is positive. Most organizations remain active and valuable to their membership. The larger international organizations remain relevant, and show no sign of stopping. The real reason for the shiny outlook, however, is due to the fact that mankind needs modeling and simulation now more than ever before.

Our ability to create complex models continues to improve, and the cost to simulate these same models decreases. Computing and storage are the cheapest they have ever been – a claim likely repeated year after year. Anyone on the Internet with a credit card can gain access to comprehensive cluster computing services. There does not yet appear to be a Moore's Law for modeling and simulation. Massive amounts of data are captured every day from myriad devices and sensors, data that is needed to make sense of our world and to monitor and the manmade and natural systems therein. As of 2014, Google had indexed 200 terabytes of data. This is estimated to be only four one-thousandths of a percent of the total Internet.[9] The quest for knowledge and understanding of our systems, both trivial and consequential, is a natural tendency of mankind.

More and more relevant questions are being asked, hard problems found, and answers and solutions sought. Modeling and simulation professionals are a commonality in a world of people seeking answers and solutions. The demand for them and their services continues to grow. This demand, in

---

9 Do You Know How Big the Internet Really Is? [Infographic], *Website Magazine*, retrieved from http://www.websitemagazine.com/content/blogs/posts/archive/2014/07/22/do-you-know-how-big-the-internet-really-is-infographic.aspx (accessed October 11, 2016).

turn, only strengthens the need for modeling and simulation societies and associations.

7.8 Summary

The value of societies and associations to the profession of modeling and simulation is immeasurable. They serve as havens where academics and professionals from both hard sciences and all industries can meet – virtually and physically – to collaborate, learn, and invent the future of the science and its applications. They provide order and process that supports vetting, instills rigor, and conserves resources. They support debate and positive conflict that results in focus, clarity, and agreement. They foster new thinking, ideas, and approaches that find their way into the body of knowledge. They help people grow professionally, helping the profession to thrive.

Without modeling and simulation societies and associations, many academics and professionals would have moved on to other concentrations simply because of a lack of peer interaction and available mentors. Without a clear professional or academic path forward, many people would find it hard to persevere; there needs to be value, beyond the intrinsic, in the effort. Societies and associations generate value by concentrating people and ideas. Members share and interact, and gain professionally and perhaps personally through the experience. Individual growth benefits the group – the society – and lifts up the profession as a whole.

References

The Center for Association Leadership. (2015) The Power of Associations: An Objective Snapshot of the U.S. Association Community, pp 5–7. Retrieved from http://www.thepowerofa.org/wp-content/uploads/2012/03/Powerof Associations-2015.pdf.

Berry, W. (2003) *The Art of the Commonplace: The Agrarian Essays* of Wendell Berry Counterpoint.

Brandin, D. (1973) The history of SIGSIM, *Simuletter*, vol. 4, no. 4, pp. 27–31.

Lopreiato, J. O. et al. (2016) *Healthcare Simulation Dictionary*, 1st edn, Society for Simulation in Healthcare, p. 32.

Padilla, J. J, Diallo, S. Y. and Tolk, S. (2011) Do we need M&S science? *SCS M&S Magazine*, n4.

8

The Uniformed Military Modeling and Simulation Professional

Rudolph P. Darken and Curtis L. Blais

Naval Postgraduate School, Monterey, CA, USA

8.1 Military Modeling and Simulation Graduate Education

At a curriculum meeting about 2002, VADM Richard W. Mayo, then Commander, Naval Network Warfare Command (NETWARCOM) asked the Commander of the Navy Modeling and Simulation Management Office (NAVMSMO[1]), CAPT Richard Bump, what the Navy's budget for modeling and simulation (M&S) related activities was at that time. He did not get an answer. It was not because CAPT Bump did not know; it was that the answer was unknowable. There was a realization at that meeting that modeling and simulation had become ubiquitous in the Navy, as it has for all the uniformed services. Because modeling and simulation has become a critical tool for the services, there is also now a realization that modeling and simulation as a technical discipline is not computer science, nor is it operations research, nor is it engineering or psychology. It is all of those things and more.

In 1996, the Naval Postgraduate School anticipated this need by creating the Modeling, Virtual Environments and Simulation (MOVES) masters degree program for uniformed officers and government civilians. This was followed by a doctoral program in 1999. Shortly thereafter, Old Dominion University and the University of Central Florida also began offering modeling and simulation degree programs. At that time, NPS, under the direction of Professor Mike Zyda, was a leader in networked virtual environments (Singhal and Zyda, 1999). NPSNET was the first freely available large-scale virtual environment and was extremely successful having been downloaded

---

1 NAVMSMO later became the Navy Modeling and Simulation Office (NMSO).

The Profession of Modeling and Simulation: Discipline, Ethics, Education, Vocation, Societies, and Economics, First Edition. Andreas Tolk and Tuncer Ören.
© 2017 John Wiley & Sons, Inc. Published 2017 by John Wiley & Sons, Inc.

and used by thousands of organizations over the years. While it could be (and was) used for nonmilitary applications, it was developed for the military and, therefore, had mostly military features and protocols as part of the core functionality. In fact, during those years, Computer Science master's students at NPS would typically create a thesis project out of an enhancement to NPSNET. Later, as virtual environment interfaces gained complexity with head-mounted displays, gloves, and other interaction devices, the research agenda broadened to include human–computer interaction issues related to large-scale networked virtual environments. Finally, it became apparent that systematic study of applications, environments, and techniques was needed. It was not enough to invent new capabilities, we needed to understand the phenomena behind why models and simulations are effective in certain situations but less effective in others. In response, the research agenda adapted again to include experimental design and data analysis.

Consequently, the initial MOVES program was a melding of computer science and experimental psychology[2] that included statistical modeling and analysis, but it has since taken on a character of its own. MOVES was created in part out of the recognition that neither a computer scientist nor an operations analyst could do what was quickly becoming the new discipline of modeling and simulation. The field has continued to evolve under new and ever increasing demands for M&S solutions to operational and analytical problems. There are currently courses in MOVES dealing specifically with simulation and training, joint combat modeling, cognitive modeling, stability operations, and simulation interoperability. None of these fit neatly into a traditional academic department.

The original program had two tracks – visual simulation and human–computer interaction. Students generally pursued either a systems-oriented program where they would build a new system or capability (these students would opt for the visual simulation track), or they would choose to study the human side of systems, either human factors or training (these students would choose the human-computer interaction track). We later realized that this was far too constraining and, therefore, created a set of concentration areas to choose from that included combat modeling, visual simulation, agents and cognitive modeling, discrete event modeling, human factors and training, and systems engineering. Even then, the program could be over-constraining. It was not unusual for a student to make a special request to modify their program to meet the needs of a unique thesis project.

---

2 Experimental psychology topics are housed in the Operations Research Department because there is no Psychology Department at NPS.

The current educational requirements have expanded further to include information assurance, wargaming, data standards, VV&A (verification, validation, and accreditation), LVC (live, virtual, and constructive simulations), social and stability models (nonkinetic), and business practices for program management. Are all of these topics still modeling and simulation? To answer that question, we need to describe how MOVES students arrive at NPS to become M&S professionals and what they do after they graduate.

Students seeking or directed to obtain graduate education at NPS typically identify their preferred curricula, but they do not actually pick their program per se; it is assigned to them by placement officers in their respective military services. In the U.S. Department of Defense, there is a quota system in place that controls how many officers can enroll in any one curriculum so that the supply will meet but not exceed the demand. Consequently, we have had MOVES students who did not state a preference for M&S at all, yet they were assigned to MOVES because that was where there was an opening. Furthermore, whether the officer requests MOVES or not, one of the unique aspects of NPS in general is that it is able to re-educate officers in new disciplines where they may not have a suitable background. It is not uncommon to have incoming MOVES students who have undergraduate degrees in history, business administration, or other nontechnical fields.

This creates a unique challenge for NPS faculty. In a typical graduate school, students are at least motivated by their own choices. While complaining about courses and professors will always be a rite of passage for graduate students everywhere, at most schools, it was the student's choice to be there. This is not always so at NPS. Furthermore, curricula at NPS have a specified length that can usually (optionally) be extended by one quarter as a "refresher" to provide a foundation for the overall instruction. MOVES is a seven quarter program (eight, counting the refresher quarter). Much of the first year is focused on advanced undergraduate material to prepare the student for the more aggressive second year that includes a research thesis. Students arrive on campus in the summer and graduate at the end of the spring quarter two years later.

8.2 Military Career Paths in Modeling and Simulation

Modeling and simulation students at the Naval Postgraduate School come from all services and numerous international partner countries. Within the Navy, the 6202 subspecialty code (P-code) defines a modeling and simulation career path, but as shown in Figure 8.1, a subspecialty is often disregarded in favor of primary Navy assignments. After one or more junior

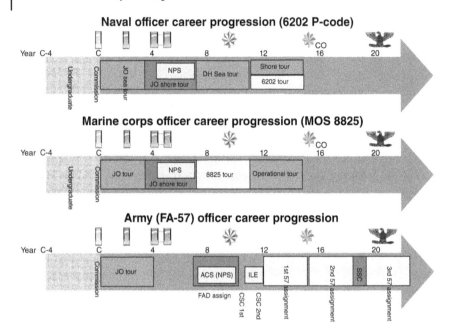

Figure 8.1 The uniformed M&S professional career progression for the services.

officer (JO) sea tours, Navy officers can ask for assignment to NPS for graduate education. However, upon graduation, they immediately return to the operational Navy for a department head (DH) tour. Only after that might they have an opportunity to pursue a 6202 modeling and simulation assignment. Consequently, Navy officers have historically had a difficult time finding assignments in modeling and simulation after graduation. However, this is a known problem that is being addressed by the Navy Modeling and Simulation Office. Most of our Navy MOVES graduates who are working in the M&S field are doing so as civilians postretirement.

This is not true, however, for the Army and Marine Corps. Upon graduation from NPS,[3] a Marine Corps officer will immediately go to a "payback" tour in the 8825 Military Occupational Specialty (MOS) in modeling and simulation. After that, they return to the operational Marine Corps but they may have a subsequent modeling and simulation assignment later on. The Marine Corps 8825 community is small but very active and they are typically placed in positions where they can immediately use the education they have just received, such as in the Marine Corps Combat Development Command (MCCDC), Marine Corps Warfighting Laboratory (MCWL), the training centers, and program offices.

---

3 All Marine Corps 8825 officers receive their graduate education at NPS.

When an Army officer becomes an FA-57 (Functional Area) Simulation Operations officer, this is considered a permanent change of occupational specialty. Those officers then serve only in modeling and simulation assignments for the remainder of their careers. It should be noted that NPS is only one of several degree programs that Army officers might be assigned to for their M&S education.[4] Furthermore, the Army has its own courses available to FA-57s. The Simulation Operations Course is their required basic course, but they also can attend the Battle Command Officer Integration Course and the Advanced Simulation Course where they receive specific training on the M&S systems that the Army employs.

Both Marine Corps and Army officers typically remain engaged in M&S activities after graduation. Because we know this will occur, while they are at NPS, we can familiarize them with actual programs they might be involved with later on. In some cases, we have had MOVES students studying topics or conducting thesis research at NPS related to programs to which they become the program officer immediately after graduation. This career progression creates a relevance that is critically important to the learning process. Furthermore, with our recent graduates running the programs that our current students will have to learn about, this forces the MOVES program to remain relevant. It is a positive feedback loop that benefits the school, the students, and the organizations that employ our graduates. In addition, the services work closely with NPS faculty and administration as curriculum sponsors to ensure continuing currency and relevance in the educational programs.

While the Air Force has not utilized the MOVES program at the master's degree level, they have sent a number of officers to NPS for doctoral degrees in modeling and simulation (see Wells (2005), McClernon (2009), Tvaryanas (2010), and Harder (2011)). Most of these officers have gone on to join the faculty at the Air Force Academy and have served in other key positions actively participating and leading Air Force modeling and simulation activities and programs. However, there is no modeling and simulation career path in the Air Force, which is somewhat counterintuitive given that the Air Force is every bit as much a consumer of M&S products as the other services, if not more so.

Internationally, NPS has had modeling and simulation students from several allied countries including Turkey, Greece, Singapore, Germany, Bahrain, Mexico, and Brazil. These students return to their home countries to assume leading roles in military modeling and simulation. While each

---

4 Old Dominion University and the University of Central Florida are others. In this chapter, when we refer to "MOVES graduates" that should be taken to mean any M&S graduate from any of these universities.

country is unique, in general, these graduates appear to remain in M&S positions long after graduation. Moreover, NPS MOVES faculty frequently collaborate with these graduates in later research projects and international modeling and simulation standards activities.

The utilization of modeling and simulation graduates is quite varied. In the Army and international allies, there is no interruption – the student transitions directly from M&S student to M&S professional. For Navy officers (who have an immediate interruption) and Marine Corps officers (who have an interruption after their "payback" tour), returning to an M&S position after serving in a non-M&S position can be problematic. Uniformed modeling and simulation professionals are unique in that they may move in and out of M&S positions, and even if they remain in M&S positions throughout their careers, these positions can call on very different sets of skills. Therefore, in military modeling and simulation education we must correctly address the breadth of knowledge our graduates are expected to have, what can be learned in an educational setting in a limited span of time, and the expected decay of that knowledge (or the utility of the knowledge) over time. As a rapidly evolving field, maintaining intellectual currency in modeling and simulation technology over time is critical and challenging.

8.3 The Modeling and Simulation Professional Landscape

One of the last courses every MOVES student takes before graduation is a capstone course intended to synthesize all of the topics that students have learned about during their prior two years of study. One of the first topics we discuss most often takes the students by surprise: "What is modeling and simulation?" and "What do modeling and simulation professionals do?" Asking those questions initially, when the students are just starting the program, seems normal. But why are we asking those same questions at the end? We have observed that after two years of intense full-time study in M&S, it is difficult to find two people who will define M&S the same way. Some students have a very tight, constrained definition that only pertains to mainstream M&S topics such as interoperability and modeling techniques. Others have a much broader definition.

We use the Modeling and Simulation Body of Knowledge (MSBoK) (Department of Defense, 2008) as a benchmark for identifying technical topics relevant to the field. Because we were early contributors to the MSBoK effort, there are still aspects of the document that mirror the MOVES program. There are two important advantages to using the MSBoK as a basic framework. First, it very clearly illustrates the breadth of topical material that can be considered "modeling and simulation." This can be very unsettling at first when students, who are about to graduate with a degree in M&S, see topics in the MSBoK that are unfamiliar to them. Second, the MSBoK includes the notion that not all M&S professionals are the same. An engineer who will be writing code is not the same as a program manager in terms of the "awareness level" required for that job. This tends to settle some anxiety when students realize that nobody is expected to master all the topics in the MSBoK.

At NPS, we have always believed that if you do not understand a technology at least at a conceptual level, you can not manage it effectively. For example, even though many of our graduates do not consider themselves to be programmers, they are taught a basic understanding of writing code and have a working knowledge of software engineering practices. An engineer describing the work plan for an M&S system in the form of UML (unified modeling language) diagrams is not intimidating to a MOVES graduate. Consequently, when a MOVES graduate, acting in a program management capacity, works with contractors and other government personnel to solve a difficult M&S problem, they are able to ask the right questions to uncover ambiguity in a specification that should result in a better product and more efficient program execution. There is a certain satisfaction when we hear that some vendors prefer to avoid having a MOVES graduate in the room when they negotiate a contract for M&S products or services. We have always believed that uniformed M&S officers, no matter what academic program they attended, need to be able to ask insightful questions in order to procure the best M&S product that solves the problem and that is programmatically sustainable. To be able to do that, M&S officers must have at least a working knowledge of technical topics that affect M&S products and services.

As part of our inquiry into what is and is not "modeling and simulation", we ask whether or not modeling and simulation is a "discipline". While there is still no formal definition of what criteria must be met before a topic area is considered a discipline, there are a few indicators that most academics agree suggest that an area has reached a level of maturity that constitutes a discipline (Krishnan, 2009). These include the following:

- A particular and articulable focus of attention
- A unique body of knowledge
- Particular and articulable methodologies and modes of inquiry
- Public recognition
- Accurate and consistent language

Some will argue that M&S is not a discipline. It is instead, a synthesis of computer science, mathematics, operations research, social science, psychology, and other traditional disciplines. They might argue M&S itself is an artifact of that synthesis and merely a tool for the purpose of study and exploration of the primary discipline. For example, Galileo created optics as a tool to study astronomy; but Galileo was considered an astronomer. An alternate argument (to which we subscribe) is that M&S is a new discipline because it has unique knowledge, tools, methods, and challenges that do not fit inside any other traditional discipline. Topics such as VV&A, LVC, semantic interoperability, human surrogation in virtual environments, and the synthesis of human and agent interactions, are all unique to M&S, and in fact require new methods of inquiry that are also unique to M&S. While we admit that M&S is an artifact or a tool that is indeed used to study and explore other phenomenon in other traditional disciplines, the artifact itself has reached a level of interest that it can now be considered a discipline of its own. Back to our example, although Galileo was an astronomer, he is now also considered the father of optics, which is widely accepted as an independent academic discipline.

Another indicator that a topic area has become a discipline is if it is capable of supporting a refereed archival journal that differentiates itself from other journals in related topic areas. There are several M&S focused journals in print today. Specific to this discussion, the *Journal of Defense Modeling and Simulation* (JDMS) published by SAGE Journals has been in operation since 2004. The journal's focus is on applications, methodologies, and technologies that promote the practice, science, and art of military and defense modeling and simulation. The contents of 12 years of issues give a reader a good idea of the research topics of specific concern to defense modeling and simulation professionals.

Early in the evolution of the MOVES program at NPS, we were asked if M&S at NPS should be the exclusive domain of MOVES. All universities deal with blurry lines between departments, but we were being asked if we thought other programs on campus should stop teaching M&S topics and instead have MOVES deliver those courses to the programs that needed them. Our response was that there are two related, but very different educational domains for M&S.

As shown in Figure 8.2, there are M&S professionals who build tools and capabilities for the constituent areas of M&S; for example: acquisition, analysis, training, test, evaluation, and planning. And there are M&S professionals in each of those areas whose expertise is in the application of M&S tools to solve their unique problems. The areas listed in Figure 8.2 happen to be the academic departments at NPS, but they could be any discipline that uses M&S. This same distinction is seen in the programmatics of military M&S program management (see Figure 8.3). Each of the primary constituents has its own unique problems. For example, the analysis community is not necessarily concerned with the same problems as the

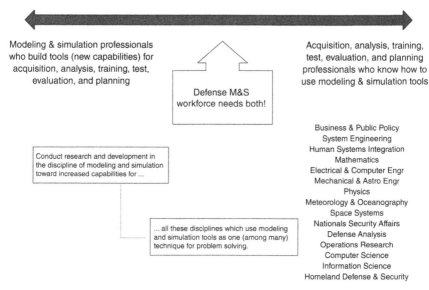

Figure 8.2 The "science" of M&S (left) and the application of M&S (right).

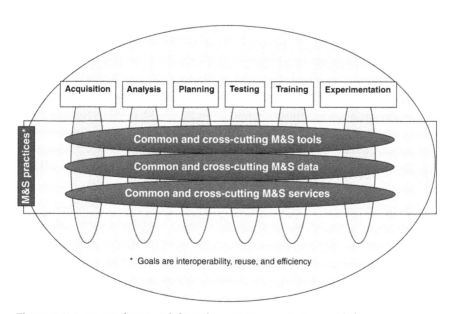

Figure 8.3 A generic framework for military M&S organizations with focus areas (application) shown vertically and cross-cutting M&S capabilities horizontally.

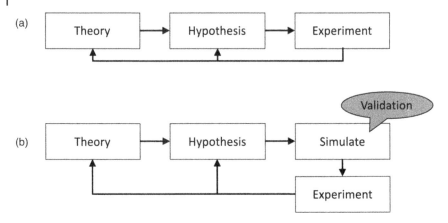

Figure 8.4 The role of simulation and validation in the scientific method.

training community. The Defense Modeling and Simulation Coordination Office (M&SCO), the Navy Modeling and Simulation Office (NMSO), and the Army Modeling and Simulation Office (AMSO) contain components that directly serve their constituents (supporting acquisition, analysis, training, etc.) while at the same time managing "crosscutting" M&S efforts that serve all their constituents. The crosscutting tools, data, and services are usually meant to address interoperability, reuse, and efficiency concerns.

In determining what M&S is and is not, we discovered that simulation has been responsible for probably the most significant change to the principles of the scientific method in hundreds of years. As shown in Figure 8.4a, the basis of the scientific method is that a theory gives rise to one or more hypotheses that test the theory. Those hypotheses are tested using sound experimental techniques that prove or disprove the hypotheses and that may or may not require modification to the theory. Simulation has become a core component of the modern idea of the scientific method. The question of the impact of simulation on the whole scientific endeavor is actively being debated in the Philosophy of Science domain (see Humphreys (2009) or Frigg and Reiss (2009) for different views on this topic). Theories still give rise to hypotheses that are tested, but they are now commonly tested in simulation, not in the real world (see Figure 8.4b). In many cases, a real-world test is not possible or so impractical that it may never be performed. Yet we still want to draw conclusions about the validity of a theory based on those experiments. This raises the critical issue of validation of the models and simulations we use for such experimentation. The well-known adage applies here: "All models are wrong. Some models are useful." (Box, 1976). How accurate does a model have to be in order to draw correct conclusions from its use? This is why the practice of VV&A is so critical.

Within the military domain, this is particularly important because so much of what we want to study cannot be analyzed in a purely real world setting. Simulation is an absolutely necessary tool. Furthermore, innovations such as LVC have allowed a mixing of the real and the virtual that also address military concerns. We often do not want to choose between an all-virtual simulation and an all-real-world exercise to study an idea or new technology.

However, if M&S is a critical tool for modern military problem solving and we require mixed mode configurations to study (a) what the future of warfare will look like, (b) how specific technologies will function under combat conditions, (c) how our military personnel will function under combat conditions, and (d) anything else that fits within the vertical topic areas in Figure 8.3, then are we not back to a definition of M&S that is so broad that it becomes meaningless? If M&S is everything, then it means nothing.

We address this concern by flipping the lens around and looking at what an M&S professional is *expected to be* from the point of view of our stakeholders – the organizations and people who employ our graduates and for whom our graduates will solve problems and produce results. In the final analysis, it does not matter what we (the educators) or our students think about what M&S is or is not. What matters is what our stakeholders expect of us. The following is what our graduates often hear:

- "You're technical, so you must know about . . ." – computers, networks, operating systems, security, devices, and displays.
- "You've studied training, so you must know about . . ." – training schedules, the science of learning, performance assessment, fidelity issues in simulators, training transfer, and skill decay.
- "You're military, so you must know about . . ." – tactics, techniques, and procedures (TTPs), best practices, standard operating procedures (SOPs), and variations to SOPs.
- "You're in government, so you must know about . . ." – the JCIDS (Joint Capabilities Integration and Development System) process, procurement, contracting, and financial rules and regulations.
- "You've got an advanced degree, so you must know about . . ." – proposal writing, strategic planning, securing resources, and inventing the future.

In the capstone course, we briefly touch on topics that none of the students would have considered to be M&S before taking the course. We learn about modern business models, intellectual property, innovation, and disruptive technology, because these officers will have to interact with industry. Therefore, they need to know how industry works to not only get what their service needs, but also to do so within a sustainable business strategy. We spend considerable time on what it means to "architect" software and how software architecture is not simply buying into a product family but rather a specification of how software and hardware components work together. This is a key prerequisite to understand how to defend your organization from product "lock in" where future choices are strictly constrained usually because of a failure to properly architect the system. We also discuss a wide variety of technologies that are actually mainstream computing topics that M&S officers may need to know about. These include wireless and mobile computing, wearable computing, social networks, cloud computing, cybersecurity, and virtualization.

The point we make is this (for example): "If you think that because you don't consider 'wearable computers' to be an M&S topic that you will not be called upon to make a recommendation pertaining to the use of wearable computers for training or operational use, you are simply wrong. If the commanding officer thinks you should know about it, then you should know about it." Since we can never be comprehensive about everything that could ever be assigned to an M&S officer, our focus is not just on presenting information, but rather on enabling graduates to constantly refresh what they know and to stay abreast of where the industry is headed.

It is important to emphasize that this is not a one-way dialog. Early on, graduates would tell us that when they would report to their first duty station post-MOVES, they would be assigned tasks such as configuring a network and troubleshooting, and this disturbed them because they saw clearly how their unique skills could help the command, but they were not being asked to participate in that way. This has improved greatly over the years, largely because of the two-way dialog that has occurred between graduates and their commanding officers who understand much more now about what a MOVES graduate can offer them. We impress upon our students that it is imperative that they seek to educate their stakeholders (superior officers and peers) about what an M&S professional is and can do for the command.

In recent years, MOVES graduates have participated in all facets of the defense M&S enterprise. The Army, having by far the largest M&S professional community, has officers embedded in science and technology, battle command, and other areas of the operational Army, as well as at the National Training Center (NTC) and other key commands. FA-57s lean more toward the training focus area within M&S than other areas, but there are analysis and acquisition components to many of these positions. The Marine Corps also places graduates in science and technology (Office of Naval Research), and also at the Marine Corps Warfighting Lab (MCWL), Twenty-nine Palms Marine Air Ground Task Force Training Command (MAGTFTC), and within the acquisition force.

These graduates participate in M&S activities that also cross the focus areas of M&S. They participate in M&S support for large-scale live or LVC exercises. They plan and design exercises. They conduct analysis for acquisition and develop new M&S tools for the acquisition workforce. They support warfare development with analysis and analytical tools that help determine the structure of the future fighting force of the United States. Quite a few M&S officers have served in M&S billets in theater in direct support of deployed forces. They also support and participate in test and experimentation functions for their service.

While the breadth of knowledge that a uniformed M&S professional might be expected to call upon for a specific job is extreme, it is clearly feasible to gain that knowledge and to employ it effectively. These graduates prove that on the job every day. But the task of educating our stakeholders will never end because the job of creating new M&S capabilities will never end. We simply do not know what new M&S products will emerge in the future that will change the reasonable expectations of what an M&S officer can do.

8.4 Preparing Military Officers to be Modeling and Simulation Professionals

Due to the unique career demands placed on our graduates that are outside their control and the extraordinary breadth of topics they might be called upon to perform or advise, one of the most important skills we can give our graduates is a strategy to combat knowledge atrophy. Even for students who excelled in the program, MOVES graduates cannot believe that the skills and knowledge that they learned at NPS will remain useful (or even be retained in their memory) up until the time pertinent skills and knowledge are needed.

We often refer to the "half-life" of an M&S graduate degree. That term was originally used by Machlup (1962) to describe the time it takes for knowledge in some domain to decay or to be superseded by new knowledge. Figure 8.5 shows a hypothetical knowledge decay curve for an engineering degree. We would argue that the decay is even steeper in M&S due to the vastness of topic areas for which graduates must be proficient. The figure shows the decay not just due to what the learner forgets, but also due to what has changed in the field.

Our students are typically surprised to see that the half-life of their MOVES degree is about one year. One year after graduation, 50% of what they learned will be either forgotten to the point of not being useful, or will have been superseded by new advances, rendering the old knowledge

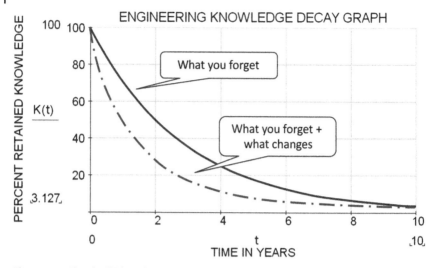

Figure 8.5 The "half-life" of an advanced technical degree.

obsolete. The point we make is that a commitment to a modeling and simulation career is a commitment to lifelong learning. But how do you do that when you are assigned to a surface ship as the operations officer with absolutely nothing to do with M&S for at least two years immediately following graduation?

The principles we give them to follow are to (1) keep your strategy simple, (2) be persistent, and (3) do something, however small, since that is better than doing nothing. There are a number of trade journals that are specific to defense M&S to which we recommend they subscribe. The online professional groups on social networks are worth subscribing to, but run hot and cold so should not be relied upon. There are also a large number of general information technology-related trade journals with which the reader can quickly become familiar with the latest trends in the industry having to do with mobile devices, wireless technologies, automation, and everything else industry is doing that might impact M&S in some way. Lastly, we recommend that our students try to attend M&S professional meetings or participate in M&S organizations that are advancing the state of the art when they can. We distinguish conferences and workshops that are more industry and product-driven, from conferences that are more academically focused.

The key message here is that if MOVES graduate do not have a plan to keep up with the pace of technological change, they will not keep up. The day will come where they will be expected to know about the latest development in some area and they will not know anything at all about it. Again, this addresses the issue of expectation. M&S officers are expected to be the "Swiss army knife" of their respective service. They need to meet that expectation. We also try to teach students how to absorb information quickly – skimming an article is often all they will have the time for, therefore, that has to have value. Lastly, we encourage our graduates to maintain contact with fellow graduates and to always reach back to MOVES and NPS as needed. Our field is just too vast for any one person to stay current on everything. But as a part of a bigger organization, it is actually quite attainable. Because the M&S communities in each service are not very large, most of our graduates know each other and serve as critical resources they depend upon for mutual support.

8.5 Conclusions

A uniformed modeling and simulation career is unique. Even after completing their graduate education, M&S officers have very little control over what area of M&S they will work in or when they will have an opportunity to have an M&S position. There can be gaps in their career where they are not expected to work in M&S at all. Taken with the sobering realization that M&S is as vast a discipline as there can possibly be, uniformed M&S professionals must be vigilant in constantly refreshing their knowledge.

Uniformed modeling and simulation professionals are engineers, technicians, warfighters, inventors, psychologists, writers, salesmen, futurists, analysts, businessmen, government officials, and all of the above. Our job as educators of uniformed modeling and simulation professionals is to raise them to a highly functional level, and to teach them how to remain there.

References

Box, G.E.P. (1976) Science and Statistics. *Journal of the American Statistical Association*, 71, 791–799.

Department of Defense (2008) Modeling and Simulation Body of Knowledge (BOK). Final Report. Available at http://www.msco.mil/msLibrary.html.

Frigg, R. and Reiss, J. (2009) The philosophy of simulation: hot new issues or same old stew?, *Synthese*, 169 (3), 593–613.

Harder, R.W. (2011) *A Quantitative Model for Assessing Visual Simulation Software Architecture*, Naval Postgraduate School.

Humphreys, P. (2009) The philosophical novelty of computer simulation methods. *Synthese*, 169, 615–626.

Krishnan, A. (2009) *What are Academic Disciplines? Some Observations on the Disciplinary vs. Interdisciplinary Debate*, ESRC National Centre for Research Methods. Available at http://eprints.ncrm.ac.uk/783/1/what_are_academic_disciplines.pdf.

Machlup, F. (1962) *Knowledge Production and Distribution in the United States.* Princeton University Press, Princeton, NJ.

McClernon, C.K. (2009) Stress effects on transfer from virtual environment flight training to stressful flight environments, dissertation, Naval Postgraduate School.

Singhal, S., and Zyda, M. (1999) *Networked Virtual Environments: Design and Implementation*, Addison-Wesley Publishing Co.

Tvaryanas, A.P. (2010) *A Discourse in Human Systems Integration*, Naval Postgraduate School.

Wells, W.D. (2005) Generating enhanced natural environments and terrain for interactive combat simulations (GENETICS), dissertation, Naval Postgraduate School.

9

M&S as a Profession and Discipline in China

Lin Zhang,[1,2] Yingnian Wu,[3] and Gengjiao Yang[1,2]

[1]*Beihang University, Beijing, P. R. China*
[2]*Engineering Research Center of Complex Product Advanced Manufacturing Systems, Ministry of Education, Beijing, P. R. China*
[3]*Beijing Information Science & Technology University, Beijing, P. R. China*

9.1 Technology System and Research Progress of Simulation Discipline

9.1.1 Technology System of Chinese Simulation Discipline

Based on the definition given by the China Simulation Federation (CSF), simulation science and technology discipline is a *comprehensive and inter-discipline* that uses modeling and simulation theories as fundamental, uses computer systems, physical effect equipment, and simulators as tools, and understands and changes the study objects by building and running models in accordance with research objectives (Chinese Association for System Simulation, 2010; Working Group for setting up the Simulation Discipline in China, 2009).

"Simulation science and technology" has been an important approach for human beings to know and transform the objective world, and plays a vital role in the critical fields that concern national strength and security, such as aerospace, information, biology, material, energy, advanced manufacturing, agriculture, education, military, transport, and pharmaceuticals.

Today, "simulation science and technology" has formed a relatively independent and comprehensive knowledge system in China, focusing on simulation modeling theory, simulation system theory, and simulation application theory. This body of knowledge consists of basic public knowledge of natural science and basic specialized knowledge of relevant disciplines, and various application fields. The knowledge foundation is integrating the professional knowledge of systems, models, and application

The Profession of Modeling and Simulation: Discipline, Ethics, Education, Vocation, Societies, and Economics, First Edition. Andreas Tolk and Tuncer Ören.
© 2017 John Wiley & Sons, Inc. Published 2017 by John Wiley & Sons, Inc.

fields, and the methodology is integrating simulation modeling on the basis of the similarity principle. Simulation system is built on the basis of system theory as well as simulation applications of whole systems, whole lifecycle, and whole orientation.

Research on simulation science and technology includes the theory and method of simulation modeling, simulation system and technology, and simulation application engineering. The theory and method of simulation modeling include similarity theory, simulation methodology and simulation modeling theory, and so on. Simulation system and technology includes simulation system theory and supporting environments of simulation systems and building and operating technology of simulation systems. Simulation application engineering includes simulation application theory, creditability theory of simulation, simulation generic application technology and simulation application in various fields, and so on.

9.1.2 Main Progresses and Achievements of Chinese Simulation Research

A multitude of key laboratories and research & development centers focusing on simulation research and application have been established by relying on universities and research institutes, including six National Key Laboratories (such as the Virtual Reality Technology Laboratory, the Beijing Simulation Center, and the Power Systems and Power Generation Equipment Control and Simulation Laboratory), two National Engineering Centers (the National CIMS Engineering Technology Research Center Simulation and Virtual Manufacturing Laboratory, and the Economic System Simulation Technology Application Research Center), and eight provincial-level Key Laboratories (such as the Advanced Simulation Technology Aerospace Technology Laboratory, the Navigational Dynamic Simulation and Control Laboratory, and the Large-scale System Simulation Center) (Web site of the Ministry of Education of PRC, 2016; China Education and Research Network, 2016).

The research on simulation science and technology discipline includes the theory and method of simulation modeling, simulation system and technology, and simulation application engineering.

- *The Theory and Method of Simulation Modeling*: Modeling is an abstract description for an entity, the natural environment, and human behavior. The theory and method of simulation modeling is the foundation for the research on simulation science and technology discipline.

  In recent years, China has achieved remarkable progress in simulation modeling system and method in the fields of system of systems, complex environment, discrete event systems, intelligent systems, and life systems.

Deep research has been made in terms of similar theory, model engineering, multidisciplinary modeling simulation language, decision theory and methods, intelligent system modeling and simulation, and complex network modeling and simulation. Chinese researchers obtained considerable results in these fields.

- *Simulation System and Technology*: Simulation system and technology includes simulation language, simulation software, and simulation supporting environments. China has gained great achievements in terms of artificial life systems, virtual prototype simulation platforms supporting multiple disciplines, complex military system integration simulation platforms, high-performance simulation computer systems, cloud simulation platforms, virtual reality development platforms, synthetic natural environment simulation platforms, and various multidomain simulators supporting engineering, training, and analysis.

- *Simulation Application Engineering*: Simulation has been widely applied in the fields of national security, national economy, advanced manufacturing, culture and education, medical care and health, sport and entertainment, and especially in national plan and engineering such as aeronautics and astronautics, cloud manufacturing, moon landing engineering, ship engineering, electronic power systems, south-to-north water diversion, synthetic natural environments, high-end medical operations, sports systems, and digital entertainment.

9.2 Situation of Simulation Education

A number of universities in various provinces, cities, and armed forces, such as Beijing, Shanghai, Shanxi, and Sichuan, have formed stable faculty staff engaged in simulation discipline education and scientific research, and owned good scientific research and laboratory conditions. By relying on "Control Science and Engineering," "Armament Science and Technology," "Computer Science and Technology," and other first-level disciplines, master and doctoral candidates are cultivated within the field of corresponding disciplines, with good cultivation quality, so that thousands of high-level simulation talents are made available to higher education institutions, research institutes, and troops. In 2016, China's Ministry of Education approved the establishment of 100 national virtual simulation experiment teaching centers, such as the Tsinghua University Digital Manufacturing System Virtual Simulation Experiment Teaching Center and the Beihang University Electric Information Virtual Simulation Experiment Teaching Center. Virtual simulation experiment teaching, by relying on virtual reality, multimedia, human–computer interaction, database and network communication technology, constructs highly simulated virtual experiment environments and objects to enable students to carry out experiments in virtual environments, so as to reach the teaching objective as required by teaching program (Web site of the Ministry of Education of PRC, 2016).

9.2.1 Undergraduate Education of Simulation Specialty

Since China has not set up a first-level discipline for simulation science and technology, the Ministry of Education has no professional simulation specialty teaching guidance committee. Simulation, however, is in high demand in various industries and specialties. As a result, some schools, especially military schools, launch undergraduate specialty in relation to simulation, such as simulation engineering of Academy of Armored Forces Engineering, Aviation University Air Force, People's Liberation Army (PLA) Information Engineering University, National University of Defense Technology, PLA University of Science and Technology, Army Officer Academy of PLA, and simulation science and technology of the Wuhan University.

Some specialties depend on existing specialties, and simulation-related specialty direction is launched, for example, the simulation cultivation direction established in the automation specialty of Beihang University, digital simulation and game design cultivation direction of computer science and technology specialty of Nankai University Binhai College, and logistics system modeling and simulation direction of mathematics and applied mathematics of Zunyi Normal College.

9.2.2 Postgraduate Education of Simulation Specialty

The statistics of the number of published doctoral papers relating to "simulation" and "simulation discipline" is made for the key universities that reveal information from January 2000 to December 2015. Such universities, by relying on first-level disciplines, cultivate 16,026 doctoral papers relating to "simulation" (remark: the underlying data come from the China National Knowledge Infrastructure, that is, that some relevant papers of some universities and military papers are not included in it), accounting for 2.82% of total students obtaining a Ph.D. degree, which reveals that Chinese key universities have owned strong capacity of talent cultivation in simulation science and technology discipline. As most of military simulation papers are not included, and military simulation is exactly one of the main forces that drive the development of the discipline, the retrieval data tend to be conservative.

According to incomplete statistics, most of postgraduates in simulation science and technology discipline have already been allocated to various universities, research institutes, and troops to engage in simulation discipline education, scientific research, and military training study. Their research area is roughly divided into 14 fields, namely, national defense and military field, aerospace field, energy field, environment field, information field, industry field, agriculture field, transportation field, ocean field, economic field, building and manufacturing field, sports culture field, public utilities field, and life medical field. The information field, aerospace field, industry field, national defense and military field, and basic science field are ranked in the top five fields of Chinese professional efforts.

According to the statistics of master and doctoral students admission catalog published by universities, there are over 40 universities that launch simulation research directions in the cultivation of master and doctoral students, including Tsinghua University, Beihang University, National Defense University, Harbin Institute of Technology, Beijing Institute of Technology, Xi'an Jiaotong University, National University of Defense Technology, University of Electronic Science and Technology of China, Xidian University, Zhejiang University, Huazhong University of Science and Technology, Tianjin University, Armored Engineering Institute Northwestern Polytechnical University, Chongqing University, North China Electric Power University, Southwest Jiaotong University, Jilin University, Harbin Engineering University, Nanjing University of Science and Technology, Hunan University, Wuhan University of Technology, South China University of Technology, Beijing University of Posts and Telecommunications, Northeastern University, Beijing Jiaotong University, Shandong University, Tongji University, and Wuhan University.

The simulation research directions are distributed in more than 100 disciplines, including control science and engineering, materials science and engineering, electric engineering, power engineering and engineering thermophysics, optical engineering, aeronautic and astronautic science and technology, chemical engineering and technology, mechanical engineering, computer science and technology, architecture, transportation engineering, military communication, mechanics, microelectronics and solid state electronics, radio physics, road and traffic engineering, hydraulic engineering, biomedical engineering, genetics, management science and engineering, vehicle engineering, electronic science and technology, sociology, sports science, environmental engineering, e-government, military equipment, weapon systems and utilization engineering, man-machine and environmental engineering, nuclear science and technology, vehicle operation engineering, digital performance, logistics engineering, and management.

Overall, more than 300 simulation research directions are involved, which cover three technology systems and directions, including the theory and method of simulation modeling, simulation system and technology, and simulation application engineering.

The research directions involving the theory and method of simulation modeling include modern simulation and virtual technology, system simulation theory and engineering technology, simulation and equipment diagnosis, high-performance simulation, man-machine engineering, design simulation, and more.

The research direction involving simulation system & technology, and simulation application engineering mainly covers the following fields and directions:

- *New Material and Optics Field*: New semiconductor device design simulation and circuit low-temperature modeling, composite material structure simulation analysis, new system photoelectric guidance and simulation technology, optical imaging guidance and simulation, optical system simulation testing technology, carbon fiber composite material forming process control and visual simulation technology.
- *Military Application Field*: Electric warfare system design and simulation, aircraft semiphysical simulation; aircraft control and simulation system; weapon system simulation, missile countermeasure simulation, micro-satellite overall design and simulation, Beidou Satellite Navigation System performance evaluation and simulation verification, ship automation control and simulation technology.
- *Virtual Reality Application*: Virtual assembly simulation and man-machine power analysis, virtual operation simulation technology, virtual welding technology, and welding process simulation.
- *Energy and Power Field*: Nuclear reactor kinetic control and simulation, electric appliance magnetic field and thermal field simulation technology, traction drive system simulation technology, power generation process control, logistics system optimization and analogue simulation, vehicle dynamic simulation and control.
- *Building and Manufacturing Field*: Refrigeration and air-conditioning system measurement and control technology and computer simulation, building simulation and control and intelligent building, heating ventilation and air-conditioning system measurement and control technology and computer simulation.
- *Health and Social Field*: Life system modeling and simulation, cardiovascular system modeling and simulation, social guarantee model recognition

and policy simulation, computer simulation and molecular modeling, big data and micro population simulation, low-carbon economy and policy system simulation.

9.2.3 Simulation Specialty Curriculum System

Beihang University provides simulation directions in automation specialty, and also launches a series of courses in relation to modeling and simulation in addition to automation specialty courses, with 240 credit hours in total. Table 9.1 shows the establishment of specific courses.

Table 9.1 Simulation courses in BUAA automation major.

No.	Course	Type	Credit	Total hours
1	Experiment Technology and Modeling	Compulsory	2.5	40
2	Designing Automation for Digital System	Compulsory	2.5	44
3	Control System Simulation	Compulsory	2	36
4	Modeling and Simulation of Aerospace Vehicle Dynamics	Compulsory	3	48
5	Virtual Instrument Technology	Elective	2	36
6	Electronic Design Automation	Elective	2	36
	Total		14	240

9.2.4 Simulation-related Teaching Materials

According to incomplete statistics, there have been over 40 simulation-related teaching books published since 1996, with teaching content covering basic theory and technology of modeling simulation, simulation evaluation, simulation tool application, and simulation application in the fields of war and defense, aircraft, robot, discrete time, system of systems, and logistics system.

9.3 Academic Exchange and Cooperation of Simulation

9.3.1 Simulation Academic Organization and Introduction

At present, Chinese simulation academic organization is under the uniform management of the China Association for Science and Technology, and is divided into first-class association, subordinate specialized committees, and local associations. The China Simulation Federation is the first-class association among Chinese simulation academic organizations (China Simulation Federation, 2016b). Other first-class associations, for example, the Chinese Association of Automation, also have simulation-related specialized committees, such as specialized committees of system simulation. The Chinese Association for System Simulation (CASS), founded in February 1989, is affiliated with Beihang University, and has been renamed into China Simulation Federation (CSF) in 2016.

Since its founding, the China Simulation Federation has held international large academic seminars for nine times, annual academic meeting of national system simulation for more than ten times, and various seminars. There are about 300 periods of journals of system simulation published in Chinese version, and in addition, English versions are being prepared. Bilateral or multilateral cooperation relations have been developed with the simulation associations of Japan, Korea, the United States, and several European countries. The China Simulation Federation established the Asia Simulation Federation with Korea, Japan, Thailand, Malaysia, and Singapore in 2010. This federation plays an important role in promoting domestic and foreign academic exchange and domestic simulation technology development.

At the time of writing this chapter, the China Simulation Federation has nearly 200 group members and over 6,000 individual members. Various provinces and cities, based on the development need of simulation, have set up local simulation associations, such as Shanghai, Guizhou, Shaanxi, Shanxi, Heilongjiang, and others.

The China Simulation Federation has five work committees, namely, the organization work committee, the education and popularization work committee, the international exchange work committee, the youth work committee, and the simulation industry work committee.

The China Simulation Federation has 19 professional committees, involving 19 research directions in the respect of the theory and method of simulation modeling, simulation systems and technology, simulation application engineering, such as modeling and simulation in terms of simulation method and modeling, simulation algorithms, simulators, aerospace system simulation, virtual technology, life systems, discrete systems, software systems, sports systems, digital entertainment, standard technology, intelligent Internet of Things (IOT) systems, electric systems, medical care, transport environment, education, and complex systems.

9.3.2 Introduction of Chinese Main Simulation Journals

The academic and organizational societies established a multitude of scientific and professional journals, reflecting the structure of simulation directions as identified. Examples for these journals are the following.

- *Journal of System Simulation*: The journal of *System Simulation*, founded in 1989, is a journal of the China Simulation Federation, and jointly organized by the China Simulation Federation and the Beijing Simulation Center. It is an authoritative and typical academic journal in the field of system simulation in China, and is aimed to report the international and domestic advanced level research achievements achieved by Chinese simulation technology, with focus on publishing innovative research papers. It is included in famous foreign and domestic retrieval institutions, and has won the title of "Chinese most influential academic journal" for two consecutive years.

- *Computer Simulation*: The journal of *Computer Simulation* is a comprehensive scientific journal in the field of simulation technology, and contains domestic and foreign research theories and new technical achievements in simulation technology. Its editorial board member is comprised of simulation expert, academician, research and professor, and so on.

- *Command Control & Simulation*: The journal of *Command Control & Simulation*, founded in 1979, is an academic/technical journal of weapon industry, focusing on C4ISR system and simulation technology, including combat system, weapon system, aerospace integrated electronic information system and equipment.

- *System Simulation Technology*: The journal of *System Simulation Technology* has the publication content involving various disciplines of simulation, mainly including simulation basic theory and key technology, simulation system design and development, and simulation new technology application. The journal reports all achievements in relation to simulation technology in the form of thesis, essay, research note, and review.

- *International Journal of Modeling, Simulation, and Scientific Computing*: The *International Journal of Modeling, Simulation, and Scientific Computing* (IJMSSC) is a peer reviewed journal that was founded by China Simulation Federation, and published by World Scientific. The members of the editorial board are from international simulation community.

9.3.3 Introduction of Main Chinese Simulation Conferences

The China Simulation Federation annually conducts and supports various simulation-related conferences, including the Asia Simulation (AsiaSim) and International Conference on System Simulation and Scientific Computing (ICSC), and the Chinese simulation conference, engaging many professional committees.

- *AsiaSim & ICSC*: AsiaSim & ICSC are formed to bring together outstanding researchers and practitioners in the field of modeling & simulation and scientific computing areas from all over the world to share their expertise and experience. AsiaSim is an annual international conference organized go-around by three Asian Simulation Societies: the China Simulation Federation, the Japan Society for Simulation Technology (JSST), and the Korea Society for Simulation (KSS) since 1999. Today, it has become a series of conferences of the Federation of Asia Simulation Societies (ASIASIM) that was established in 2011. ICSC is a prolongation of the Beijing International Conference on System Simulation and Scientific Computing (BICSC) sponsored by CSF since 1989. The AsiaSim & ICSC 2012 is organized by Chinese Association for System Simulation (CASS) and Shanghai University.

    The 2016 International Simulation Multi-Conference has been a joint conference of the 16th Asia Simulation Conference and the 2016 Autumn Simulation Multi-Conference, organized by the Society for Modeling and Simulation (SCS). The AsiaSim/SCS AutumnSim 2016 Multi-Conference focuses on the theory, methodology, tools and application for M&S of complex systems and provides a forum for the latest R&D results in academia and industry. Together with the conference, the International Simulation Expo 2016 is conducted in Beijing, China.

- *Chinese Simulation Conference*: The Chinese Simulation Conference is a national annual simulation conference organized by China Simulation Federation, with attendees from scientific works and members in simulation science and technology field at home and abroad. The number of annual attendee is over 300. Different topics are developed annually according to current simulation research hotspot. The topic of the conference was "Modern Modeling and New Progress in Simulation Technology" and "Simulation and Modeling in Information Era" in 2015 and 2014, respectively.

- *Other Simulation Conferences*: Apart from the simulation conference organized by first-class associations, there are also the simulation conferences organized by the subordinate professional committees of first-class association, such as the International Conference on Life System Modeling and Simulation organized by the Life System Modeling and Simulation Professional Committee of China Simulation Federation, the Chinese Intelligent IOT System Conference organized by the Intelligent IOT System Modeling and Simulation Professional Committee of China Simulation Federation, the International Seminar on Complex Management System Modeling and Simulation organized by the Scatter Simulation Professional Committee of China Simulation Federation, the Chinese Transport Modeling and Simulation Academic Conference sponsored by

Chinese Association for System Simulation and conducted by the Modeling and Simulation Professional Committee of China Simulation Federation and Tsinghua University Future Transport Research Center, and the Academic Conference for Chinese System Simulation Technology and Application organized by the System Simulation Professional Committee of Chinese Association of Automation and Simulation Technology Application Professional Committee of China Simulation Federation.

9.4 Development of Simulation Industry

9.4.1 Application of Simulation Technology

China carried out the research and application of simulation technology for several decades already, so the Chinese simulation technology has been developing rapidly.

Since the 1950s, simulation technology has been first applied in the realm of automation control. Equation-oriented modeling and mathematical simulation that uses analogue computers have been widely applied. In parallel, the semiphysical simulation for automatic flight control systems that uses self-developed triple-axis simulation rotation tables has begun to be applied in the development of aircraft and missile engineering.

In the 1960s, the search for the simulation of discrete event systems, such as transport management and enterprise management, started while continuous system simulation continued to be carried out.

In the 1970s, Chinese training simulators were developed rapidly, and a number of Chinese self-designed simulators were developed successively. Examples are aircraft simulators, ship simulators, thermal power units training simulation systems, chemical process training simulation systems, locomotive training simulators, tank simulators, and automobile simulators. These simulators and training systems play a significant role in operator training.

In 1980s, China constructed a batch of semiphysical simulation systems on high level and large scale, such as a radio frequency (RF)-guided missile semiphysical simulation system, an infrared (IR)-guided missile semiphysical simulation system, fighter engineering flight simulators, fighter semiphysical simulation systems, destroyer semiphysical simulation systems, and more. These semiphysical simulation systems play a key role in the development of weapon model.

In 1990s, China began to engage in the research on the advanced technology and its application, such as distributed interactive simulation and virtual reality, and carried out large-scale, complex system simulation from single weapon platform performance simulation to multiweapon system warfare simulation in war fighting environment.

Currently, simulation technology has been widely applied in military, industry, agriculture, and water conservancy (China Simulation Federation, 2016a). In the respect of water conservancy, simulation technology plays a vital role in the engineering of "diverting the water of the Yellow River into Shanxi," Dianchi Lake water pollution governance, and South-to-North Water Diversion, which provides strong support for engineering preliminary decision-making and management scientization. It could be said that simulation technology greatly accelerates the development of Chinese water conservancy from traditional water conservancy to modern water conservancy, and makes positive contributions to water control and water supply security that could be observed since simulation technology was applied.

Simulation system technology is broadly applied in various aspects of Chinese national defense as well as in industries such as aerospace, weapon, national defense electronics, ship, electric power and putrescence. Especially in the demonstration, development, production, utilization, and maintenance process of modern high-tech equipment simulation technology is widely applied. Today, simulation system industry has been an industry representing national key technology and core scientific competitiveness, and owning a large scale. Until now, the overall scale of global computer simulation market has exceeded hundreds of billions of U.S. Dollars, and Chinese computer simulation market has exceeded 70 billion Renminbi (RMB), revealing the enormous development potential of computer simulation industry in the future.

In terms of products, the Chinese market also makes various types of simulation models, data, simulators, simulation software, simulation computers, and visual engines available. The investment in research and development (R&D), promotion and sale of such products account for 30% around of market, which is approximately RMB 21 billion, and the expenses arising from promotion, implementation, and maintenance of such products account for approximately 70%, which are about RMB 49 billion.

In terms of engineering projects, Chinese aerospace, ship, and electric power group and high-tech enterprises are the leader of simulation industry, with around RMB 5 billion annually in scientific research, product development, and output scale in simulation engineering projects.

In terms of service, the numerous medium and small-sized companies play an important support role in the implementation of simulation product and engineering, apart from national large companies. As Party B or Party C of national groups or enterprises, they focus on the application, implementation, and maintenance of simulation products, and the simulation service has the market profit space of around RMB 1 billion annually.

In terms of application of simulation products, service and engineering, the use of high-tech innovation products brings a lot of new opportunities and market demands, including digital entertainment, high-end medical surgery simulation, moon landing plan simulation, military equipment simulation test, and so on, and further gives rises to new technology competition and product demand, with market demand of around USD 15 billion annually.

Nowadays, with the implementation of Chinese strategic objective "Rejuvenating the country through science and technology" and the upgrading from downstream "Made in China" to "Created in China", the simulation technology is enjoying high concern and policy support by the State as an important tool and means for scientific research, and Chinese simulation industry is facing the good opportunity of fast development. During the period of the 12th Five-year Plan, the compound growth rate of domestic simulation industry is expected above 26%. If the growth rate were 20%, the total market scale would exceed RMB170 billon by 2015.

9.4.2 Demand of Simulation Talents

Professional simulation talent is in demand in enterprise research, development, production, as well as social, economic, and financial activity. Various industrial and national important scientific research project and engineering project are inseparable from the participation of professional simulation talent, especially in national defense equipment research and army combat training.

- *Enterprise and Industry Demand*: Chinese military research institutes set up simulation centers (laboratory) with dozens of personnel around, which are in urgent need of simulation talent. The employment of simulation talent by manufacturing enterprise, civil aviation, and logistics enterprise will fundamentally change the existing research & development design process system, and integrate simulation technology into research & development design process. Only with simulation talent, the simulation training system in power plant could be put into practice. Simulation science and technology has been widely applied in typical industries, such as manufacturing, military, electricity, agriculture, mechanics, digital pharmaceutics, and medical care industry.
- *Research Demand of Large-Scale Science and Engineering Complicated Problem*: Quite a few cases of simulation applications are made available in Chinese aerospace field, navigation field, and large engineering projects such as Changjiang Gorges five-level ship lock project, and South-to-North Water Diversion Project. Equipment manufacturing enterprise

needs the talent who has the background of automation control, computer application, electronics or similar background, is familiar with MATLAB partial differential equation solution, model recognition and neural network, and is capable of process modeling and simulation. Engineering installation and building industry needs the simulation talent who has a good command of C, C++ language and finite element analysis software ANSYS or other specialized simulation software. Urban design and planning are in need of the talent who has experience in urban and road intersection design, and could use software tool for planning simulation.

- *Demand of Military Training*: With the modernization of weapon equipment and the development of modern war space scales, the military confrontation in nonwar condition is inseparable from simulation, including demonstration, development, production, shaping and use of equipment training simulator, large-scale combat simulation training, and war fighting deduction research at the level of strategy, battle, tactic and combat, which all need to cultivate a batch of military simulation talents at all levels in the aspects of research, design, development and application of simulation system.

9.5 Conclusion

In this chapter, we have introduced modeling and simulation as a profession and discipline in China. The technology system and research progress of simulation discipline has been presented first. "Simulation science and technology" has formed a relatively independent and comprehensive knowledge system in China. A number of key laboratories and research & development centers focusing on simulation research and applications have been established at universities and research institutes. Lots of achievements on simulation-related research have been obtained in China.

The situation of simulation education in China has been introduced. The undergraduate and postgraduate programs on simulation have been set up at a number of universities. China have formed stable faculty engaged in simulation discipline education and scientific research, and owned good scientific research conditions. Many universities have simulation program in automation specialty, and also launch a complete curriculum in relation to modeling and simulation. There have been over 40 simulation-related teaching books published since 1996.

China is active on academic exchange and cooperation with simulation societies of other countries. China Simulation Federation (formerly named as Chinese Association for System Simulation, CASS) initiated the Asia

Simulation Conference with Japan and Korea and promoted the establishment of the Federation of Asian Simulation Societies (ASIASIM). There are several peer-reviewed journals on simulation in China, with an increasing number published in English and open to international researchers to publish their research in them.

The computer simulation market of China has exceeded RMB 70 billion, which reveals the enormous development potential of simulation industry in the future. In summary, China is serious in contributing to the profession of modeling and simulation nationally as well as internationally.

References

China Education and Research Network, (2016) http://www.edu.cn.

China Simulation Federation, (2016a) Research report on Chinese Simulation Industry Strategy.

China Simulation Federation, (2016b) http://cass-sim.buaa.edu.cn/.

Chinese Association for System Simulation, (2010) Report on the progress of simulation science and technology discipline, science and technology of China Press.

Web site of the Ministry of Education of PRC, (2016) http://www.moe.gov.cn/

Working Group for setting up the Simulation Discipline in China, (2009) Suggestion on the setup of first-class discipline simulation science and technology.

10

Modeling and Simulation for the Enterprise: Integrating Application Domains for the M&S Professional

Steve Swenson,[1] Robert M. Gravitz,[2] and Gary M. Lightner[2]

[1]Aegis Technologies, Newport, RI, USA
[2]Aegis Technologies, Orlando, FL, USA

10.1 Introduction

Modeling, simulation, virtual, and augmented reality have influenced our lives in more ways than most understand or can even comprehend. Indeed, models and simulations have become ubiquitous necessities in planning, analysis, engineering, materiel development, and training. Typically, the application of modeling and simulation (M&S) is narrowly targeted to particular scientific and/or business domains and acquisition phase. The results of this narrow targeting are principally reduced productivity and incoherent results.

This chapter provides a vision for an enterprise-level view of M&S. We first outline the challenges and constraints of applying M&S, and then define an Enterprise M&S Strategy and how it maps to business strategy and practice. Finally, we outline extant technologies that may be brought to bear to realize the vision of an enterprise M&S strategy. In the end, the picture that will emerge is of a practice that is multidisciplinary, enabling new and creative ways to view our world and address – and solve – issues and problems that cannot be adequately addressed using a domain stovepiped approach to M&S.

As this chapter has been written by practitioners of the field with many years of experience, it may be different from the chapters produced by academia. In particular, the use of references may be sparser than scholars are used to. However, instead of pure literature research, we used the life lessons learned to enumerate what industry currently expects from educators and researchers, and also how we can contribute to solutions. If scholars can back these claims up with more research, we already

The Profession of Modeling and Simulation: Discipline, Ethics, Education, Vocation, Societies, and Economics, First Edition. Andreas Tolk and Tuncer Ören.
© 2017 John Wiley & Sons, Inc. Published 2017 by John Wiley & Sons, Inc.

accomplished one of our objectives: to give practically relevant inputs to the community of M&S professionals.

10.2 Outline of Challenges and Constraints of Applying M&S

The development and use of models and simulations occupy an influential position in all of their respective application domains. Modeling and simulation is integrally involved in a wide variety of business, human performance, scientific, and engineering programs, and within those programs, M&S employment and results are strongly affective. Many such programs depend entirely on M&S-based representations, and even when they use complements of other techniques, the continuing improvement of M&S makes program outcomes progressively dependent upon the consequences of M&S employment. As M&S becomes part of the critical path of a program, risk increases with the attendant increase in programmatic consequences should a risk be realized. This has the unintended consequence of driving the M&S perspective inward. This stovepipe hardening – manifested in reduced or eliminated sharing across pipelines – means that programs take on a larger percentage of the M&S work increasing unnecessary duplication and expense. Further, as programs limit their use of talent available in other parts of the business enterprise, application of enterprise-wide skill is traded for programmatic safety. In the end, the stovepipes result in higher cost, longer development time and, ironically, higher risk.

10.2.1.1 Our Thought Experiment

As we venture into our discussion about an enterprise perspective on modeling and simulation, we'll draw upon an example business case to codify the principles, both good and bad, to help the reader develop a practical grasp of the topic. Photon Motors has developed a solar-powered automobile concept and they've received a significant injection of venture capital funding to bring their vehicle into production. The founders of Photon grew up in the information age and intend to advertise their car's capabilities in a consumer-grade video game. They have formed their company around two business units. The first is responsible for the design and manufacture of the Photon automobile and will use modeling and simulation to get their product to market quickly. The second business unit will be developing the video game and modeling and simulation forms the very core of their product. Admittedly, our thought experiment is thoroughly contrived. Nonetheless, it provides a reasonable foundation upon which to understand challenges faced by a divided M&S culture.

10.2.1.2 All Models are Wrong; Some are Useful

That's an expression you'll often hear from the more conservative elements in the modeling and simulation community. While bleak, the statement has more than just a whiff of truth about it. Models are, by their very nature, approximations of that which they represent. These approximations are guided by the use of the model; the assumptions about what is important for that use; cost and time constraints which dictate how well the important things need to be represented; and, the available knowledge about how best to represent the thing modeled. An automobile's dynamics in a consumer video game can be rather vague because the principal motivation of a video game is to entertain human players. On the other hand, an automobile manufacturer levies far stricter requirements upon their vehicle dynamics model to reflect, for example, characteristics such as body flex and the dynamic coefficient of friction of the tires.

10.2.1.3 Stovepipes Often Involve Incompatible and Competing Objectives

Customers drive needs and a business enterprise marketing to multiple customers must effectively and profitably respond to a constellation of customer needs to maintain a competitive advantage. Profit and loss is often assessed at the business unit level, and, in turn, reward is assessed on profit and loss. Managers of individual business units, thus, generally have limited incentive to help another business unit perform better. Nonprofit organizations, including civil and government organizations, may not be guided by profit and loss but nonetheless suffer a similar fate as organizational components within those enterprises have varying performance criteria. Individual business units establish processes, protocols and technologies to best meet the demands of their respective customers. This fine-tuning results a culture that is specifically aligned with one set of objectives. Each culture forms its own understanding of success with technical and programmatic decisions aligning with that understanding. The video game developer and the automobile manufacturer both, at the highest level, define success in terms of the profits their respective products generate. However, the video game culture believes the best way to sell games is to make them fun. The manufacturer culture believes the best way to sell cars is to make them fun to drive, safe, and efficient.

10.2.1.4 Stovepiped Cultures Become Technical Monocultures

As the stovepiped culture develops processes, protocols, and technologies to maximize its success with their customer(s), employees that are aligned in skill and interest tend to flourish while those who aren't tend to fall away. Additionally, hiring practices naturally follow organizational objectives. Over time, the stovepiped organization tends toward becoming a technical monoculture. On initial inspection, this makes for a highly tuned, focused organization and one, frankly, that's hard to argue against. However, the technical monoculture tends to gravitate toward a "local maxima" with respect to technical solutions, optimizing for the individual business unit at the expense of enterprise performance. Again, from our example, Photon's video game business unit will tend to focus on things like visual interest (e.g., are the texture maps eye-popping) and storytelling (e.g., the semantic motivation for advancing through levels) – ingredients for a successful game. On the other hand, the manufacturing unit will tend to focus on things like physics (e.g., the static and dynamic coefficient of friction of the tires). Video game developers, of course, consider physics in their design, but the value of the physics lies in and is subordinate to visual interest and gameplay.

10.2.1.5 Stovepiped Cultures are Often Uninformed about Capabilities Available in Other Parts of the Enterprise

Intentional human-to-human discourse has a cost, and if the perceived value of the discourse is less than the perceived cost, discourse doesn't happen except, maybe, accidentally around the water cooler. Within the stovepipe, the cost for communications is relatively small compared to the value. But cost of communication between stovepipes is high and the value is often very unclear. If one of the video game developers has an epiphany that her portion of the video game would benefit from a more robust model of the vehicle's tires, where does she go to get that robust model? It may or may not be immediately obvious to a game developer that automobile manufacturers are, indeed, interested in finely tuned tire models. It may be or may not be obvious with whom, in the manufacturing stovepipe, the game developer should speak about tire models. It may or may not be obvious how the game developer is going to employ this newly found and retrieved tire model. It's very difficult for the game developer to reasonably assess cost and value of a proposed game developer to tire modeler colloquy.

10.2.1.6 Leaders of Stovepipes are Ultimately Disincentivized to Cooperate

For all of the reasons discussed above, leaders of stovepipes are disincentivized to foster cooperation with other stovepipes. And here's where it gets interesting. Leaders of stovepipes are building their business units to serve their customers, and, any attempt on senior management's part to push enterprise-value of modeling and simulation down onto the stovepipes is easily met with definitive and unassailable logic to frustrate the push. From a technical perspective, the enterprise-value, as the argument goes, is greatly exaggerated; to assess the value of using modeling content, each business unit must be allowed to determine usefulness in light of customer needs and objectives. The video game, so the leader of the game developer stovepipe complains, is not greatly improved by the introduction of a robust implementation of the static coefficient of friction; and, indeed, the cost is so high (which leads to the business argument against) that its required implementation represents a barrier to profitability. What is enterprise-level leadership to say to this? They are, after all, entrusting their stovepipe leads to maximize customer satisfaction and profitability.

10.2.1.7 Business Practice can Represent a Further Challenge to Sharing Among Stovepipes

Large customers paying for the development of complete, or nearly complete, products are unlikely to support efforts that benefit their provider's enterprise writ large. Capability shared across stovepipes, could, in fact, be used to advance a competitor's interest. Indeed, customers will often write contracts that stipulate establishing a "firewall" around their product's development.

Some business structures fail to address, or otherwise limit, inside sales. In the Federal Government, for example, organizations do not earn money from their customers. Rather, program budgets are allocated through a well-defined political process and that process is largely inviolable. Funding is allocated against program requirements that have been clearly delineated in program documentation. As such, programs are generally not permitted to expend allocated funding to solve another program's challenges. This effectively removes any consideration of solutions that have value for generality's sake. Within the Department of Defense construct, the problem is more acute as one program office is not responsible for the entire lifecycle of the product. System analysis, development, deployment, training, and maintenance are typically handled by different organizations. Therefore, a model and/or simulation used for the development of a system is different from the model and/or simulation used to train personnel on the system.

Finally, there are manifold technical challenges attendant to moving modeling and simulation capability from one organization to another. Identifying and acquiring the right data for the task is more challenging than one might expect. Data are typically gathered for a particular purpose and the fidelity, content, and rigor may or may not correspond with the new target application. Even harder can be getting the one responsible for collecting the data that are authoritative and valid for the intended task. Repurposing models/simulations poses additional constraints. Verification and validation (V&V) of a modeling capability is most effectively guided by intended use. Leveraging V&V from another source requires careful analysis before applying it to new use cases. Furthermore, in the perfect case, model developers rigorously document model assumptions but those assumptions may be limited to those most significant for the original use. Owners of models for their own system may be reluctant to share the model for fear of the receiver misusing the model to highlight faults in the system.

10.3 An Argument for Overcoming the Challenges and Constraints

The divided organization potentially leads to inefficient optimization of capability and individual business units settling on local maxima along the enterprise value vector. An enterprise perspective enables enterprise-level assessment of value across the organization and through time. Modeling and simulation is particularly promising in its relevance to an enterprise[1,2] operations precisely because of the relative size, scope, and strategic significance of enterprise initiatives, and because of its breadth and inter-related complexity of prevailing enterprise environments their contexts[3]. Several circumstances motivate the trend toward collaborative M&S operational practice, some with technical foundations, others with institutional or business practice groundings, but each having its own peculiar influence on likely credibility, confidence, and uncertainty of model and simulation results.

From the relatively technical perspective, the increasing need for subject matter expertize beyond that of the M&S professional himself (assuming those roles to be differentiated) can affect consequent M&S quality. Multiple agents with distinct roles necessitate collaboration, admit to diffusion of responsibility, and can result in lack of coherent, persistent, and accessible assumptions and modeling conceptualizations. Meanwhile, computational complexity and scope increases with emerging representational challenges. Under these circumstances, a broader scope of election of representational techniques bears on eventual relevance of results. The fundamental difficulty of M&S programs can affect achievement of correctness of representations, particularly among disparate representation domain component elements.

---

1 En'ter.prise - n. 1. An undertaking, esp. one which involves activity, courage, energy or the like; an important or daring project; a venture (K Dictionaries Ltd, 2005).
2 Enterprise – n. One or more organizations under common control. Generally refers to the broadest scope of organizations and operational process relevant to the subject discussion rather than to individual components thereof (North Atlantic Treaty Organisation: Research and Technology Organisation, 2012).
3 Enterprise context – n. The operational or environmental context within which enterprise considerations, agents, relationships, and transactions are relevant.

10.4 Definition of an Enterprise M&S Strategy and How It Maps to Business Strategy and Practice

Because of the diversity, technical sophistication, and "abstractness" of M&S, a purely top-down modeling and simulation strategy – by pressing policy down on midlevel management and technical staff – often results in initial but reluctant compliance that translates to inefficient implementations and thus fails the business case. On the other hand, a purely bottoms-up, grassroots strategy often produces technically strong and efficient results but with organizational incoherence.

An enterprise M&S strategy seeks to address the social and economic barriers to achieving the full potential of modeling and simulation by melding top-down vision with bottoms-up technical how-to. An enterprise strategy takes a broad organizational view of modeling and simulation that transcends individuals and individual programs yet recognizes and encourages grassroots contributions of individuals and individual organizations to improve the whole.

We must first define what we intend by the use of the word "enterprise." The term is purposefully vague; by it we intend to imply that an enterprise transcends individuals and individual products or services. However, as an economic entity, an enterprise should not be so broadly considered as to not have a clear socioeconomic intent. The United States, we suppose, could be considered an enterprise; an enterprise so broadly defined sheds little social and economic clarity within that to develop a modeling and simulation strategy. So, where should enterprise lines be drawn? Our preference would be to draw the enterprise line around the largest possible economic unit that aligns goals, objectives, funding, and authority. But, we also understand that circumscription of too large an enterprise boundary may limit success by introducing too many voices and watering down clear authority.

The critical requirement for drawing boundaries around the enterprise bears repeating – goals, objectives, funding, and authority need to be aligned. Experience tells us that when the critical requirement has not been met, mixed results, at best, are achieved. Within the Department of Defense, for example, the military services and the Office of the Secretary of Defense each established their own modeling and simulation offices. Roles and responsibilities varied among them as did their funding levels. Cost to comply with the directives that came out of those offices was, in large part, borne by other organizations. The M&S offices had seed funding but would never have been sufficient to achieve a wholesale focusing of M&S efforts within their purview. Rather, they were required to show how compliance with their directives was advantageous to the organizations actually providing the funding. As one can imagine, success ebbed and flowed as a function of the personality and vision of the one making the case. The goals, objectives of the Military Service modeling and simulation offices were not aligned with attendant funding and authority; despite heroic efforts, the DoD-wide M&S vision never achieved its full potential.

An enterprise strategy for modeling and simulation embraces the notion that models and simulation are systems and that the tenets of a well-developed systems engineering discipline are ripe for application to modeling and simulation.

First, the practice of systems engineering – including system development, evaluation, and life-cycle management – is a widely practiced discipline, with a familiar body of knowledge, and a generally appreciated suite of protocols for establishing and justifying the quality of the subject system and, consequently the confidence that might reasonably be held in its expected performance. Second, if models and simulations are a-kind-of system, then the practices associated with systems life-cycle management ought to be inherited by model and simulation life-cycle management.

Support for the premise that simulations are a kind of system includes invoking analogies between systems and simulations. Systems are certainly created and used with *a priori* intention[4] – even though alternative applications may be discovered. Systems by definition[5] consist of elemental components, each of which has its own identity, classification, attributes, and methods. Any system depends for its own nature strongly upon the relationships among its constituent's components and is created largely by the conjunction and interoperation of those components. All these descriptive properties apply to models and simulations in useful and evocative ways. Naturally, models and simulations being inheritable from the class of all systems, the relationship is asymmetric with respect to class inclusion – that is, all simulations are systems, but not all systems are necessarily simulations. Also, reasonable care needs to be taken to preserve discrimination between the simulation, which is a system and the mission space world which is the simulation's representational target system.

10.4.1.1 Modeling and Simulation in an Enterprise Context

While modeling and simulation can be a relatively private activity, pursued by individual practitioners and conducted in such a way that only the

---

4 System – "... an integrated set of elements that accomplish a defined objective." (INCOSE, 2000)
5 System – A system is a construct or collection of different elements that together produce results not obtainable by the elements alone (INCOSE, 2016).

determinations and decisions pursuant M&S exercise is disclosed, such close-hold behavior is yielding with considerable alacrity to a more social form of business practice. Modeling and simulation is particularly auspicious in its relevance to enterprise operations precisely because of the relative size, scope, and strategic significance of enterprise initiatives, and because of the breadth and interrelated complexity of prevailing enterprise environments and their contexts. As discussed above, several circumstances motivate the trend toward collaborative M&S operational practice, some with technical foundations, others with institutional or business practice groundings, but each having its own peculiar influence on likely credibility, confidence, and uncertainty of model and simulation results.

From the relatively technical perspective, increasing need for subject matter expertise beyond that of the M&S professional himself (assuming those roles can be differentiated) can affect consequent M&S quality. Multiple agents with distinct roles necessitates collaboration, admits to diffusion of responsibility, and can result in lack of coherent, persistent, and accessible assumptions and modeling conceptualizations. Meanwhile, increasing computational complexity and scope grow with emerging representational challenges. Under these circumstances, broader scope of election of representational techniques bears on eventual relevance of results. Additionally, fundamental difficulty of M&S programs can affect achievement of correctness of representations, particularly relationships among disparate representation domain component elements.

From the more institutional point of view, several factors incumbent upon the quality of M&S may be applicable. In enterprise contexts, M&S challenges often arise in the context of broader and more far-flung systems engineering and business practice concerns. Under those conditions, organizational interface discontinuities may inhibit prompt and effective M&S development and usage. Further, over broad-based organizational agendas, intended-use ambiguity is certainly possible, and the justification of alternative-use or reuse relevance is sometimes questionable. The temporal sustainment and project or product life-cycle management relevance of M&S in enterprise environments has its own concomitant entailments. Diversity required across a program of product life-cycle management (PLM) domains may be difficult to manifest in a single set of M&S assets; consistency of alternative representations so employed may be difficult to sustain; and temporal and organizational custodial persistence, stability/currency/consistency must of necessarily be administered explicitly, often across organizational boundaries with their distinctive concepts of operations and M&S practices. Requirements for developmental and operational resource investment is challenging in any case for large or complex M&S efforts; but in fully enterprise-wide contexts, labor, equipment, and (most of all) data resources may be constrained in such a way as to affect M&S artifact and results quality. Finally, enterprise intentions to have M&S influence be manifest broadly across enterprise scope establishes more and more rigorous demands for demonstration of asset and results quality; and it introduces potentially inconsistent or antithetical stakeholder requirements for credibility, confidence, and uncertainty management.

Altogether, modeling and simulation practice altogether occurs in a socioeconomic environment. A brief account of that practice, as understood by the authors, is posited so that the subsequent exposition of plausibly motivated M&S evaluation process, practice, and techniques may be most clearly appreciated.

Naturally, we can do no more than roughly distinguish scopes of consideration for each of these perspectives. In this spirit, we can only suggest the processes and consequent product artifacts under consideration. In addition, we defer any detailed consideration of the evaluation of M&S artifacts, composites, or the consequences of their uses to a later section of the text for heuristic convenience. Nevertheless, we consider the identification and discrimination of these commonly conflated domains of M&S activity to be helpful to the reader by being suggestive in further classifying the pragmatic influences on M&S credibility, confidence and uncertainty and in serving as a basis of consideration for the M&S stakeholders precisely what influences among those possible warrant attention, effort, and capture as evidence of M&S credibility, confidence, and uncertainty.

10.5 Extant Technologies That Should Be Brought to Bear to Realize the Vision of an Enterprise M&S Strategy

10.5.1 Verification, Validation, and Accreditation

As has been implied, the technical roots of an enterprise modeling and simulation strategy are grounded in the credibility of the modeling and simulation components therein. As discussed in Chapter 3 of this book in more detail, the need for trustworthy evaluation of assumptions, capabilities, and component applicability are unassailable. A trustworthy workforce bounded by a declaration of ethical practice is undeniably necessary, but it is not sufficient. The workforce must be bounded by a rigorous set of processes for coherently evaluating and assessing modeling and simulation components to establish reasonable grounds upon which to reasonably apply those components to a particular problem set. In the modeling and simulation world, that class of processes is circumscribed by the terms Verification, Validation, and Accreditation (VV&A).

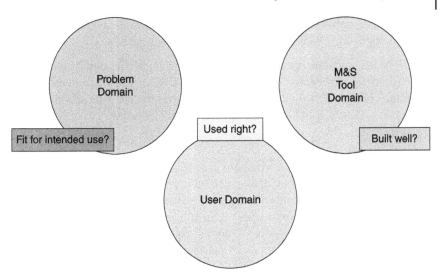

Figure 10.1 The M&S solution space.

Enterprise guidance for VV&A is important as it assures that everyone in the enterprise is doing it and documenting it the same way. In the sections that follow, we address a few processes that we feel are generally relevant to the management of VV&A and that are of particular value given the nature of pursuing model and simulation credibility across the enterprise.

Ideally, the design, development, and application of models and simulations should be defined using a systems engineering, multidimensional view of the solution space[6,7]. Credibility of M&S may then be defined using those domains that encompass the important attributes of the assessment. This systems engineering solution space addresses the major elements of a simulation system that results in the following three dimensions depicted in Figure 10.1:

- The Problem Domain, which addresses how well the M&S fits the intended use (Intended Use or IU).
- The M&S Development Domain, which addresses how well the M&S is built (Built Well or BW) with respect to the intended use.

---

6 The solution space is defined by the dimensions of the solution that match the demands of the problem being addressed and represents choice options within those dimensions where customer and problem heterogeneities matter in satisfying a particular case.

7 These domains were first noted in Appendix A of NASA-STD-(I)-7009 (NASA, 2009), Standard for Models and Simulations, an interim standard released for evaluation using pilot programs in December 2006.

- The User/Analyst Domain, which addresses how well the M&S is used (Used Right or UR) with respect to the intended use.

Any M&S solution must meet the requirements and design elements established by the Problem Domain, M&S Tool Domain, and the User Domain. The intersection and degree of overlap is the solution space for the application. Let us consider each of these domains.

10.5.1.1 Problem Domain

This domain addresses the intended use of the M&S resource, wherein systems engineering efforts define and document the foundation of the simulation design, the issues, trades, and associated development and use risks associated with its application (e.g., training, analysis, or experimentation). The overarching systems engineering challenge in this domain is ensuring the simulation as conceived is fit for its intended use. The domain addresses the appropriateness of the M&S resource to be used for a specific application.

10.5.1.2 M&S Tool Domain

This domain addresses the M&S design and development, wherein M&S resources and tools are developed and managed for support of the problem domain activities and the associated mechanization of its key attributes, be it live, virtual or constructive; continuous or discrete; stochastic or deterministic, and so on. The overarching systems engineering challenge in this domain is ensuring the simulation as is (designed, development, integrated and tested, and delivered) is built well enough to support its intended use.

10.5.1.3 User Domain

This domain addresses the M&S operators, analysts, subject matter experts, wherein users successfully make use of the M&S resource, their capability to define the use (e.g., specific training events, analyses, or experiments) being supported by the M&S, their capacity to actually apply the M&S resource (e.g., setting-up the training scenario, analyses, or experiment), and their ability to execute the M&S, generated and capture the relevant M&S results and draw the correct conclusions. The overarching systems engineering challenge in this domain is ensuring the user is adequately trained and sufficiently proficient in the execution of the M&S resources to ensure it is used right (correctly).

10.5.1.4 Integrated Solution Space

Some of the challenges and the associated criteria within each of these domains are depicted in Figure 10.2.

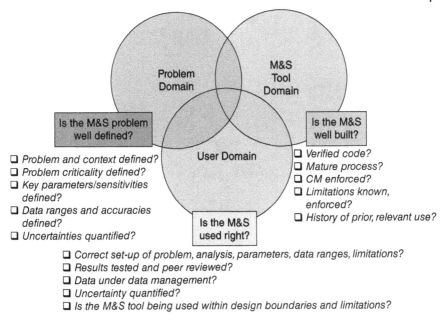

Figure 10.2 The integrated solution space.

These domains are not orthogonal and in a perfect world (the solution) would intersect one another. Also, this figure captures the essence of several significant credibility factors (criteria) within the solution space that are aligned with the associated domain.

For example, within the Problem Domain:

- Is the system/problem well defined?
- Are the system parametric data defined in common terms with the other domains (common ontology)?

For example, within the M&S Tool Domain:

- Is the M&S tool well built?
- Does it have verified and validated requirements, models, and parameters?
- Was it built using a mature development process?
- Does the M&S match the intended use such that the analysis parameters are synchronized to the system/problem parameters (common ontology)?

For example, within the User Domain:

- Is the user's execution of the M&S done well?
- Is the M&S resource correctly set up and executed properly?
- Are the system parametric data defined using terms in common with the other domains (common ontology)?

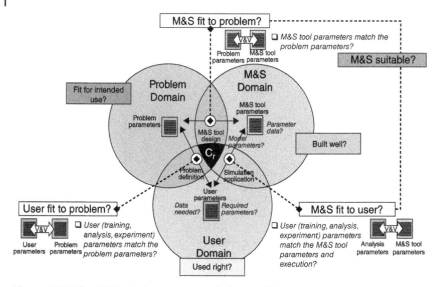

Figure 10.3 The M&S solution space correlation coefficient.

10.5.1.5 Solution Space Correlation Coefficient

As noted in the Venn diagram depicted in Figure 10.3, the actual solution space for a simulation can be represented by the intersection of these domains and the degree of correlation that exists. The Venn diagram portrays the intersections between these domains.

The interface between the Problem Domain and the User Domain is the User's Problem Definition. The interface between the User Domain and the M&S domain is the application of a selected simulation for use. The interface between the M&S domain and the Problem Domain is an M&S tool design that represents some selected and defined aspect of the referent system and/or its operation in the System/Problem Domain. Credibility of M&S Results (Cr) is then achieved when all three of these criteria sets are simultaneously satisfied. This is a unique metric we have labeled as correlation coefficient – or the overlapping section C_r – shown in the center of the figure.

This M&S credibility then is the defined intersection across all three of these domains in the solution space. Ideally, the User's needs (parameters) intersect the M&S resource parameters within the context of the defined Problem. This goodness-of-fit statistic represents how well an M&S resources fits a specific user and specific problem. As illustrated in Figure 10.4, a measure of this overall goodness of fit summarizes the discrepancy between what exists in the solution; for example, M&S Space, User Space, or Problem Space and what is desired for the M&S resource in question.

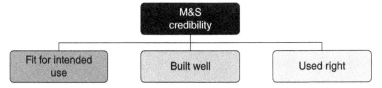

Figure 10.4 M&S credibility is a function of the goodness of fit.

10.5.1.6 Credibility Criteria

Dendritic analysis is an effective way of decomposing these credibility criteria (Critical Operational Issues, or COIs) to the point where actual V&V data requirements and test measurements can be identified for each domain.

Measurement of M&S results credibility is not practical at the domain level, even though it is a useful category for summarizing the key types of credibility obtained by lower level criteria. Thus, each of the three domains is further hierarchically decomposed into a set of lower level, measurable criteria, as depicted in Figure 10.5.

A representative set of example criteria are illustrated in Figure 10.6, which graphically portrays in one dimension these three domains of the solution space along with the associated criteria and scale for measuring their credibility.

Using this technique, these COIs can be successively decomposed into subordinate requirements in a root-like structure. In this approach, objectives are used to clearly express the broad aspects of the simulation's usage to the Problem, M&S and User domains.

Measures of Performance (MOP) or Measures of Evaluation (MOE) can be developed as subsets within each of these domains and are designed to treat specific and address issues exclusive to each. The specific criteria selected for each domain are influenced by the M&S intended use. Each MOE or MOP is traceable and a direct contributor to the assessment and, through it, is identifiable as a direct contributor to addressing a COI. Each issue can be linked to one or more MOPs or MOEs that may be quantitative or qualitative measures that are in turn are tied to specific V&V activities and data elements. Data elements are observations and/or measurements under specified conditions.

10.5.1.7 Enterprise Solution Manifold

What emerges then is an ability to conceptually represent where in the solution space a particular M&S resources resides if we apply a uniform scale across the enterprise.

M&S credibility domains

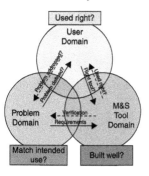

Dendritic identification of credibility criteria

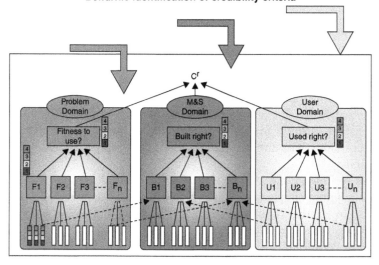

Figure 10.5 Solution manifold.

For example, the National Air and Space Agency (NASA) established a mechanism enabling this across their enterprise with its release of NASA Technical Standard 7009, *Standard for Models and Simulations* (NASA, 2009)[8]. This standard defined and applied a unitary scale to meet guidance of the NASA Chief Engineer to: "Include a standard method to assess the credibility of the models and simulations presented to the decision maker

---

8 This standard was published by NASA to provide uniform engineering and technical requirements for processes, procedures, practices, and methods that have been endorsed as standard for models and simulations developed and used in NASA programs and projects, including requirements for selection, application, and design criteria of an item.

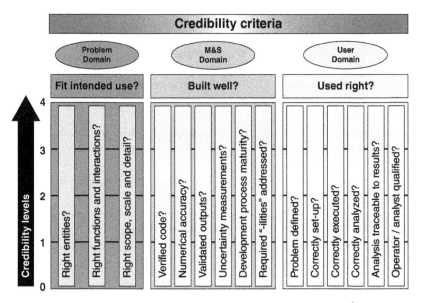

Figure 10.6 M&S credibility assessment domains and example criteria.[9]

when making critical decisions (i.e., decisions that affect human safety or mission success) using results from models and simulations."

As a consequence, the M&S results used in critical decisions within NASA may be assessed using a Credibility Assessment Scale (CAS) with five levels as illustrated in Table 10.1 for key factors like M&S development.

Level 1 is the minimal evidence that must be available for any credit to be given. Level 2 gives credit to comparison with referent data from expert opinion or other M&S. Level 3 can only be achieved if several key factors (Validation, Input Pedigree, and Results Uncertainty) are based on experimental data. Level 4 sets a high bar: there is minimal chance of model implementation errors, numerical errors are insignificant, validation has been accomplished against the real system in its real environment, input data are validated against data from the real system in its real environment, uncertainty estimates are nondeterministic and based upon real data, sensitivities to the most sensitive variables are known, the M&S has a strong track record, the management processes are strict, and the personnel are highly qualified.

---

9 These criteria were first noted in Appendix A of NASA-STD-(I)-7009, Standard for Models and Simulations, an interim standard released for evaluation using pilot programs in December 2006 (NASA, 2009).

Table 10.1 Key elements of NASA's Credibility Assessment Scale.[a]

Level	M&S development	
	Verification evidence	Validation evidence
Level 0	Insufficient evidence	Insufficient evidence
Level 1	Conceptual and mathematical models verified	Conceptual and mathematical models agree with simple referents
Level 2	Unit and regression testing of key features	M&S results agree with experimental data or other M&S on unit problems
Level 3	Formal numerical error estimation	M&S results agree with experimental data for problems of interest
Level 4	Numerical errors small for all important features	M&S results agree with real world data

a) This table only illustrates scoring for M&S development, which is 1 of the eight factors detailed in the NASA standard. As detailed in Appendix B of the standard, additional factors to be considered include: M&S Operations, which includes Input Pedigree, Results Uncertainty, and Results Robustness; and Supporting Evidence, which includes Usage History, M&S Management, and People Qualifications (NASA, 2009).

Achieving a level 4 rating on the scale may be technically feasible, albeit difficult. Obviously, level 4 can only be achieved across the board for a system that is in the operations phase of the life cycle. Lower levels on the scale are more appropriate targets for earlier phases of the life cycle. More details on using this kind of scalar approach to summarizing credibility findings across an enterprise are provided in the NASA Technical Standard 7009.

Using a unitary scale across an enterprise is useful if the user (decision maker) has multiple M&S resources that can potentially be used to address a need. Each simulation can then be scored using a common scale as notionally illustrated in Figure 10.7. Subsequently, trade-offs across the enterprise can be conducted in selecting one M&S over another M&S for use.

This evaluation can be applied uniformly across an enterprise and is notionally depicted in Figure 10.8, which is a manifold that yields a view of the solution space using a three-dimensional array and a color-coded incremental scale representing unacceptable (red), marginal (yellow), acceptable (green), and superior (blue) assessments of the fundamental questions (criteria) for the M&S Tool, User, and Problem domains.

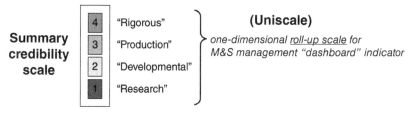

Figure 10.7 Unitary scale[10].

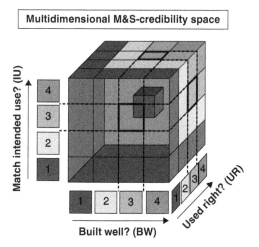

Figure 10.8 Solution manifold[11].

The uniform application of this approach across the enterprise enables decision makers to

- compare many M&S that may be available and evaluate the trades associated with selection of one M&S for use over other available M&S;
- make better decisions based on the use of M&S data; and
- do trade-off analysis that support investment in future M&S development/refinement.

---

10 This unitary scale was first introduced in Appendix A of NASA-STD-(I)-7009, Standard for Models and Simulations, and continues in use with its successor, NASA Technical Standard 7009 (NASA, 2009).
11 This multidimensional solution space was first introduced in Appendix A of NASA-STD-(I)-7009, Standard for Models and Simulations (NASA, 2009).

10.5.1.8 Caveats

The illustration just provided, while correct from a big picture standpoint, is actually very simplistic from an M&S practitioner's perspective, owing to the following properties of modeling and simulation:

- *Nonlinear:* The domains are depicted as occurring in a strict linear space. This is not typical; there is a high degree of healthy interaction between them. This interaction often takes the form of iteration (do it until you get it right!) and recursion (keep doing it until you are done!).
- *The M&S Tools, Users, and Problems are Not Static:* Each can be quite dynamic. For example, problems may be defined, redefined or rescoped so they can be addressed with the M&S as it is. Similarly, M&S resources mature and evolve in response to technology insertion, user needs, and new problems. Users may move on and be replaced by new personnel that are unfamiliar with the problems and the resources being used.
- *Complex System Model:* The solution space was depicted as operating in a simple environment. In practice, for nontrivial systems, the solution space consists of many layers and levels. Throughout this layered structure, multiple M&S resources are being applied to a number of problems simultaneously.

So, a lot of things are going on at the same time in the real world and are being done by many different users and M&S tools. Given this, it's all too easy to actually make chaos out of order!

10.5.2 Standards

An enterprise modeling and simulation strategy seeks to establish coherence across modeling and simulation conception, development, maintenance, and application. Standards provide the base upon which to build that coherence. In NASA's 2004 report "A Renewed Commitment to Excellence," there is an obvious attention to the fundamental need for full lifecycle modeling and simulation standards. The report recommends that NASA:

"Develop a standard for the development, documentation, and operation of models and simulations.

a) Identify best practices to ensure that knowledge of operations is captured in the user interfaces (e.g. users are not able to enter parameters that are out of bounds).
b) Develop process for tool verification and validation, certification, reverification, revalidation, and recertification based on operational data and trending.

c) Develop standard for documentation, configuration management, and quality assurance.

d) Identify any training or certification requirements to ensure proper operational capabilities.

e) Provide a plan for tool management, maintenance, and obsolescence consistent with modeling/simulation environments and the aging or changing of the modeled platform or system.

f) Develop a process for user feedback when results appear unrealistic or defy explanation." (NASA, 2004).

The report demonstrates the breadth of regard when considering standards for modeling and simulation at the enterprise level. We would add two other categories for consideration: interoperability standards and data standards. These two categories recognize that an enterprise view necessitates sharing of capability across individual and program boundaries. Building capability and data according to *a priori* designated standards, whether officially recognized or simply developed by the enterprise itself, will greatly reduce the cost, time, and risk of integrating disparate capabilities.

The Department of Defense has led the way in defining simulation interoperability standards and has borne much of the expense of vetting those standards through internationally recognized standards development organizations. There are several simulation interoperability standards worth considering and an enterprise need not settle on one but can designate particular interoperability standards for specific uses. Two prominent standards are explained briefly here.

Distributed Interactive Simulation (DIS) was developed in the early 1990s and consists of a set of predefined messages in which to encode simulation state. These messages are called Protocol Data Units (PDUs). DIS is a very structured approach to simulation interoperability and is primarily targeted at entity-level (i.e., airplane, ship, and soldier) simulations. DIS is recognized by the Institute of Electrical and Electronics Engineers (IEEE) as IEEE 1278 (IEEE, 2012, 2015).

High-Level Architecture (HLA) was developed in the mid 1990s in recognition of some of the limitations of DIS. HLA consists of (1) an interface specification, (2) an object model template, and (3) a set of rules with which simulations must comply. Unlike DIS, HLA does not mandate content for representing simulation state. Rather, such content is explicated by simulation developers in object models. HLA continues to evolve to account for advances in Information Technology and application demands. IEEE also recognizes HLA as IEEE 1516 (IEEE, 2010a, 2010b, 2010c, 2003).

The Public Health Data Standards Consortium defines data standards as "documented agreements on representations, formats, and definitions of common data." (Public Health Data Standards Consortium, 2016) The enterprise context is replete with potentially common data; structuring, defining, and documenting that common data are critical for effective reuse. Furthermore, domain-specific data standards are readily available to the public at-large. For example, the Federal Geographic Data Committee endorses nearly one hundred geospatial information standards and specifications. Through standards, data are more readily shared within the enterprise and between the enterprise and the rest of the world. This has the potential to significantly reduce cost and risk and significantly increase coherence across M&S application spaces. Data sharing, however, has its potential pitfalls that can become particularly acute in an enterprise context. First, garbage-in invariably leads to garbage-out. Poorly chosen data can have catastrophic consequences and sharing poorly chosen data multiply the effect. Data should be selected and applied with the same care as when selecting and applying modeling and simulation components as described under VV&A above. Second, data must be treated securely. In the commercial sector, organizations need to be cognizant of the proprietary nature of the data in question. And as data are shared across an enterprise, special care is called for as the consequences of disclosure are typically most acutely felt with the originator and the sense of the consequences tends to diminish as the data travel farther from the source. Additionally, as data become more complex (i.e., data within data), it must be remembered that the consequence of disclosure of the whole is at least as high as the greatest consequence among the parts.

A robust effort to identify, select, inculcate, and use standards as broadly as practicable, requires investment with little to no apparent impact to customers, users, and shareholders in the short term. Automobile sales brochures will highlight a car's safety; but few, if any, will bother to draw attention to a manufacturer's use of SAE J1102 (Society of Automotive Engineers, 2016) (Mechanical and Material Requirements for Wheel Bolts). On the other side of failure, the value of standards can be abundantly clear. Poor fiscal decisions based on incoherent financial models, over-budget not-invented-here programs, and customers frustrated by inconsistent human–machine interfaces, all bring standards into sharp focus. But, the *a priori* business case for standards is almost always difficult to make; and, it's often more social than economic. To use standards is to stand upon giant's shoulders. A lot of effort went into developing DIS as a standard to get two simulators to "talk" to one another. To use DIS is to leverage the expertise, time, energy, and expense of many talented engineers and scientists.

10.5.3 Other Techniques and Technologies That Should Be Considered

While not as critical as VV&A and standards for the successful implementation of an enterprise M&S strategy, there are several other categories of technologies and best practices that are worthy of consideration.

Free and easy communication among members of the M&S community will promote serendipitous information exchange, facilitate broad collegiate problem-solving, and foster establishment of a unified culture. Point-to-point communication is personal and powerful but presupposes the one with the problem knows who might have the answer. Broadcast communication gets the information to the most people in the shortest time but inhibits valuable interchange. In the software world, where monitor and keyboard are always close at hand, Instant Messaging provides a valuable low-cost bridge between the two. As there are many products available, the enterprise should carefully standardize on the most advantageous. Finally, for distributed teams where personal contact is sparse, use of online video group chat is enormously valuable to bringing the personal connection (along with the attendant relationships and emerging culture) to otherwise cold text, voice communication.

As implied in the previous paragraph and elsewhere above, a common culture goes a long way to building the enterprise M&S capability. Models and simulations ultimately rely on development, testing, and maintenance of software. Initial movement toward standard software development tools and practices can be met with intense, passive, and active, resistance. New graduates learned their trade in their school's preferred environment. Journeymen practiced their trade in their microculture's preferred environment. To be sure, when journeymen from different microcultures and students from different schools gather together to formulate a standard development strategy, strong opinions will abound. The options must be carefully studied, similarities and differences noted, and vision for standardization developed and brought to bear on a skilled software community. Ultimately, everyone in the enterprise will need to comply either willingly or by policy; however, diligent bottoms-up study will yield a qualitatively better solution and also one that's more quickly embraced by those most affected.

The last area we'll offer for consideration is a physically or logically central repository for shared modeling and simulation capabilities, data and documentation. The one-stop-shop is a relatively simple concept, only slightly less simple to implement and with a clear benefit to the enterprise M&S community. Enterprise IT specialists are well aware of network threats and vulnerabilities and their expertize should be used liberally.

10.6 Concluding Thought

An enterprise perspective on modeling and simulation is one that represents exceptional value to an organization but its ultimate realization is challenging. Top-down vision and controlled, thoughtful guidance provides the direction. Bottoms-up technical knowhow and enthusiasm provides the optimal solution for the particular enterprise. Don't be fooled into thinking that top-down and bottoms-up are separated by an intellectual chasm. They are not. Bottoms-up can inform and even initiate cogent vision and guidance. Likewise top-down can bring business and technical experience to the detailed discussions as appropriate. As systems grow ever-more complex, an enterprise M&S strategy is critical to the successful and profitable application of this important discipline.

Acknowledgment

The authors wish to acknowledge the mentorship and leadership provided by our esteemed colleague and friend, William F. (Bill) Waite. Bill served as the Chairman, Chief Technology Officer, and was cofounder of The AEgis Technologies Group. He was a substantial contributor to the M&S community, workforce, industry, and market. Over a professional career spanning five decades, Bill was instrumental in the invention and evolution of M&S technologies, practices, and standards impacting a broad spectrum of M&S programs and activities, including simulation technologies evolution, simulation systems development, simulation verification, validation, and accreditation, simulation-based studies and analyses and systems engineering, and the development of hardware and software products supporting modern M&S practice. The M&S community had no stronger advocate, his colleagues' had no better friend, and his employees at AEgis had no better mentor.

References

IEEE (2010a) *IEEE Std 1516–2010 - IEEE Standard for Modeling and Simulation (M&S) High Level Architecture (HLA)– Framework and Rules*, IEEE, Piscataway, NJ.
IEEE (2010b) *1516.1–2010 - IEEE Standard for Modeling and Simulation (M&S) High Level Architecture (HLA)– Federate Interface Specification*, IEEE, Piscataway, NJ.

IEEE (2010c) *1516.2–2010 - IEEE Standard for Modeling and Simulation (M&S) High Level Architecture (HLA)– Object Model Template (OMT) Specification*, IEEE, Piscataway, NJ.

IEEE (2003) *1516.3–2003 - IEEE Recommended Practice for High Level Architecture (HLA) Federation Development and Execution Process (FEDEP)*, IEEE, Piscataway, NJ.

IEEE (2012) *IEEE 1278.1–2012 - IEEE Standard for Distributed Interactive Simulation–Application Protocols*, IEEE, Piscataway, NJ.

IEEE (2015) *IEEE 1278.2–2015 - IEEE Standard for Distributed Interactive Simulation (DIS) – Communication Services and Profiles*, IEEE, Piscataway, NJ.

INCOSE (2000) *INCOSE Systems Engineering Handbook Version 2. 0*, John Wiley and Sons, Inc.

INCOSE (2016) What is Systems Engineering, INCOSE. A system is a construct or collection of different elements that together produce results. (accessed December 28, 2016).

K Dictionaries Ltd (2005) *Random House Kernerman Webster's College Dictionary*, Random House, Inc.

NASA (2004) A Renewed Commitment to Excellence: An Assessment of the NASA Agency-Wide Applicability of the Columbia Accident Investigation Report, NASA, Greenbelt, MD.

NASA (2009) *Standards for Models and Simulations*, NASA, Cape Canaveral.

Public Health Data Standards Consortium (2016) Promoting Standards Through Partnership. Available at http://www.phdsc.org/standards/health-information/d_standards.asp. (accessed December 28, 2016).

Society of Automotive Engineers (2016) Mechanical and Material Requirements for Wheel Bolts(STABILIZED Nov 2016), SAE International, 2016. Available at http://standards.sae.org/j1102_201611/. (accessed December 28, 2016).

Part IV

Application

11

A Complexity and Creative Innovation Dynamics Perspective to Sustaining the Growth and Vitality of the M&S Profession

Levent Yilmaz

Auburn University, Auburn, AL, USA

11.1 Introduction

The National Academy of Engineering report (NAE, 2006) urges that leadership in innovation is essential to prosperity and security. In a global, knowledge-driven economy, technological innovation – the transformation of new knowledge into products, processes, and services – is critical to competitiveness, long-term productivity growth, and the generation of wealth. To raise awareness about the role of and significance of simulation in advancing science, engineering and technology, the National Science Foundation (NSF) convened a panel on Simulation-Based Engineering Science (SBES, 2005) that highlighted the value of simulation modeling in advancing the computational science and engineering fields.

This chapter aims to promote the recognition of modeling and simulation (M&S) as its own discipline, which has a core research domain with implications to the broader context of science and engineering. This view is in contrast to the more traditional view of simulation as simply a "research tool or instrument" to support research in a variety of diverse fields. The distinction between these views and the growth and sustainment of M&S as a discipline is critical for the growth and expansion of the field.

11.1.1 Background on the Scientific and Societal Role of Simulation

M&S is a unique discipline in that it interacts strongly with other disciplines and draws much of its vitality from serving the needs of other diverse communities of science. Dramatic improvements in the capability and capacity of computational models in empirical and normative understanding

The Profession of Modeling and Simulation: Discipline, Ethics, Education, Vocation, Societies, and Economics, First Edition. Andreas Tolk and Tuncer Ören.

of artificial and natural information processes are enabling the creation of vast archives of knowledge of significant value to scientists. For example, the particle physics community is considering computational virtual experiments on supersymmetric particles, which may shed light on the dark matter problem of cosmology. Similar opportunities are emerging in medicine, where the promise of artificial organs is being explored via computation-based simulation of natural and artificial organ functions. Numerical simulations are providing new modes of computational science where supercomputers act as the central power plants in a scientific Grid. The increasing prominence of simulation and data-driven science is also driving experimental science by introducing and integrating new apparatus into the scientific problem-solving process. For instance, the earthquake engineering community is deploying telepresence capabilities that allow participants to remotely design, execute, and monitor experiments without traveling to the actual experiment facilities. The Network for Earthquake Engineering Simulation (NEESGrid) is using Grid technologies to link earthquake engineering across the United States with shared engineering research equipment, data resources, and leading edge computing resources. Other resources that focus on the role of simulation in engineering and science include the Presidents Information Technology Advisory Committee (PITAC) Report (Benioff and Lazowska 2005), which emphasizes the development of computational science for national competitiveness; the SCiDAC (2000) and SCaLeS reports (Keyes *et al.*, 2003), which emphasize opportunities for scientific discovery at the high-end of today's simulation capabilities and the need for a new scientific culture of interdisciplinary teamwork to realize them; and the Cyberinfrastructure report (Atikins 2003), which outlines a diverse program of interrelated research imperatives stretching well beyond simulation into communication and computational virtual experimentation.

11.1.2 Motivation

This chapter is based on the observation that the significant potential benefits and advances in the realm of simulation-based engineering and science require explicit understanding of basic and applied research in the scientific components of M&S itself. Physics, biology, medicine, as well as human and social sciences have well-refined public explanations of their research processes and modes of scientific inquiry and knowledge production. These explanations provide guidance as to what constitutes good and innovative research.

Despite its broad impact on science and technology, M&S does not yet have this sort of explicitly delineated, well-understood, and communicated guidance. However, if M&S is going to transform twenty-first century science, engineering, and technology, and continue to be relevant, then the characterization of the science and technology of modeling and simulation must also be explicitly delineated and documented. Understanding research strategies, guiding principles, and criteria for evaluating good and creative innovations in M&S will not only help scientists in developing next-generation simulation technologies and designing research plans using computational models but will also facilitate explaining the nature and the importance of M&S research to other engineers and scientists at large.

11.2 Sustaining the Growth and Vitality of M&S

What is necessary for a scientific field to form and develop? What are its underlying guiding principles, research strategies, and its criteria for evaluating good and creative innovations? From basic research and concept formulation to diffusion of innovations, scientific fields rest on fundamental strategies that not only provide guidance to scientists but also provide explanations for the society and institutions that have stakes in the produced knowledge. The theme and the issues addressed in this chapter are grounded on a generic model (Gardner, Csikszenthmihalyi, and Damon 2001) of creative research depicted in Figure 11.1.

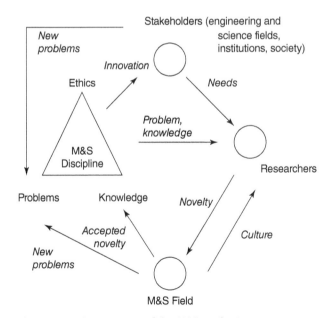

Figure 11.1 Components of the M&S profession.

The elements of the framework are the researchers, the field, the discipline, and the stakeholders. The field is comprised of experienced scientists, who decide on the novelty of contributions, as well as other contributors and apprentices. The discipline has three dimensions (problems, ethics, and knowledge) that depict the knowledge, skills, practices, and values of the professional realm. Stakeholders refer to other engineering and scientific fields that use simulation, as well as general society and institutions, which benefit from the scientific and technical production of the M&S field. The relations between the elements of the model pose challenging issues. The dependency between the domain and researchers in M&S suggests exploring the following questions pertaining to nature of scientific production:

• What kinds of issues (methods for analysis, generalization, feasibility, empirical understanding, normative understanding, theory generation etc.) are interesting and worthy of allocating limited resources in scientific, technological, and applied M&S research and why?
• What research methods can help attain these results (abstraction, empirical models, analytical models, procedure, or technique)? What are strengths and limitations of existing methods and how can existing weaknesses be mitigated?

Considering the novelty relations between the researchers, the field, and the discipline, as well as the competitive nature of M&S research:

• What types of evidence are required to demonstrate the credibility and novelty of research results in M&S (e.g., persuasion, analysis, experience, evaluation, examples)?

The social conditions that stimulate research are captured through the culture imposed by the field on the researchers, who are practicing to grow and extend the scope of the M&S body of knowledge. Hence:

• What are the desirable characteristics of the M&S field to stimulate community-building and alignment of the expectations of the stakeholders and the discipline, the field, and the researchers?

As pointed out in SBES (2005), as well as the National Research Council report on Defense Modeling, Simulation, and Analysis, the changing landscape in science and engineering (e.g., industrial and defense application, medicine, predictive homeland security, energy and environment) introduces new types of problems and challenges into the M&S domain. In light of these emergent needs:

• How can M&S stay relevant as new critical fields such as global climate change mitigation, energy restructuring, genetic engineering impacts on society, universal healthcare, and so on emerge and come into prominence? Surely, the systems point of view and the tools that M&S brings to the table are key to these new directions. So, how does M&S play its rightful role in these?

As depicted in the model, a discipline and its domain need to contain more than just knowledge and problems. In addition, there has to be an ethical dimension reassuring that knowledge and skills will not be used against the common interest shared by the stakeholders. Given the significance of ethics and institutional support for the sustained growth of a scientific field:

• What are the ethical concerns that are peculiar to M&S and its relation to society?

The issues above suggest exploring the M&S discipline under the umbrella of four major dimensions. The first dimension focuses on the scientific discourse by exploring the nature of scientific and technical knowledge production and application. The second dimension recognizes the significance of evaluation of knowledge leading to its certification, which is the result of a process of competition. The next two dimensions acknowledge M&S profession as a sociocultural practice. Hence, desirable characteristics of the social and institutional organization, as well as ethics in M&S, are considered as important levers to sustain growth and further development of the professional realm of M&S.

11.3 Alignment of the Discipline, Field, and Individual Researcher in M&S

Recognizing the significance of the proper alignment of the discipline, the field that represents the social context, expectations of the stakeholders, and personal backgrounds of individuals, we take a close look at each one of these components and raise questions pertaining to their impact on the M&S profession. The discipline component as shown in Figure 11.1 represents a symbolic system that indicates the codification of specialized knowledge, practices, and norms for transmission to new practitioners. Besides the ethics dimension, a discipline embodies ideas, practices, knowledge, as well as new problems worthy of exploration. Unless the M&S profession convinces its stakeholders that its practices (procedures) and ethical standards are valuable, support for its advancement is likely to find resistance.

The M&S field constitutes the social context of the professions by enacting the symbolic system depicted by the M&S discipline. More specifically, the field constitutes the engineers and scientists that practice the skills and knowledge embodied by the discipline. Great majority of the field is represented by experts. Vitality of the M&S profession requires a second group, called the apprentices or students, to facilitate growth and continuity of the profession. A small minority of the members of the field are categorized as evaluators that decide which contributions are worthy of expanding the domain of the discipline. Finally, the scientists and engineers who opt to enter into profession represent the individuals who receive training to be enculturated in the norms and practices of the discipline to make contributions to further extend its depth, as well as scope.

The alignment of the field, the culture represented by the discipline, and the expectations of the stakeholders are critical to the growth of the profession. Furthermore, to sustain the vitality of the profession, the field has to pay attention to emergent challenges and, if necessary, revisit assumptions regarding the discipline, field, and practitioners. Tables 11.1–11.3 suggest issues that are worth considering to examine the desired characteristics of discipline, field, and practitioners to sustain the relevance and vitality of the M&S research and profession, respectively.

Reasoning and addressing the issues raised above indicate a systemic view of the interactions among disciplines, fields, individuals, and the domain. Moreover, the interaction aspect suggests a network-centric view that appears to be effective in sustaining complexity toward novelty and resilience observed in socioecological systems, including innovation systems,

Table 11.1 Issues Pertaining to the influence of discipline and domain on M&S profession.

1. How is knowledge stored and disseminated?
   Permanency and ease of access of knowledge facilitates assimilation of knowledge, as well its future growth

2. How is knowledge protected? Does new copyright mechanisms such as Open Access Data Protocol improve diffusion of innovations?
   Accessibility improves participation in M&S research

3. How differentiated is the discipline?
   Differentiation improves specialization, which then enables advancement

4. Is the knowledge within the discipline sufficiently integrated so that advances in one subdomain become relevant for the overall discipline, while avoiding rigid or tight integration that inhibits innovation?

5. Is the discipline open enough to facilitate interdisciplinary as well as transdisciplinary research to make the profession vital during rapid advancement of science and technology for emergent new application domains?

Table 11.2 Issues pertaining to the influence of context on M&S profession.

1. Does the M&S field encourage innovation?

2. What is the extent of mobility and internal conflict in the field?

3. Is the field balanced in terms of integration and differentiation to facilitate generation of innovations, as well as their adoption.

4. To what extent is the field independent of its stakeholders (e.g., institutions, other science and engineering fields)?

5. Is there an agreement and consensus on the what constitutes novelty? Too much or lack of agreement may inhibit innovation and meaningful growth

6. What is the extent of institutionalization in the field?

7. Is the field open and supportive of change? Higher field receptivity in conjunction with bold individual contributions is likely to result in innovations

Table 11.3 Issues pertaining to the influence of personal qualities on M&S profession.

1. Does the practitioner exhibit the talent and skills needed to succeed?

2. What is the impact of curiosity and intrinsic motivation on generating novelty?

3. Do practitioners exhibit divergent, yet meaningful and useful thinking that is conducive to discovery?

4. What are the requisite personality traits relevant to successful research? Do certain personality traits bring advantage in generating novel contributions?

that promote the significance of balancing connectivity, symbiosis, and decentralization.

11.4 Managing the Complexity of the M&S Innovation Ecosystem

The M&S innovation ecosystem is comprised of individuals, groups, communities, and societies that conduct research and development in the M&S discipline.

11.4.1 M&S Innovation EcoSystem

A diverse group of agencies and communities compete and cooperate to produce new knowledge and novelty in the M&S domain (Ören, 2016). The interaction among these communities in terms of joint, cosponsored workshops, as well as researchers from different communities working on joint projects, facilitates exchange of information, thus generating interdependencies that self-organize through autonomous and self-motivated individuals. The following analogy illustrates the aspects of scientific communities and disciplines that exhibit behavior that is akin to self-organizing behavior that is observed in complex adaptive innovation systems. Specifically, one can view the innovation ecosystem in a way similar to the self-organizing energy–metabolism dynamics observed in ecological systems where energy is converted into resources utilized by the individuals in the environment to perform actions.

In this view, each scientific community can be construed as having a phenotype that defines its domain or discipline that is comprised of norms, practices, and skills that are deemed to be critical to collective creativity within the field of study. Each community is comprised of individuals that relate to members of species in ecological models. A discipline accommodates multiple communities that interact and coevolve with each other through processes of learning, transformation, and mobility. Individuals migrate across communities and disciplines that serve as cognitive niches to individuals seeking environments conducive to creative problem solving. Individuals within scientific communities will have the ability to change and modify their environment as a result of their development within, and interaction with, the environment. Individuals and scientific communities are associated with a scalar health or fitness measure indicating success in their environment. Scientific communities undergo stages of coalescing, growth, stability, and renewal that may affect their behavior. The external environment (e.g., funding agencies) influences decisions of individuals and communities by altering the availability and distribution of resources. Knowledge production converts human, financial, and knowledge capital into resources (e.g., open problems, skills), which are then transformed into solutions and products.

11.4.2 Sustaining the M&S Field as a Complex System

As a self-organizing innovation ecosystem, the M&S field exhibits the characteristics of a complex adaptive system. Complexity science involves the study of structural interdependencies and the emergent system-wide effects they produce. The emergent behavior of the innovation ecosystem is driven by two interrelated variables: heterogeneity and interdependence. Heterogeneity refers to the level of variation of certain key characteristics and behavior among the system's components (e.g., societies, individuals, communities of practice), whereas interdependence refers to the way the components interact with each to affect or influence each other's behavior.

11.4.2.1 Heterogeneity

The role of heterogeneity and diversity within the M&S innovation landscape can be examined from different perspectives, including but not limited to the following:

• How do variations and differences among communities of innovation help define the emergent behavior of the field and the state of the discipline?
• How can right amount of diversity can be advanced so that the sociotechnical innovation system can influence its adaptive capacity to respond to emergent problems and needs generated by the stakeholders.

The main advantages of heterogeneity include its benefits to improving the ability of the discipline and the field to overcome three major constraints or obstacles involved in experimentation: (i) bounded rationality, (ii) replication, and (iii) premature convergence.

Bounded Rationality

Individuals, communities of practice, and societies often make decisions with limited information or without full knowledge about an issue at stake. Heterogeneous scientific fields with groups that have distinct objectives and missions can have better opportunity to explore a wide range of experiments. On the other hand, in homogeneous fields, the communities will have similar priorities, resulting in an exploration that is bounded by the common perspective.

Replication

In the lack of heterogeneity and variation, a strategy explored by a society can produce information that is available at little or no cost to other groups, hence communities of practice and innovation have an incentive to "replicate" each other's innovation experiments. One of the key concerns of replication is that it leads to less experimentation. Also, the societies that conduct experiments and explore new practices for knowledge generation and dissemination may choose to reveal or share less information to prevent others' ability to replicate. This deprives the M&S field of critical data and observations that can help move the discipline forward. Heterogeneity offers a solution that mitigates such inadvertent consequences of replication. Because of variety in priorities and perspectives, strategies and solutions discovered by one community may not be optimal or even applicable in a different society within the M&S field. Hence, some communities of the field are less likely to copy the strategies available to others, and they are expected to explore and experiment under the constraints that their members are experiencing.

Convergence by Information Contagion

When members of the M&S field all arrive to the same strategy, which may be suboptimal, the result is a premature convergence that puts the overall field in a less than desirable state that is not congruent with the needs of its stakeholders. Premature convergence is often the result of contagion and risk aversion. When the members of the field stop relying on the local information available to them and instead leverage or mimic the decisions made by others, the result is an information cascade that facilitates diffusion and adoption of policies and technical strategies that are less than optimal. Also, risk aversion results in communities to avoid experimenting with specific initiatives that they perceive as likely to fail. Heterogeneity reduces the risk of risk aversion, for the initiative that works for one or more groups may not necessarily be relevant to a community that has a different priority and perspective within the M&S domain.

While the advantages offered by heterogeneity in relation to the above three criteria provide incentives for achieving a reasonable level of variety, we also need to recognize the downside of heterogeneity to help mitigate its undesirable consequences. For instance, heterogeneity can be a source of conflict between competing societies and communities. Furthermore, extensive variety can limit the ability of the field to internalize knowledge and practices due to lack of coherence, increased cost of information transfer, and limitations of collective action. Highly complex initiatives require participation of multiple communities of innovation and practice with centralized goal setting and coordination.

11.4.2.2 Interdependence

Heterogeneity is only part of the puzzle in managing the complexity of the M&S innovation landscape. Information interdependence and innovation diffusion play critical role in opening communication channels and allowing members of the discipline to engage in new avenues of exploration. Within the context of interdependent societies and communities of innovation, the number, strength, and topology of connections help determine the overall innovation behavior of the field. The likelihood that communities will exchange information, as well as the accuracy, frequency, and type of information are influenced by the collaborative network structure into which the field is embedded.

The network structure has influence on discovering the equilibrium for optimal performance of a discipline by managing a balance between explorative and exploitative activities. Exploitation of existing solutions to move the discipline forward brings stability and predictability while minimizing risk to existing communities of practice. However, exploitation can result in stagnation and may hinder the ability of the M&S discipline to compete with other scientific and engineering domains. On the other hand, exploration facilitates maintaining diversity of initiatives and solutions and hence increases the ability of the field to experiment. However, extensive degree of exploration is costly and may result is a field that loses its ability to leverage benefits of convergence and stability.

Interdependence and network connections among the members of a field can nudge the behavior of the field toward either exploration or exploitation. The notions of strong and weak ties are critical to this connection. Strong ties that characterize dense networks that include connections among a large number of constituent members facilitate fast spreading of information, whereas networks with weak ties reduce the pace of information aggregation due to slower diffusion based on limited number and scarcity of ties. With strong ties, the field is pushed toward exploitation-based strategies and homogeneity. On the other hand, weak times incentivize exploration, for information about other experiments is limited. However, experimentation with new initiatives requires managing information deficits, in which case a central and focal society can serve as a clearinghouse or information portal.

11.5 Conclusions

This chapter aims to raise awareness about the nature of scientific knowledge production in M&S, as well as its application as a scientific instrument in the broader context of science and engineering. As such, the first part of the chapter contributes to both the development of the M&S profession and to its recognition within the scientific/technical community at large. A better understanding of the M&S discipline based on the types of problems, the research methods adopted by the members of the discipline, and the evaluation criteria used by the field should help researchers design research plans and report results clearly, while helping to explain the character of modeling and simulation research to other scientists and engineers.

The second part of the chapter is devoted to characterizing the role of heterogeneity (diversity) and interdependence as critical factors that influence the performance of the M&S field. The complexity perspective is leveraged to examine the positive and negative implications of the identified critical variables and provided a framework to make recommendations for managing the overall system toward a performance equilibrium in terms of balance in exploration and exploitation of knowledge.

References

Atikins, D. (2003) Revolutionizing Science and Engineering through Cyberinfrastructure. Technical Report, National Science Foundation Blue Ribbon Panel, National Science Foundation.

Benioff, M. and Lazowska, E. (2005) Computational Science: Ensuring Americas Competiveness. Technical Report, Presidents Information Technology Advisory Committee (PITAC).

Gardner, H., Csikszenthmihalyi, M., and Damon W. (2001) *Good Work: When Excellence and Ethics Meet*, Basic Books, New York, NY.

Keyes, D., Colella, P., Dunning, T., and Gropp W. (2003) A Science-Based Case for Large-Scale Simulation: Volume 1. Technical Report, Department of Energy.

NAE (2006) *Assessing the Capacity of the U.S. Engineering Research Enterprise*, National Academy Press, Washington DC.

Ören, T.I. (2016) Societies and Organizations in Modeling & Simulation. Available at http://www.site.uottawa.ca/~oren/links-MS-AG.htm

SBES (2005) Simulation-Based Engineering Science: Revolutionizing Engineering and Science through Intelligent Simulation. Technical report, National Science Foundation Blue Ribbon Panel on Simulation-Based Engineering Science.

SCiDAC (2000). Scientific Discovery through Advanced Computing. Technical Report, Department of Energy Office of Science Strategic Report.

12

Theory and Practice of M&S in Cyber Environments

Saurabh Mittal[1] and Bernard P. Zeigler[2]

[1]*The MITRE Corporation, McLean, VA, USA*
[2]*University of Arizona, Tucson, AZ and RTSync Corporation, Rockville, MD, USA*

12.1 Introduction

William Waite, to whom this book is dedicated, strongly advocated for the acceptance of modeling and simulation (M&S) as a discipline, practiced by professionals skilled in its concepts, methods, and tools. This chapter affirms this point of view and argues that it is even more critical to adopt it now that we have entered the era of Cyber environments. The World Wide Web, Internet of Things, Information Technology, and the advancing Cyber environment are transforming the practice of engineering sociotechnical systems as well as system engineering's handmaiden, M&S. Cyber environments are open systems, continually evolving with uninhibited interactions of users at various levels causing behaviors to emerge that impact everyday lives in unforeseen ways. M&S has traditionally required two people: a domain expert and an M&S expert. However, the situation is different with Cyber environments. M&S in Cyber environment will require working with professionals from multiple disciplines. Many times, the domain subject matter experts (SMEs) fail to recognize the contribution by M&S expert as they believe the modeling and simulation are only tools, a piece of code, where they will apply their science to, and nothing more. However, as we will see in this chapter, creating a M&S tool that is founded on M&S theory is a nontrivial task. A successful collaboration between multiple domain experts and an M&S expert begins by acknowledging M&S as a scientific discipline and a profession. This chapter recognizes that contemporary model-based system engineering must be robustly supported by M&S professionals armed with theory, concepts, and tools up to the challenges of Cyber

The Profession of Modeling and Simulation: Discipline, Ethics, Education, Vocation, Societies, and Economics, First Edition. Andreas Tolk and Tuncer Ören.

environments replete with multiple SMEs in any given scenario. It will also establish the need for formal methods and tools to build M&S workbenches to create reusable results, have a common, understandable, and reproducible way to solve problems with M&S.

Modeling and simulation using traditional systems engineering practices was tractable as a systems engineering process that aids the development of model-based systems engineering and simulation-based systems engineering. This was because systems built using these approaches had well-defined boundaries and it was relatively easy to understand their limitations and/or plan their extensions/modifications when migrating to a different environment. On the other hand, cyber-complex adaptive systems have soft boundaries, are highly reconfigurable, and must be available anytime and anywhere to serve the purpose at hand. To develop a system engineering process for cyber-complex adaptive systems, we must first strive to understand how to model them in a repeatable way. This will eventually lead to the cyber systems engineering discipline.

Classical Systems Theory is based on two broad concepts: the relation between a system's structure and its behavior, and closure-under-composition. Systems Theory coupled with Control Theory provides mechanisms for instituting control at various levels of systems specifications. Much of the advance in systems architecting as applicable to various marvels of modern society, for example, automobile, aircraft has been hinged on developing closed systems so that the composite system's behavior is closed under composition. However, in contemporary society, any system engineering endeavor also requires the application of Network Theory to address issues such as scalability, geographical separation, and black-box integration. Further, the open Cyber environment does not allow behavior to be engineered and predicted while constrained within a meld of Classical Systems and Network principles. Consequently, we begin our exploration with complex adaptive systems (CAS) (Holland, 1992). Mittal (2014) defined Cyber CAS (CyCAS) as CAS manifesting in a cyber environment and discussed eight characteristics of CyCAS. He presented various metrics that could be used to perform model engineering and performance evaluation for CyCAS, which we later refine in the chapter. The CyCAS engineering must be accompanied by a test and evaluation (T&E) methodology that incorporates near-real experimentation environment, such as one available through Cyber Ranges. We will examine the eight CyCAS Views and how they can be implemented in a Cyber Range T&E environment. Cyber M&S using Cyber-Range Events emphasized the use of Logical Ranges to determine the "right level" of abstraction required to balance a mix of actual assets and systems with limited and representative asset and system models for simulations (Damodaran and Couretas, 2015). We will extend these ideas and augment them with modeling and simulation-based engineering methodology that must be brought in for Cyber M&S. A Cyber M&S endeavor is a complex endeavor and requires a set of skills and a design and engineering process. Another important aspect in any M&S endeavor is the validation and verification of models. While model validation is user-faced, verification is implementation-specific. In Cyber M&S, verification becomes extremely hard as Cyber-Range comprises of Live, Virtual, and Constructive (LVC) assets, fundamentally, are a lot of black boxes (Ashby, 1956) for which complete information is seldom available. This makes V&V for Cyber M&S an area of further opportunity and research.

The Cyber environment will always be open to emergent behavior as it is a human-in-the-loop system, an open system (Mittal, 2012; Mittal and Zeigler, 2014). New approaches such as activity-based system of systems engineering (Zeigler, 2012; Mittal and Zeigler, 2014) are required that align multilevel effects for SoS' sustainability in a cyber environment. Without a formal M&S-based engineering approach, Cyber systems are always prone to cascaded failures as cyber CAS are highly networked at multiple levels. In the early days of Internet, such cascaded failures caused widespread router outages (Coffman et al., 2002). However, a cyber outage that is exponentially large-in-magnitude will have far reaching impact on the society and various interconnected critical infrastructures. This leads to the role of M&S as a discipline to deal with the complex challenges for systems deployed in Cyber environments. The proposed theoretical constructs provide the basis for robust continued growth of M&S as a discipline and a profession in a challenging cyber environment.

While critical in any undertaking related to "cyber," the subjects of Security and Resiliency engineering will be only briefly considered and will not be the focus of Cyber M&S in this chapter.

The chapter is organized as follows. Section 2 provides an overview on Cyber Complex Adaptive Systems (CyCAS) framework and its various characteristics. Section 3 describes the cyber and Cyber-Range environment. It also describes the concept of a Logical Range. Section 4 presents the fundamental M&S theory and introduces formal Discrete Event Systems (DEVS) (Zeigler et al., 2000). Section 5 discusses Cyber M&S with DEVS at length. It presents the modeling, simulation-platform requirements, experiment design, application of DEVS M&S theory and security and resilience elements that need to be specified for a Cyber M&S environment. Section 6 deals with the verification and validation of cyber models with a simulation-based engineering process. Finally, a discussion and conclusion closes the chapter in Section 7.

12.2 Cyber Complex Adaptive Systems (CyCAS) Framework

Complex adaptive systems are systems that display strong emergent behavior, have positive- and negative-feedback loops, and have large number of adaptive agents (Mittal, 2014). Examples include ant colonies, the biosphere, business enterprise, the brain, the immune system, communities, social systems, stock market, financial systems, High Frequency Trading systems, and so on. CyCAS are CAS that occurs in contemporary society where computational systems interact with both live human agents and software agents (Mittal, 2013c, 2014). These software agents may control physical assets connected through the network. Essentially, cyber-physical system (CPS) is a more generally accepted term that does not explicitly mention software as an integral part. While CPS emphasize cyber and physical, CyCAS emphasizes cyber and software abstraction layers that must be analyzed accordingly as the control layer rests at the software layer between the cyber and physical. With software-defined networking, cloud computing, and virtualization technologies becoming the norm today, cyber environment indeed is a human and software environment with a transparent physical environment. It is termed a CAS as it manifests all the properties of a CAS, such as strong emergent behavior, positive- and negative-feedback loops, and large number of heterogeneous adaptive agents.

Mittal (2014) discusses the following eight characteristic properties of CyCAS:

1) *Human-in-the-System*: A critical component of the system. Various human behavior modeling approaches and context quantification (Mittal and Zeigler, 2014) need to be incorporated from fields like cybernetics, cognitive psychology, and agent-based modeling. The human behavior models must conform to the model configuration management for managing the complexity within these models.

2) *Multiagent System*: Agents in CyCAS can be homogeneous, heterogeneous, deliberative, or communitive in varying degrees and combinations in a live and virtual environment. Canonically, such systems are multilevel algorithms subject to multilevel control in a real-time environment.

3) *Control and Communication in a Netcentric Environment*: This is the cyber aspect. Spatiotemporal profile of communications is the source of all emergent behavior. The links establish positive and negative feedback loops, and it is critical to determine who is controlling what structure and which behavior. Netcentric environment allows standards to be applied for communication protocol but they alone are not sufficient to manage the structure of the resulting system.

4) *Resource-Constraints and Economy of Scale*: Both software and physical assets have finite computational resources and being networked, have interdependencies. These resources provide services when available and their dependencies dictate if their failure has cascaded effects. Scalability and resourcefulness need to be addressed in a formal manner in CyCAS.

5) *Emergent Attention and Second-Order Cybernetics*: Various analytical and audit (observers) models motivated by the pragmatics of SoS need to exist to identify new correlations and causations within the manifested CyCAS behavior. This must conform to second-order cybernetics philosophy where in a single observer is not the only evidence. A consistent set of observations from multiple observers is required to ensure the observed emerging patterns are categorized correctly by subject matter experts. This will eventually result in development of new feedback loops that help risk mitigation.

6) *Phase Transition*: CyCAS, fundamentally, being an SoS, manifests emergent behaviors. Due to the black-box nature of constituent systems, the emergent behavior has surprise effects on these systems. This results in macrolevel behaviors getting amplified (positive feedback) or desired behaviors vanishing (negative feedback) by actions taken by the constituent systems. These phase transitions must be studied through computational techniques.

7) *Structure of Knowledge*: Each black-box software or physical asset maintains its own knowledge-base. While syntactic interoperability is easily achieved, semantic interoperability is an NP-hard problem (Diallo, 2016). The fundamental issue with semantic interoperability is the multiple interpretations that different systems attach to the same data object. Issues like knowledge sharing, transformation, evolution, computational representation (ontology), interoperability, context switching according to emergent attention, and new knowledge synthesis must address multiple operational levels (Mittal, 2013c).

8) *Resilient or Antifragile*: CyCAS undergoes a lot of dynamic reconfiguration due to their computational nature. It needs to be formally determined if the engineering CyCAS is fail-fast, resilient/robust, or antifragile where it continues to function, albeit evolving, in unknown ways.

Based on these characteristics, CyCAS Framework, Mittal (2014) also proposed eight views and the associated metrics that would allow their specification and evaluation. These views are explored ahead in the chapter. This concept of CyCAS was also explored for evaluating the agility of adaptive command and control network (Tran et al., 2015). It focused on the core CyCAS concepts for C2 domain but did not go all the way to use the

CyCAS Framework. Likewise, the core concepts were also used for service-oriented modeling and simulation for system of systems engineering (Zeigler and Zhang, 2015). In a moment, we will relate the views and metrics to operational elements of modeling and simulation such as experimental frames and modeling formalisms.

12.3 Cyber Experimentation and T&E Environment

In this day and age, almost all the life support (e.g., electricity, water, telephone, banks) and mission-critical (defense, nuclear, stock-market, etc.) systems have a cyber component, that is, they share information and can be software controlled remotely. Any weakness at the cyber level can bring these infrastructures to a halt (US-Canada Power System Outage Task Force, 2004). While the information sharing can add to ease of maintaining systems, a simple compromise of Personally Identifiable Information (PII) can quickly turn this into a financial and legal liability (Damodaran and Couretas, 2015). The Defense Science Board (2013) is keen to ensure that military systems are cyber hardened and have T&E. Since training is an essential part of cyber systems, and military systems in particular, M&S is a prime contender to provide the needed platform. Before it can be used for training and T&E (Arnwine, 2013), it must have the needed fidelity.

12.3.1 Cyber ranges and Logical ranges

To provide a realistic T&E environment, a cyber range is needed. A cyber range is a virtual environment that is used for cyber training and cyber-technology development and testing. In comparison to traditional physical ranges where weapon testing and assets change slowly, a cyber range is a place where assets change rapidly. Cyber ranges are used to evaluate effectiveness of cyber defenses and to train cyber warfighters, along with the common goals of tactics, techniques, and procedures (TTP) development and mission rehearsal. Following cyber ranges are available (Damodaran and Couretas, 2015; OSD, 2015; Defense Science Board, 2013; Hansen, 2008):

- National Cyber Range (NCR) managed by the Test Resource Management Center (TRMC) which is part of the Office of the Under Secretary of Defense for Acquisition, Technology, and Logistics (OUSD(AT&L)).
- DoD Cyber Security Range (DCSR) managed by Defense Information Systems Agency.

- Command, Control, Communications, and Computers Assessments Division (C4AD) managed by Joint Staff J6.
- Joint IO Range (JIOR) managed by Joint Staff J7.

TRMC oversees the National Cyber Range (NCR) that supports a wide variety of different types of cyber events including the following:

1) R&D testing
2) Product evaluation
3) Training
4) Mission rehearsal
5) Risk reduction activities
6) Architecture analysis
7) DT&E
8) OT&E
9) Malware analysis
10) Forensic analysis

Each NCR event has the following stages (Damodaran and Couretas, 2015; OSD, 2015):

1) *Event Preplanning and Planning*: Discussions and use-case development.
2) *Event Design*: Goals, objectives and assumptions, environment design, outputs, and data collection plan.
3) *Event Development*: Red team operations, environment build, and verification.
4) *Event Execution*: Conducting tests, obtaining data, review results, and adapt as needed.
5) *Event Completion*: Data analysis, reporting, briefing, next event planning.

The event workflow is semiautomated and the infrastructure is largely virtualized and deployed in a secured datacenter. Figure 12.1 shows the process.

The deployment process includes the definition of a logical range. A logical range is the actual secured set of network connections that interconnect the various participating entities (e.g., cyber ranges) in a particular event. These logical ranges can separate many cyber events as many such events can happen concurrently. A logical range is typically documented in a manner that enumerates the software/hardware components used to support the event.

Conceptually, a logical range masks the geographical location of various physical assets and live players. The logical range abstraction is created during the planning phase, is self-contained and is isolated from surroundings. This isolation is deliberate and serves two purposes

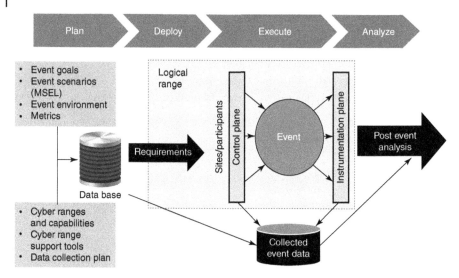

Figure 12.1 Cyber event process. (*Source*: Damodaran 2015. Reproduced with permission of Society of Computer Simulation International.)

(Damodaran and Couretas, 2015): First, the technologies within the logical range do not escape into the real-world causing damage. Second, this isolation prevents unintended exposure of the contents of the logical range. Within the logical range, the event plane (denoted by circle in Figure 12.1) consists of red and blue teams that participate in a cyber wargame. Red team is the attacker side and blue team is the defending side. The control plane is used for controlling the operating environment and may have live assets as green team that supports white and gray teams. Gray team generates traffic. White team exercises authority over the complete logical range.

Cyber ranges provide a LVC simulation environment for cyber wargaming. In the cyber event context

1) live simulation denotes actual assets, systems, operators, network devices, full-stack protocols, hardware, nonemulated software, and live attackers (but no real live enemy);
2) virtual simulation denotes actual assets may interact with representative system models and vice versa; and
3) constructive simulation denotes limited or representative asset models interact with limited or representative system models. The entire model is at a lower fidelity than the live simulation.

12.3.2 Multilevel Effects

The objective of a cyber LVC simulation is to observe the effects in a cyber M&S "sandbox." Depending on the configuration of the event (aspect of L, V, and C), the effects may vary greatly. An "effect" is defined as a change in the Event plane because of interaction of red, blue, and gray teams. These effects may be predefined anticipated changes or new ones that still need to be defined. One of the primary requirements of a good experiment is the background network traffic. In an isolated logical range, sometimes synthetic traffic may not be enough. In that situation, actual prerecorded traffic from real world can be injected into the logical range for realism and to reproduce effects. In other scenarios, effects such as "site-down by fire" might be injected in virtual environment to examine the resiliency of the system.

As the cyber LVC simulation environment provides opportunity to experiment at multiple levels of specifications for hardware, protocols, software, and people (agents), the analysis can be conducted at multiple levels. For example, in a pure constructive simulation, mathematical models could be used that are more abstract and have more variability. Additional application of constraints to these abstract models lead to incorporating emulators that may model full protocol stack for high-fidelity interaction between a constructive asset and virtual system. Further application of constraints will lead to incorporation of actual assets and systems. This capability allows rapid feedback for model validation (in constructive simulation) so that model can be used for large-scale simulation without losing correctness. Consequently, an optimized logical range can be constructed that depicts the right cyber effects.

12.4 Theory of M&S for System of Systems

In systems theory as formulated by Wymore, systems are defined mathematically and viewed as components to be coupled together to form a higher level system (Oren and Zeigler, 2012). As illustrated in Figure 12.2, systems theory of Wymore (1967) mathematically characterizes the following:

- *Systems:* Well-defined mathematical objects characterizing "black boxes" with structure and behavior.
- *Composition of Systems*: Constituent systems and coupling specifications result in a system, called the resultant, with structure and behavior emerging from their interactions.
- *Closure under Coupling*: The resultant is a well-defined system just like the original components.

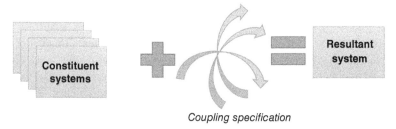

Coupling specification

Figure 12.2 Wymore's system composition.

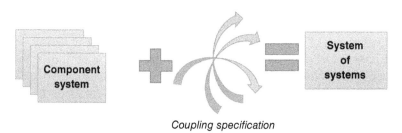

Coupling specification

Figure 12.3 System of systems.

As illustrated in Figure 12.3, a System of Systems (SoS) is a composition of systems, where often component systems have legacy properties, for example, autonomy, belonging, diversity, and emergence (Boardman and Sauser, 2006). In this view, a SoS is a system with the distinction that its parts and relationships are gathered together under the forces of legacy (components bring their pre-existing constraints as extant viable systems) and emergence (it is not totally predictable what properties and behavior will emerge.) Here in Wymore's terms, coupling captures certain properties of relevance to coordination, for example, connectivity and information flow. Structural and behavioral properties provide the means to characterize the resulting SoS, such as fragmented, competitive, collaborative, and coordinated.

12.4.1 Discrete Event Systems Specification (DEVS) Formulation of SoS

The DEVS formalism by Zeigler *et al.* (2000), based on Systems Theory, provides a framework and a set of modeling and simulation tools to support systems concepts in application to SoSE (Mittal et al., 2008; Mittal and Martin, 2013a). A compilation of state-of-the-art in M&S for SoSE is also available in Rainey and Tolk (2015). A DEVS model is a system-theoretic concept specifying inputs, states, outputs, similar to a state machine.

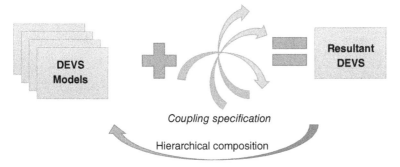

Figure 12.4 DEVS formulation of systems-of-systems.

Critically different however, is that it includes a time-advance function that enables it to represent discrete event dynamic systems, as well as hybrids with continuous components, in a straightforward platform-neutral manner. DEVS provides a robust formalism for designing systems using event-driven, state-based models in which timing information is explicitly and precisely defined. Hierarchy within DEVS is supported through the specification of atomic and coupled models. Atomic models specify behavior of individual components. Coupled models specify the instances and connections between atomic models and consist of ports, atomic model instances, and port connections. The input and output ports define a model's external interface, through which models (atomic or coupled) can be connected to other models.

As illustrated in Figure 12.4, based on Wymore's systems theory, the DEVS formalism mathematically characterizes the following:

- DEVS Atomic and Coupled models specify Wymore Systems (Boardman and Sauser, 2006).
- Composition of DEVS models and coupling result in a Wymore system, called the resultant, with structure and behavior emerging from their interaction.
- Closure under coupling property ensures that the resultant is a well-defined DEVS just like the original components.
- Hierarchical composition based on closure under coupling property enables the resultant coupled models to become components in larger compositions.

12.4.2 Hierarchy of System Specifications

Concepts for organizing models and data for simulation based on systems theory (Zeigler et al., 2000) and implementable in Model-Based Systems

Table 12.1 Levels of system specifications.

Level	System specification	Description
4	Coupled	Hierarchical system with coupling specification. System of systems is defined at this level
3	Atomic	State space with transitions. Behavior and internal structure are specified at this level
2	I/O function	State space with defined initial state. Component temporal behavior with respect to initial state is defined at this level
1	I/O behavior	Collection of input/output pairs defining temporal behavior is defined at this level
0	I/O frame	Defines input/output variables with associated ports over a time base

Engineering (Mittal and Martin, 2013a) are a necessary background for discussing the modeling and simulation framework (MSF). The system specification hierarchy provides an orderly way of establishing relationships between system descriptions as well as presenting and working with such relationships. Pairs of system can be related by morphism relations at each level of the hierarchy. A morphism is a relation that places elements of system descriptions into correspondence. For example, at the lowest level, two observation frames are isomorphic if their inputs, outputs, and time bases respectively, are identical (Table 12.1). In general, the concept of morphism tries to capture similarity between pairs of systems at the same level of specification. Such similarity concepts have to be consistent between levels. When we associate lower level specifications with their respective upper level ones, a morphism holding at the upper level must imply the existence of one at the lower level. The morphisms are set up to satisfy these constraints. The most fundamental morphism, called homomorphism, resides at the State Transition level (Level 3 in Levels of Systems Specification shown in Table 12.1).

12.4.3 The M&S Framework (MSF) and its Entities

MSF presents entities and relationships of a model and its simulation as background. The basic entities of the framework are: source system, model, simulator, and experimental frame. As illustrated in Figure 12.5, the basic entities in M&S are the actual system (the "Source System"), the "Model,"

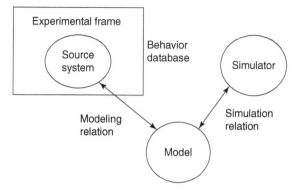

Figure 12.5 M&S entities and their relationships (Zeigler et al., 2000).

and the mechanism for executing the Model (a "Simulator") when the model generates a description of events over time. It is important to understand the relationship between the Model and the Source System (the "Modeling Relation") and the relationship between the Model and the Simulator (the "Simulation Relation").

The adequacy of the model must be judged with respect to the context of use, which includes the domain of input values, the range of output values, and the intent of the user. The experimental frame was originally introduced to operationalize this contextual dependence of adequacy as an object in a status equal to real system, model, and simulator. In the simulation theoretic usage we employ here, the MSF separates models from simulators as entities that can be conceptually manipulated independently and then combined in a relation which defines correct simulation. In addition, the Experimental Frame (EF) defines a particular experimentation process for model input, state, and outcome measurements in accordance with specific analysis objectives. **The EF formally recognizes that the intended use (IU) of a model is a fundamental determinant of its validity with respect to the source system.** Modular reuse, validity, and executability of simulation compositions are common aspirations among enterprises regularly relying on M&S of System of Systems (SoS) throughout their lifecycles. Such enterprises invest significantly not only in development and experimentation but also in verification and validation (V&V). The MSF helps clarify many of the issues involved in such activities.

The *source system* is the real or virtual environment viewed as a *source of observable data*, in the form of time-indexed trajectories of variables. The data that has been gathered from observing or otherwise experimenting with a system is called the *system behavior database*. This concept of system is a specification at level 0 and its database is a specification at level 1 and 2

(Table 12.1). It is at this level the conceptual states and I/O trajectories are identified. This data is viewed or acquired through experimental frames of interest to the modeler. In *data rich* environments, such data is abundant from prior experimentation or can easily be obtained from measurements. In contrast, *data poor* environments offer meagre amounts of historical data or low-quality data (whose representativeness of the system of interest is questionable). The modeling process can direct the acquisition of data to those areas that have the highest impact on the intended uses of the M&S.

In its most general guise, a *model* is a system specification at any of the levels of the Hierarchy (Table 12.1). However, in the traditional context of M&S, the system specification is usually done at levels 3 and 4. Thus, the most common concept of a simulation model is that it is a set of instructions, rules, equations, or constraints for generating I/O behavior. In other words, we write a model with a state transition and output generation mechanisms (level 3) to accept input trajectories and generate output trajectories depending on its initial state setting. Level 3 usually have a state-space. Such models form the basic components in more complex models that are constructed by coupling them together to form a level 4 specification. The definition of model in terms of system specifications has the advantages that it has a sound mathematical foundation and it has a definite semantics that everyone can understand in unambiguous fashion.

Table 12.2 characterizes the level of system specification that typically describes the entities. The level of specification is an important feature for distinguishing between the entities, which is often confounded in practice. Based on this framework, the basic issues and problems encountered in

Table 12.2 Defining the basic entities in M&S and their usual levels of specification.

Basic entity	Definition	Related system specification level
Source system	Real or artificial source of data	Known at Level 0
Behavior database	Collection of gathered data	Observed at Level 1 and 2
Experimental frame	Specifies the condition under which the system is observed or experimented with	Constructed at Level 3 and 4
Model	Instructions/algorithms for generating data	Constructed at Level 3 and 4
Simulator	Computational device for generating behavior of the model	Constructed at Level 4

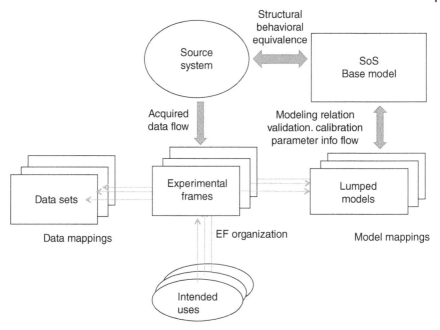

Figure 12.6 Architecture for SoS V&V based on MSF.

performing M&S activities can be better understood and coherent solutions developed.

Figure 12.6 shows the application of the framework to V&V for M&S for SoS. Given the importance of the experimental frame concept to V&V, we review it in more depth based on detailed developments found in Zeigler and Nutaro (2015). Roughly, an experimental frame (EF) is a specification of the conditions under which the system is observed or experimented with. As such experimental frames are the operational formulation of the IUs that motivate a modeling and simulation project. Many experimental frames can be formulated for the same system (both source system and model) and the same EF may apply to many systems. There are two equally valid views of an EF. In the first **data storage view**, a frame is a definition of the type of data elements that will be stored in the database to be tagged with the frame for later extraction. In the second, **data acquisition view**, a frame is a system that interacts with the SoS to obtain the data of interest under the specified conditions that are part of the frame. In this view, the frame is characterized by its implementation as a measurement system or observer. In this implementation, a frame typically has three types of components: *generator* that generates input segments to the system; *acceptor* that monitors an experiment to see the desired experimental conditions are met; and *transducer* that observes and analyzes the system output segments.

12.4.4 Intended Uses (IUs) and Experimental Frames (EFs)

Balci (2012) emphasized the role of "intended uses" in directing the construction of an M&S application and consequently how it should be subjected to V&V. He provided by way of illustration a hierarchy of intended uses for a particular M&S application but did not attempt a general formulation of such a hierarchy. Recently, an organization with a heavy reliance on M&S for SoS, the US Missile Defense Agency, is establishing a **standard IU specification** that contains a comprehensive listing of specifics in relation to the analysis problem that the model is intended to address (Zeigler and Nutaro, 2015). Indeed, this specification significantly expands the set of elements that characterize the objectives of the user. The fundamental elements of an IU include pertinent analyst tasks, model inputs and outputs, experimental designs, calibration methods and data, test objectives, and concepts of operations (CONOPS). In addition, the specification requires characterization of key attributes, that is, aspects and values that identified stakeholders and developers agree on such as Focus (from narrow consideration of a component, such as a specific radar, to the broad scope of end-to-end SoS evaluations), Simulation Type (Constructive, Virtual, or Live), Fidelity, Uncertainty Quantification, Interoperability, Level of detail and relation to operator training or exercise experience.

Our formulation based on such a specification is that once the IU is known, suitable experimental frames (EFs) can be developed to accommodate it. There is *1:N* relationship between an IU and EFs. Such frames translate the IU elements into more precise experimentation conditions for the SoS Base Model and its various abstractions and aggregated models called Lumped Models. A model developed for an application is expected to be valid in each frame associated with the IU specification that formalizes that application. An IU specifies a focus, fidelity, and a level of detail to support the problems and tasks it concerns. Different foci, fidelities, and levels of details may both require and allow different models that exploit these factors to enable optimal set up and run time attributes. **The basic concept in Figure 12.6 is that IUs act as keys to all data and models that have been acquired and developed thus far**. In the storage process, a new data set or model is linked to the IU that motivated its development. In retrieval, given an application of interest to the user, the system supports formulating a representative IU and finding the closest IU matching the newly formulated IU. If the user is unsatisfied with the match, or wishes to explore further, the system supports synthesizing a composite IU using available lattice-like operations (upper and lower bounds, decomposition, etc.).

A Cyber Event (Section 12.3) can be directly mapped to an IU and accordingly, can address multiple EFs.

12.5 Cyber M&S with DEVS

Cyber Systems are multidomain systems. A sufficiently high fidelity cyber model has to be a multiparadigm model where the model is expressed in discrete, continuous, and/or hybrid paradigm. The model specification language adds another layer of complexity in developing a unified cyber model. Assuming that the model specification language is supported by a model transformation engine that transforms the user-specified model into a computational language, another challenge that props up is the execution of the model in a computational environment. This model execution, when added with various experiments and "what-if" conditions, is often called simulation. However, it is a nontrivial extension. An incorrect execution of model, invalidates the model, even though it was a valid model, that is, a useful model serving an IU. To solve the above challenges and apply it to Cyber M&S requires the integration of Cyber event (Section 12.3) with the core M&S Framework (MSF).

12.5.1 Modeling a CyCAS

Modeling a CyCAS essentially begins with modeling a cyber-physical system (CPS) at a fundamental level. A CPS is an integration of physical subsystems together with computing and networking (Lee, 2008). The "adaptive" aspect of CyCAS is not considered in CPS modeling, and may be broadly incorporated in computing. A CPS model is a hybrid system model made up of both continuous and discrete systems. A continuous system (CS) is one that operates in continuous time and in which input, state, and output variables are all real values. A discrete (dynamic) system is one that changes its state in piecewise constant event-based manner (which also included discrete-time systems as they are a special case of discrete event systems) (Lee et al., 2015). A typical example of a hybrid system is a CPS in which the computation subsystem is discrete and a physical system is CS. A CyCAS in LVC environment also qualifies as a CPS, with live systems as CS, constructive systems as discrete and virtual systems as hybrid, containing both continuous and discrete. At the fundamental level, there are various ways to model both timed and untimed discrete event systems (Figure 12.7), all of

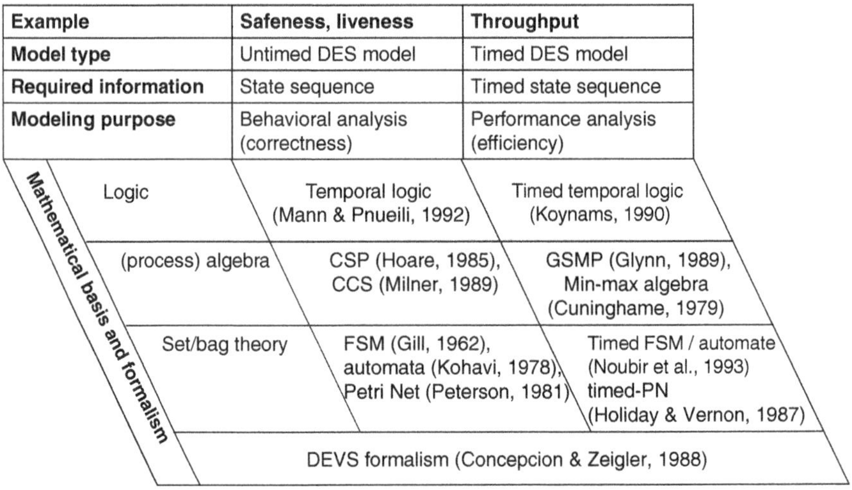

Example	Safeness, liveness	Throughput
Model type	Untimed DES model	Timed DES model
Required information	State sequence	Timed state sequence
Modeling purpose	Behavioral analysis (correctness)	Performance analysis (efficiency)

Figure showing a parallelogram labeled along the left side "Mathematical basis and formalism":

Logic	Temporal logic (Mann & Pnueili, 1992)	Timed temporal logic (Koynams, 1990)
(process) algebra	CSP (Hoare, 1985), CCS (Milner, 1989)	GSMP (Glynn, 1989), Min-max algebra (Cuninghame, 1979)
Set/bag theory	FSM (Gill, 1962), automata (Kohavi, 1978), Petri Net (Peterson, 1981)	Timed FSM / automate (Noubir et al., 1993) timed-PN (Holiday & Vernon, 1987)

DEVS formalism (Concepcion & Zeigler, 1988)

Figure 12.7 Math formalisms for discrete event modeling (communication sequential process (CSP), calculus of communicating system (CCS), generalized semimarkov process, finite-state machine (FSM), and petri nets (PN). (*Source:* Lee *et al.* (2015). Reproduced with permission of ETRI Journal.)

which can be transformed to, and studied within, the formal DEVS theory (Vangheluwe, 2000; Zeigler et al., 2000).

Note that model types are characterized by two types of questions: those relating to Behavioral Analysis and Performance Analysis. These give rise to experimental frames that, for example, relate to studying the safeness/liveness of a system or its production throughput. In other words, as Figure 12.7 shows, untimed models are typically developed to accommodate correctness frames while timed models typically accommodate efficiency frames. To model CyCAS, Table 12.3 elaborates on the nature of models for various CyCAS views. It associates each CyCAS view typical questions it concerns, expressed as experimental frames and metrics and with the applicable modeling paradigm (or domain), discrete or continuous nature and the platform the model may represent.

As can be seen from the column on model domain, a CyCAS model borrows concepts from multiple domains that come with their own modeling paradigm. Consequently, a multiparadigm modeling (Vangheluwe et al., 2002) effort is needed that would align different paradigms with their corresponding modeling formalisms and implementation types toward a composite model capable of exchanging information across various abstraction levels. Developing a modeling workbench for multiparadigm modeling environment is a nontrivial exercise and two solutions exist. Given a set of modeling formalisms that a CyCAS model needs, the first option requires the use of a Formalization Transformation Graph (Vangheluwe, 2000) that

Table 12.3 Multiparadigm modeling for CyCAS views.

ID	CyCAS view	Model domain	Model type	Platform represented
1	Human view	Cognitive psychology, human factors, estimation theory, identity access management	Discrete	Software
2	Multiagent view	Complexity science, distributed artificial intelligence (DAI)	Hybrid	Hardware/ software
3	Control and communication view	Network theory, communication theory, cybernetics, systems of systems	Discrete	Hardware/ software
4	Resource and constraints view	Logistics, supply-chain, network theory, operations research	Discrete	Hardware/ software
5	Emergence view	Complexity science, network theory, second-order cybernetics	Discrete	Software
6	Phase-transition view	Complexity science, network theory, systems of systems	Hybrid	Software
7	Knowledge view	Ontology, network theory	Discrete	Software
8	Resilience view	Complexity science,	Discrete	Hardware/ software

transforms a relatively simpler formalism to be transformed into a more rigorous formalism. Such was an approach used by AtoM3 (Vangheluwe et al., 2002). However, one then must understand each of the other formalisms and their semantics to ensure that there are no leaky abstractions and the mapping is correct. The other approach is to transform these modeling formalisms to hybrid super-modeling formalism that can model both discrete and continuous systems at the fundamental level, for example, DEV&DESS (Cellier, 1979; Praehofer, 1991; Mittal and Martin, 2013a; Zeigler et al., 2000). This would ensure that the appropriate abstractions from each of the formalisms are integrated at the state-event and time-event levels and are simulated in a mathematically verifiable way, as implemented in the DEVS formalism.

In modern times, the model-driven engineering community, led by Eclipse platform[1] has made a lot of progress in creating editing workbenches for domains that lacked a comprehensive integrated development environment (IDE). Using the ANTLR context-free grammars, as a basis for creating new modeling languages, frameworks like Xtext[2] allows creation

---

1 Eclipse Framework. Available at http://www.eclipse.org
2 Xtext Framework. Available at: http://www.xtext.org

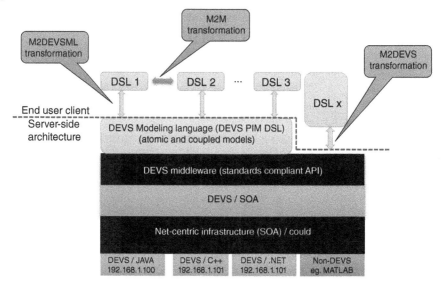

Figure 12.8 DEVS modeling language stack.

and transformation of these modeling languages to a chosen target modeling formalism. Efforts made in an Xtext-based DEVS Modeling Language (Mittal and Martin, 2013a; Zeigler and Sarjoughian, 2013) and the DEVSML Stack (Figure 12.8) (Mittal and Martin, 2013a, 2013b; Mittal et al., 2007; Mittal and Douglass, 2012) allows the transformation of multiple modeling formalisms to the DEV&DESS environment. Figure 12.8 shows the following three transformations that could be attempted to bring in various DSLs to the DEVS ecosystems:

- Model-to-Model (M2M)
- Model-to-DEVSML (M2DEVSML)
- Model-to-DEVS (M2DEVS)

12.5.2 Simulation of CyCAS

Having established that a multiparadigm modeling is required for CyCAS modeling, let us now focus on the execution of these models, that is, simulation. Table 12.4 enumerates some of the tools that are currently used in a particular domain. Some tools are language dependent (e.g. C++, Java, LISP, DSL) and/or some are platform dependent (e.g. Windows, Linux, Mac). Some are discrete event, some are continuous and closed-form. These tools have their own software architecture, subscribe to a scientific theory

Table 12.4 Domains and their tools (not a complete list).

S. No.	Domain	Modeling tool/architecture
1	Cognitive psychology	ACT-R (Adaptive Character of Though-Rational (ACT-R, 2016)), jACT-R[a], SOAR[b]
2	Network science	Gephi[c], NetworkX[d], Igraph[e], Statnet[f], Pajek[g]
3	Human factors	JACK[h], Kinemation[i]
4	Communication systems	Opnet[j], Omnet[k], NS-3+[l], MATLAB/Simulink[m]
5	Ontology	Protégé[n], TwoUse[o], NeOn[p], FlexViz[q]
6	Complexity science	NetLogo[r], RePast/Symphony[s], DEVS (Zeigler et al., 2000), R[t]
7	Supply-chain	MS Excel, Arena, SAS
8	Power systems	GridLab-D[u]

a) Java ACT-R. Available at http://jact-r.org/.
b) SOAR. Available at http://soar.eecs.umich.edu/.
c) Gephi. Available at http://gephi.org.
d) NetworkX. Available at https://networkx.github.io/index.html#.
e) Igraph. Available at http://igraph.org/redirect.html.
f) Statnet. Available at https://statnet.csde.washington.edu/.
g) Pajek. Available at http://mrvar.fdv.uni-lj.si/pajek/.
h) JACK Available at https://www.plm.automation.siemens.com/en_us/products/tecnomatix/manufacturing-simulation/human-ergonomics/jack.shtml.
i) Kinemation. Available at http://www.cs.cmu.edu/~german/research/HumanApp/humanapp.html.
j) Opnet Modeler. Available at https://www.riverbed.com/products/steelcentral/opnet.html Omnet.
k) Omnet. Available at https://omnetpp.org.
l) NS3. Available at https://www.nsnam.org.
m) MATLAB/Simulink. Available at https://www.mathworks.com/products/simulink.html.
n) Protégé. Available at http://protege.stanford.edu.
o) TwoUse. Available at http://semanticweb.org/wiki/TwoUse_Toolkit.html.
p) NeOn Toolkit. Available at http://semanticweb.org/wiki/Neon_Toolkit.html.
q) FlexViz. Available at https://sourceforge.net/projects/flexviz/.
r) NetLogo. Available at https://ccl.northwestern.edu/netlogo/.
s) RePast/Symphony. Available at https://repast.github.io/repast_simphony.html.
t) The R-Project. Available at https://www.r-project.org.
u) GridLab-D. Available at http://www.gridlabd.org.

and sometimes the software is proprietary. Figure 12.9 shows A, B, C, D, E as sample architectures. It also shows a layered M&S architecture addressing the pragmatic, semantic, and syntactic levels of interoperability. For more details, see Mittal (2014) and Mittal *et al.* (2008).

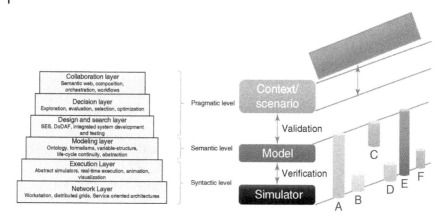

Figure 12.9 M&S with V&V and Testing and Evaluation (T&E) (Mittal, 2014).

Consider, for example, a complex system comprising of Electric Grid, thousands of smart homes, and data-communication network, would require modeling to be done for the following:

1) Continuous system for the power system using GridLab-D.
2) Continuous system for building simulation using LabView.
3) Discrete system for the smart home behavior using model-predictive controllers in Generic Algebraic Modeling Systems (GAMS).
4) Discrete system model for data communication network using OmNet++.

Examples of such a hybrid system demonstrate how a large-scale hybrid modeling could be attempted. The first example at Oak Ridge National Lab, USA (Nutaro et al., 2008) developed a complex system with item #1 and #4. The second example at National Renewable Energy Lab (NREL), USA (Mittal et al., 2015; Ruth et al., 2015) developed a system comprising of item #1, #2, and #3. Both efforts integrated different modeling paradigms and ran simulations on high-performance computing (HPC) environment in virtual (as-fast-as-possible) and real (wall-clock) time. The NREL effort also integrated air-conditioner (A/C) hardware with the simulation exercise for a real-time 7-day scenario (Mittal et al., 2015).

The above examples are not yet a CyCAS as some of the analysis requiring Complexity Science and Network Theory needs to be done. For example, Emergence View and Resilience View were absent from the above works.

Without the DEVS super-formalism as a foundation, this would have become a software engineering exercise wherein multiple tools would be stitched to perform as a complex system. It would resemble engineering a multithreaded software program with no verifiable inter-thread communication protocols to guarantee timeliness and accurate concurrent execution

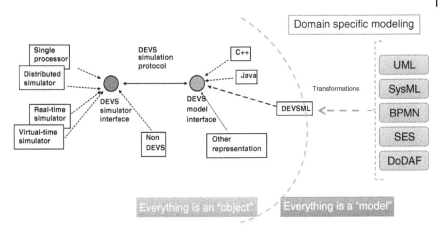

Figure 12.10 MSF and transformation for multiple domains using model driven engineering (MDE) (Mittal and Martin, 2013a).

in an SoS setting. The task of integrating various simulators to perform together as a composite simulation is termed as co-simulation. This involves weaving the time series behavior and data exchanges accurately, failure of which, will yield inaccurate simulation results. Every such hybrid system would require a dedicated effort to build a co-simulation environment. When the simulator code is open-source or available, the architecture of the tools could be understood and interventions can be made to weave the external tool's input. However, many times these simulation tools are proprietary architectures with no access to the code-base. In these cases, there is truly no means to verify the model's execution at the simulator level. In that situation, the only solution possible is to wrap the tool in a DEVS wrapper (Figure 12.10). Sometimes, this is also done to avoid the labor required to make the tool update as understanding large code-bases is time and cost prohibitive.

In our second solution, as shown in Figure 12.10, a DEVS middleware is created that would take any modeling (after M2DEVS or M2DEVSML transformation) to the DEVS simulation environment. A DEVS simulator can be executed in a local, distributed, virtual-time, real-time, and high-performance environments (Mittal and Martin, 2013a; Zeigler et al., 2000; Wainer and Mosterman, 2011; Li et al., 2015).

Earlier work on DEVS-Bus (Kim and Kim, 1998), the netcentric SOA simulation infrastructure (Mittal and Martin, 2013a) and recent work in building cyber-physical simulation environments (Lee et al., 2015), along with a multiagent toolkit MECYSCO (Camus et al., 2016) provide a solid foundation to execute a DEVS model in a parallel distributed co-simulation environment producing accurate results. The MSF underlies the DEVS

Simulation Protocol (Mittal and Martin, 2013a; Zeigler and Sarjoughian, 2013) that provides provably correct simulation execution of DEVS models thereby obviating commonly encountered sources of errors in legacy simulations.

These DEVS engineering efforts need to be supplemented with analysis pertaining to Complexity Science and Network Theory to incorporate Emergence and Resilience Views.

Mittal (2014) stresses the fundamental difference between software-based discrete event simulation and systems-based discrete event simulation. While the former is strictly based on object-oriented software engineering paradigm (Schmidt, 2006; Volter et al., 2006), the latter enforces Wymore's System Theory on the object-oriented discrete event simulation engine. Since cyber complex adaptive systems are multiagent adaptive systems at the fundamental level, there are many agent-based modeling (ABM) tools available to represent them. Unfortunately, due to their software-based object-orientation, the large majority of these tools do not conform to Wymore Systems Theory's closure under composition principle. In contrast, a DEVS-based agent has the notion of a system attached to it and is built on formal semantics that adheres to Wymore's Systems Theory. Such an approach makes it possible to develop a simulator, a simulation protocol and a distributed high performance engine for agent/system model's execution that ensures that closure of coupling is not violated and holistic behavior as a result of multilevel interactions is accurately manifested (Mittal, 2012). Moreover, DEVS formal specification allows it to interface with model-checking tools to supplement simulation with formal verification and validation, a critical feature of model engineering (Zeigler and Nutaro, 2015).

Mittal and Martin (2016) describe packaging all these functionalities in a netcentric DEVS Virtual Machine (VM) that provides agent-execution environment in Cloud-setting within the DEVSML Studio[3] (Mittal and Martin, 2016). Several M&S environments exist that support the DEVS-based methodology just described, including DEVS-Suite, CD++, DEVSim++, JAMES II, Python DEVS and VLE (see the list at DEVS Standardization Group (DEVS, 2016) for descriptions). The M&S environment MS4Me[4] was developed as the first in a commercial line of DEVS products.

DEVS enables new frameworks for application domains, especially those that feature hybrid systems. Some examples such as production flows in the food industry, building energy design, quantum key distribution (QKD) systems, and agent-based transportation evacuation are presented in Table 12.5 in terms of

---

3 DEVSML Editor. Available at http://duniptechnologies.com/jm/downloads.html
4 Ms4Me. Available at http://ms4systems.com

Table 12.5 Some DEVS-enabled frameworks.

Application area	Components	Novel feature	Unique capability
Development of DEVS models for food industry production lines	processing units, conveyor belts	New framework for carrying out simulations of continuous-time stochastic processes	Keep track of parameters related to the process and the flowing material (temperature, concentration of pollutant) is also considered. Since these parameters can change over time in a continuous manner, the possibility to transmit those laws as functions is introduced in the model
Development of DEVS models for building energy design	occupants, thermal network points, windows, HVACs	Allow different professions involved in the building design process to work independently to create an integrated model	Results indicate that the DEVS formalism is a promising way to improve poor interoperability between models of different domains involved in building performance simulations
Quantum key distribution (QKD) system with its components using DEVS	Classical pulse generator, polarization modulator, electronically variable optical attenuator	DEVS assures the developed component models are composable and exhibit temporal behavior independent of the simulation environment	Enable users to assemble and simulate any collection of compatible components to represent complete QKD system architectures
DEVS framework for transportation evacuation integrating event scheduling into an agent-based method	Vehicles, agents	This framework has a unique hybrid simulation space that includes a flexible-structured network and eliminates time-step scheduling used in classic agent-based models	Hybrid space overcomes the cellular space limitation and provides flexibilities in simulating evacuation scenarios. Model is significantly more efficient than popular multi-agent simulators. Keeps high model fidelity and the same cognitive capability, collision avoidance, and low agent-to-agent communication cost

their novel features and unique capability offered when compared to existing approaches. DEVS also supports tools for simulating such models. For example, a compiler that employs DEVS to execute models expressed in the well-known simulation language, Modelica.

Some of the main reasons for basing the architecture on DEVS are as follows:

- DEVS formalizes what a model is, what it must contain, and what it doesn't contain (e.g., experimentation and simulation control parameters are not contained in the model).
- DEVS represents a System of Interest (SoI) using well-defined input and output interfaces. This is critical because composing models requires respecting such boundaries for the constituent referent SoIs.
- DEVS is *universal* and *unique* for discrete event dynamic system models in the sense that any system that accepts events as inputs over time and generates events as outputs over time is equivalent to a DEVS in a strong sense: that is, its behavior and structure can be described by such a DEVS.
- A DEVS model is a system-theoretic concept specifying inputs, states, outputs, similar to a state machine. Critically different however, is that it includes a time-advance function that enables it to represent discrete event systems, as well as hybrid systems with continuous components in a straightforward platform-neutral manner.
- DEVS-compliant simulators execute DEVS-compliant models correctly and efficiently. DEVS defines what's necessary to compose modularized models that will be executable by a compliant simulator.
- DEVS models can be executed on multiple different simulators, including those on desktops (for development) and those on high-performance platforms, such as multicore processors.
- DEVS supports model continuity that allows simulation models to be executed in real-time as software by replacing the underlying simulator engine.

12.5.3 From CyCAS Views to M&S

CyCAS modeling is a multidisciplinary exercise that requires input from multiple professionals. It requires weaving multiple abstraction levels using semantic events (Mittal, 2013c; Sarjoughian and Mayer, 2011) to achieve semantic interoperability (Figure 12.9) within the scenario-context. Ontology harmonization is another way (Zeigler and Hammonds, 2007). This issue becomes more pronounced if there are competing or contrasting theories at different abstraction levels (Roca et al., 2014; Winsberg, 2010). Further, without the context, the problem of semantic interoperability is proven to be NP-Hard (Diallo, 2016).

The problem of multiparadigm modeling has been recognized in Grand Challenges in M&S (Taylor et al., 2012). DEVS Theory has been shown to support multiparadigm modeling (Zeigler, 2013). Incorporating the scenario for CyCAS' intended use allows us to formulate various metrics that can aid model development. This is very much needed as it would help build the Phase Transition and Resilience View in the set of CyCAS views. Table 12.6 adapted from Mittal (2014) associates various metrics with CyCAS views, model representations and their intended uses.

12.5.4 Security and Vulnerability inclusion in CyCAS Views

There has been little work done in the area of security modeling meta-model, partly because securing systems is highly domain dependent. Every domain has conducted a security assessment in some form or another and yet they are incomplete, vulnerabilities exist, and this eventually leads to cyber attacks. Part of the problem is the lack of availability of structured domain knowledge (e.g., in ontological format) that would aid the integration of security models with the larger system models. Recent works like Cyber Security Modeling Language (CySeMoL) (Sommestad et al., 2013) use metamodeling approaches and create their own ontology of attack vectors and systems. Their approach could very well need to be integrated with MITRE's ATT&CK Framework (MITRE, 2014) for a comprehensive knowledge base that could participate in integrated modeling for CyCAS. Much of the work in security and vulnerability assessment deals with developing attack-networks through various available attack-surfaces in the system and matching them with the user profiles. This section warrants a full chapter of its own and is left for future work as this is not the focus of this chapter. The rich body of knowledge available in Security engineering, Mission Assurance and Vulnerability assessment is needed as a metamodel so that it can be used for integrated modeling in the CyCAS framework.

12.6 Validation and Verification in Cyber M&S

Figure 12.9 shows the relationship between validation and verification (V&V) and the MSF elements: model, simulator, and scenario context (experimental frame). The subject of V&V is a critical aspect in developing any theory or the modeling thereof. Without valid models, the theory cannot be tested. Without verified models, model's correctness cannot be ensured.

Table 12.6 Model representation, intended uses (IUs), and experimental frames (EFs) for CyCAS views.

ID	CyCAS view	Model representations	Intended uses (IUs)	Experimental frames and metrics
1	Human view	Inclusion of cognitive, social, and technical abilities of agents for incident scenario interactions	Improved ability to account for agent motivations and predict novel cyberattacks and unforeseen responses	Vulnerability metrics and vulnerability counts for both static and dynamic situations, behavior quality, spread and quantification, quantized context
2	Multiagent view	Sophistication of linguistic levels at semantic and pragmatic levels for agent–agent interaction	Improved accounting for formation of cooperative/collaborative versus competitive teams and subgroupings	Small-world and other graph-based metrics for leader/follower behaviors, group cohesion, observability spectrum
3	Control and communication view	Lumped and system emulators to represent communication and security protocols and stacks	Analysis of information technology (IT) stacks for protocol evaluation. Command and control strategies across various partitions of the system deployment	Degree of control (from centralized and totalitarian to completely decentralized and autonomous), interactivity graphs, feedback loops, heat maps
4	Resource and constraints view	Resource utilization, availability, limitations and affordance, constraint network graphs	Accounting for the limits placed on performance objectives placed by resource availability and cost	Conventional utilization and waiting time metrics employed by operations research
5	Emergence view	Multilevel dynamic structure and behavior coupling,	Inclusion of computational and repeatable emergent behavior	Emergence observers designed to detect novelty in behaviors and structures, measures of degree

Table 12.6 *(Continued)*

ID	CyCAS view	Model representations	Intended uses (IUs)	Experimental frames and metrics
		validation and recognition		of agent adaptation in dynamic environments
6	Phase transition view	Processes to identify quantum changes in manifested behavior by the CyCAS, transitions to novel, albeit identified, holistic properties of CyCAS	Ability to identify steady state of the CyCAS in presence of threshold crossing at multiple levels, cascaded failures or amplification of positive feedback loops resulting in new functional networks	Multilevel transition probabilities, credit assignment, new behavior detection, and encoding in a live system or at runtime in a computational environment, credit assignment, dynamic role switching, runtime behavior encoding
7	Knowledge view	Semantic network, semantic validity through subject matter expert (SME) and keyword-rank	Ability to manage universal truth (and semantic event) at multiple abstraction levels.	Ontology harmonization, knowledge propagation effects, spread and detection of new word-groups in a multilayer network or ontology. Identification of semantic gaps between multiple-levels
8	Resilience view	Processes to recover from catastrophic events to restore normal operation	Ability to assess recovery success and times after catastrophic events	Degree of robustness at various levels and across levels, threat vectors, attack surfaces across multiple levels

12.6.1 Validity and Partial-Order Relations

The basic modeling relation, *validity*, refers to the relation between a *model*, a *system*, and an *experimental frame*. Validity is often thought of as the degree to which a model faithfully represents its system counterpart. However, it makes much more practical sense to require that the model faithfully captures the system behavior only to the extent demanded by the objectives of the simulation study. In our formulation, the concept of validity answers the question of whether it is impossible to distinguish the model and system *in the experimental frame of interest*. The most basic concept, *replicative validity*, is affirmed if, for all the experiments possible within the experimental frame, the behavior of the model and system agree within acceptable tolerance. Thus, replicative validity requires that the model and system agree at the I/O relation level 1 of the system specification hierarchy.

Stronger forms of validity are *predictive validity* and *structural validity*. In predictive validity, we require not only replicative validity but also the ability to predict as yet unseen system behavior. To do this the model needs to be set in a state corresponding to that of the system. Thus, predictive validity requires agreement at the next level of the system hierarchy, that of the I/O function level 2 (see Table 12.1). Finally, structural validity requires agreement at level 3 (state transition) or higher (coupled component). This means that the model not only is capable of replicating the data observed from the system but also mimics in step-by-step, component-by-component fashion, the way that the system does its transitions.

The term *accuracy* is often used in place of validity. Another term, *fidelity* (Ruth et al., 2015) is often used for a combination of both validity and detail. Thus, a high fidelity model may refer to a model that is both high in detail and in validity (in some understood experimental frame). However, when used this way there may be a tacit assumption that high detail alone is needed for high fidelity, as if validity is a necessary consequence of high detail. In fact, it is possible to have a very detailed model that is nevertheless very much in error, simply because some of the highly-resolved components function in a different manner than their real system counterparts.

Besides the two fundamental relationships, there are others that are important for understanding modeling and simulation work. These relations have to do with the interplay and orderings of models and experimental frames (Figure 12.11). The inescapable fact about modeling is that it is severely constrained by complexity limitations.[5] Complexity is measured typically on resource usage in time and space relative to a particular simulator, or class of simulators.[6] However, properties intrinsic to the

---

5 As technology advances, such constraints may continue to be loosened. Nevertheless, fundamental limitations to computation are well established in computer science. Moreover, we conceive of human comprehension of models as involving mental simulation and therefore cognitive computational constraints are also included in such limitations.

6 Human understanding of models formulated as a class of cognitive simulators is included, per the previous footnote.

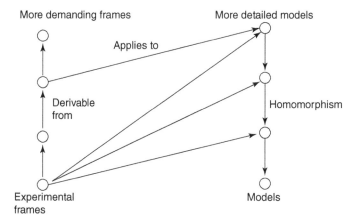

Figure 12.11 Fundamental ordering relations for data architecture.

model are often strongly correlated with complexity independently of the underlying simulator. Successful modeling can then be seen as *valid simplification*. We need to *simplify*, or reduce the complexity, to enable models to be executed on resource-limited simulators. But the simplified model must also be *valid*, at some level, and within some experimental frame of interest. As in Figure 12.12, there is always a pair of models involved, call them the base and lumped models. Here, the base model is typically "more capable" and requires more resources for interpretation than the *lumped model*. By the term "more capable," we mean that the base model is valid within a larger set of experimental frames (with respect to a real system) than the lumped model. However, the important point is that within a **particular frame of interest** the lumped model might be just as valid as the base model. The concept of morphism, introduced above, affords criteria for judging the equivalence of base and lumped models with respect to an experimental frame.

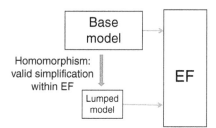

Figure 12.12 Validity of base and lumped models in EF.

12.6.2 V&V within Cyber M&S

Kavak *et al.* (2016) lists broad scenario elements categorized under two groups: *cybersystems* (including data and wide network infrastructure), as observed in LVC systems and *actors* (attacker, user, and system security personnel). Cyber Resiliency Engineering Framework (Bodeau and Graubart, 2011; Bodeau et al., 2014) elaborate on various phases of a resiliency engineering process such as: anticipate, withstand, recover, and evolve. This will need to be incorporated in the Resiliency View and further work is needed to develop a M&S framework for T&E of resilience in simulation experiments. Further, MITRE's Adversarial Tactics, Techniques and Common Knowledge (ATT&CK 2016) (MITRE, 2014) Framework can aid the development of attacker behavior models.

For CyCAS models, the validation aspect answers the question: How useful a model is for a particular cyber scenario or a Cyber event (Section 3)? The verification aspect falls-back to the co-simulation of various tools and paradigms that need to be brought in for an accurate simulation of a cyber model. The entire simulation experiment, the model, and the simulation-infrastructure must be automated through a model-base repository and transparent simulation framework. These three aspects can be represented using a System Entity Structure (SES) framework (Mittal and Martin, 2013a; Zeigler et al., 2000; Zeigler and Hammonds, 2007).

The SES contains the decomposition, taxonomies, coupling specification, and constraints of a given Internet infrastructure. The modeler prunes the SES to extract a desired network structure and the network component models, the attacker models, and analyzer models expressed in DEVS stored in the model base. As illustrated by the SES in Figure 12.13, a scenario can contain the following:

- Attackers
- Users that the attacker targets initially
- A cyber system and data that the attacker targets
- System security personnel that detect the incident
- Interactions between the attacker, users, the system and the security personnel
- A wide network infrastructure that facilitates connection between cyber systems and people

Kavak *et al.* (2016) found a large number of studies focusing on attacker and security personnel modeling only their technical abilities. We illustrate how the absence of cognitive and social abilities in the current formulations of security scenarios can be represented using specializations in the SES. In particular, in Figure 12.13, presence is pervasive specialization for agents. It

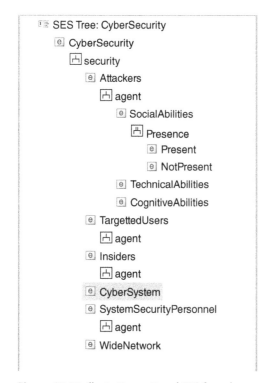

Figure 12.13 Illustrative notional SES for cybersecurity scenarios.

enables choice of inclusion or not of social, technical, or cognitive abilities using pruning operations that select *Present* or *Not Present* from the specialization.

Increasingly diverse and ingenious hacking techniques on the Internet infrastructure allow a variety of access paths (You and Chi, 2009). Such paths have grown exponentially in number with Internet of Things (IoT) developments that connect myriads of unsophisticated computational elements performing sensing and control tasks under remote networked guidance. You and Chi (2009) use the SES/MB (System Entity Structure/Model Base) framework (Figure 12.14). It illustrates DEVS potential to support quantitative evaluation of cybersystem vulnerability, assessing vulnerable resources and determining defense policy to protect these resources. Moreover, such an environment deals with dynamically changing (time-dependent) vulnerabilities in contrast with only static (time-independent/scanning-based) vulnerabilities of given network components.

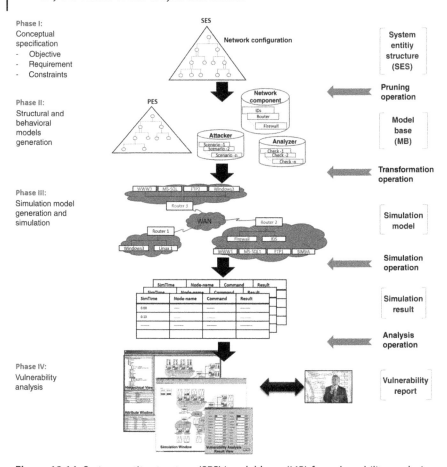

Figure 12.14 System entity structure (SES)/model base (MB) for vulnerability analysis.

The simulation model is constructed by integrating the dynamic models in the model-base along with the pruned network structure enabling simulation execution and generation of behavior in the specified EF and its vulnerability metrics. The simulation results can be analyzed using the metrics so that the vulnerabilities of each network component is evaluated. This environment illustrates the paradigm discussed in the section of DEVS-based hierarchical and modular M&S able to construct security models including network vulnerability metrics in quantitative manner. Although illustrative of DEVS-based capabilities, still more advanced modeling and simulation methodology is needed for identifying vulnerabilities that may result from cyber-attacks that can occur within the broader framework of Kavak *et al.* (2016).

12.7 Discussion & Conclusions

Inspired by William Waite's broad perspective, we affirm that M&S is an authentic scientific discipline and an important new profession. It is deemed as a US National Critical Technology (H. Res 487) along with multiple resolutions in the M&S Legislation by J. Randy Forbes (2007). Indeed, such a perspective underlies any successful collaboration between multiple domain experts and M&S professionals. Recognizing that cyber complex adaptive systems have soft boundaries, are highly reconfigurable, and must be 100% available anytime, we must first strive to understand how to model them in a repeatable way. Thus, contemporary model-based system engineering must be robustly supported by M&S professionals armed with theory, concepts, and tools up to the challenges of Cyber environments replete with multiple SMEs in any given scenario. This also led us to emphasize the need for formal methods and tools, such as those enabled by DEVS, to build M&S workbenches that create reusable results, and proffer a common, understandable, and reproducible way to solve problems with M&S. All this will eventually lead to the cyber systems engineering discipline.

As for the M&S professionals that are needed to support successful cyber system engineering we stress that a cyber systems modeling effort requires interdisciplinary contributions and strong relationships between domain experts and modelers. Also needed is a person with expertise in modeling multiple domains who can provide inputs on the domain interfaces and knowledge exchanges between various constituent domains to ensure event-semantics are preserved when multiple disciplines are brought together. We have seen that CyCAS models need to be validated in LVC setting through Cyber-events in a Cyber Range. Management and automation of a Cyber Range event is another complex endeavor. Another level of complexity is the execution of virtual environment with co-simulation infrastructure such as simulators and emulators that are also part of the LVC infrastructure. All of the M&S infrastructure engineering has to ensure that the hybrid nature of the system (i.e., both discrete and continuous nature) is implemented faithfully in a testable manner and the LVC simulation does not manifest any unintended emergent behaviors. The testing and evaluation using a Cyber event also require a library of models and experimental frames that support the intended uses of the CyCAS. As indicated, we believe that DEVS, and its associated M&S framework, enable the fundamental modeling and simulation theory that incorporates the above aspects required for robust Cyber M&S.

Finally, validation and verification of a CyCAS model is the hardest nut to crack, primarily because, a CyCAS, either a model or a real system, will always contain black-boxes (either as systems or simulators) for which there is not enough information about the inside architecture. This will always lead to inaccurate simulation results when the manifested behavior of the black-box is not consistent within the scenario-context, and if deemed correct, is incorrectly executed through the complex Cyber Range LVC simulation infrastructure. Validation is the only aspect that can be ensured in CyCAS M&S. Verification must be ascertained using statistical and stochastic methods.

This chapter has reviewed the state of the art in Cyber M&S. Barely covering the critical security modeling and resilient systems engineering practices here, we look forward to incorporating them in our modeling and simulation framework in the near future.

Acknowledgements

The authors would like to acknowledge the book editors and MITRE peer-reviewers for their suggestions and feedback that enhanced the quality of the chapter.

Disclaimer

'The author's affiliation with The MITRE Corporation is provided for identification purposes only, and is not intended to convey or imply MITRE's concurrence with, or support for, the positions, opinions or viewpoints expressed by the author(s). Approved for Public Release: Case Number 16–4658.

References

Adaptive Character of Though-Rational (ACT-R) . http://act-r.psy.cmu.edu/. Last accessed: Decemeber 16, 2016.

Arnwine, M. (2013) Joint Mission Environment Test Capability (JMETC) Improving T&E with distributive test capabilities, *NDIA Systems Engineering Conference*.

Ashby, W.R. (1956) *An Introduction to Cybernetics*, Chapman & Hall.

Balci O. (2012) A life cycle for modeling and simulation. *Simulation*, 88 (7), 870–883.

Boardman J. and Sauser B. (2006) System of Systems – The meaning of "of". *IEEE/SMC International Conference on System of Systems Engineering, Los Angeles*.

Bodeau, D.J. and Graubart, R. (2011) Cyber Resilience Engineering Framework, MITRE Technical Report, MTR 110237.

Bodeau, D.J., Graubart, R., and Laderman, E.R. (2014) Cyber resiliency engineering: Overview of the architectural assessment process. *Conference on Systems Engineering Research, Redondo Beach, CA.*

Camus, B., Paris, T., Vaubourg, J., Presse, Y., Bourjot, C., *et al.* (2016) MECYSCO: a multi-agent DEVS Wrapping Platform for the Co-simulation of Complex Systems (Research Report) LORIA, UMR 7503, CNRS.

Cellier F.E. (1979) Combined continuous/discrete system simulation languages-usefulness, experiences and future development.methodology in systems modeling and simulation, pp 201–220.

Coffman, E., Ge, Z., Misra, V., and Towsley, D., (2002) Network Resilience: Exploring cascading failures within BGP, *Allerton Conference on Communication, Control and Computing.*

Concepcion, A.I. and Zeigler, B.P., (1988) DEVS Formalism: A Framework for Hierarchical Model Development, *IEEE Trans. Softw. Eng.*, vol. 14, no. 2, 1988, pp. 228–241.

Cuninghame-Green, R. (1979) *Minimax Algebra, Lecture Notes in Economics and Mathematical Systems 166*, New York, NY, USA: Springer-Verlag New York.

Damodaran, S. and Couretas, J. (2015) *Cyber Modeling and Simulation for Cyber-Range Events*, Summer Computer Simulation Conference, Chicago, IL.

Defense Science Board (2013) Resilient Military Systems and the Advanced Cyber Threat. Washington DC: Office of the Under Secretary of Defense for Acquisition, Technology and Logistics, 146.

DEVS (2016) DEVS Standardization Group http://cell-devs.sce.carleton.ca/devsgroup/?q=node/8

Diallo, S. (2016) *On the complexity of Interoperability, Proceedings of Symposium on M&S of Complexity in Intelligent, Adaptive and Autonomous Systems*, Society of Computer Simulation International.

Forbes, J.R. (2007) Modeling and Simulation Legislation. Available at: http://forbes.house.gov/biography/mslegislation.htm

Gill, A. (1962) *Introduction to the Theory of Finite-State Machines*, New York, NY, USA: Mc-Graw Hill.

Glynn, P.W. (1989) A GSMP Formalism for Discrete Event Systems, *Proc. IEEE*, 77(1), pp. 14–23.

Hansen, Andrew P. (2008) Cyber Flag - A Realistic Cyberspace Training Construct. Wright Patterson Air Force Base: AFIT.

Hoare, C.A.R. (1985) *Communicating Sequential Processes*, Upper Saddle River, NJ, USA: Prentice Hall.

Holland, J.H. (1992) *Complex Adaptive Systems*, 121 (1), 17–30.

Holliday, M.A., Vernon, M.K. (1987) A Generalized Timed Petri Net Model for Performance Analysis, *IEEE Trans. Softw. Eng.*, 13(12), pp. 1297–1310.

Kavak, H., Padilla, J.J., and Vernon-Bido, D. (2016) A characterization of cybersecurity simulation scenarios. *Proceedings of the 19th Communications & Networking Symposium, Society of Computer Simulation International.*

Kim, Y.J. and Kim, T.G. (1998) A heterogeneous simulation framework based on the DEVS BUS and the high level architecture. *In Proceedings of Winter Simulation Conference. Vol. 1.*

Kohavi, Z. (1978) *Switching and Finite Automata Theory*, 2nd ed., New York, NY, USA: McGraw-Hill.

Koymans, R., Specifying Real-Time Properties with Metric Temporal Logic, *Real-Time Syst.*, vol. 2, no. 4, Nov. 1990, pp. 255–299.

Lee, E.A. (2008) Cyber physical systems: design challenges, *IEEE International Symposium on Object Oriented Real-time Distributed Computing*, Orlando, FL, USA.

Lee, K.H., Hong, J.H., and Kim, T.G. (2015) System of systems approach to formal modeling of CPS for simulation-based analysis. *ETRI Journal*, 37, (1) 175–185.

Li, B.H., Chai, X., Li, T., Hou, B., Qin, D., Zhao, Q., Zhang, L., Hao, A., Li, J., and Yang, M. (2015) Research on High Performance Modeling and Simulation for Complex Systems, in *Concepts and Methdologies for Modeling and Simulation: A Tribute to Tuncer Oren*, (ed. L. Yilmaz,), Springer, UK.

Mann, Z., and Pnueli, A., *The Temporal Logic of Reactive and Concurrent Systems*, New York, NY, USA: Springer-Verlag New York, 1992.

Milner, R. (1989) *Communication and Concurrency*, Upper Saddle River, NJ, USA: Prentice Hall

MITRE (2016) Adversarial Tactics, Techniques and Common Knowledge (ATT&CK) Framework. Available at: https://attack.mitre.org/wiki/Main_Page

Mittal, S. (2012) Emergence in stigmergic and complex adaptive systems: a discrete event systems perspective, special issue on stigmergy. *Cognitive Systems Research*, 21, 22–39.

Mittal, S. (2013) Netcentric complex adaptive systems, in *Netcentric System of Systems Engineering with DEVS Unified Process*, (eds S. Mittal and J.L.R. Martin), CRC Press, Boca Raton, FL.

Mittal, S. (2014) *Model Engineering for Cyber Complex Adaptive Systems*, European Modeling and Simulation Symposium, Bordeaux, France.

Mittal, S. and Douglass, S.A. (2012) DEVSML 2.0: the language and the stack, *Spring Simulation Multi-Conference, Orlando, FL.*

Mittal, S. and Martin, J.L.R. (2013) *Netcentric system of systems engineering with DEVS unified process*, CRC Press , Boca Raton, FL USA, p. 712 ISBN: 9781439827062.

Mittal, S. and Martin, J.L.R. (2013) Model-driven Systems Engineering in a netcentric environment with DEVS unified process, invited paper to *Winter Simulation Conference, Washington DC.*

Mittal, S. and Martin, J.L.R. (2016) DEVSML Studio. A framework for integrating domain-specific languages for discrete and continuous hybrid systems into DEVS-based M&S environment, *Summer Computer Simulation Conference, Montreal, Canada.*

Mittal, S. and Zeigler, B.P. (2014) Context and attention in activity-based intelligent systems, *ITM Web of Conferences*, 3, DOI: 10.1051/itmconf/20140303001

Mittal, S., Martin, J.L.R., and Zeigler, B.P., (2007) DEVSML: Automating DEVS simulation over SOA using transparent simulators, *DEVS Symposium, Spring Simulation Multi-Conference.*

Mittal, S., Zeigler, B.P., Martin, J.L.R., Sahin, F., and Jamshidi, M. (2008) Modeling and simulation for system of systems engineering, in *System of Systems Engineering for 21st Century*, (ed. Mo Jamshidi,) John Wiley and Sons, Inc.

Mittal, S., Ruth, M., Pratt, A., Lunacek, M., Krishnamurthy, D., and Jones, W., (2015), A system-of-systems approach for integrated energy systems modeling and simulation, *Summer Computer Simulation.*

Noubir, G., Stephens, D.R., Raja, P. (1993) Specification of Timed Finite State Machine in Z for Distributed Real-Time Systems, *Proc. IEEE Workshop Future Trends Distrib. Comput. Syst.*, Lisbon, Portugal, pp. 319–325.

Nutaro, J., Kuruganti, P.T., Shankar, M., Miller, L., and Mulle, S. (2008) Integrated modeling of the electric grid, communications, and control. *International Journal of Energy Sector Management*, 2 (3), 420–438.

Oren, T. and Zeigler, B.P. (2012), System theoretic foundations of modeling and simulation: a historic perspective and the legacy of A Wayne Wymore, *Simulation*, 88 (9), 1033–1046.

OSD (2015) National Cyber Range Overview. Available at: http://www.acq.osd. mil/dte-trmc/docs/20150224_NCR%20Overview_DistA.pdf

Peterson, J.L. (1981) *Petri Net Theory and the Modeling of Systems,"* Upper Saddle River, NJ, USA: Prentice Hall.

Praehofer H. (1991) System theoretic formalisms for combined discrete-continuous system simulation. *International Journal of General System*, 19 (3), 226–240.

Rainey, L. B., and Tolk, A. (eds). (2015) *Modeling and Simulation Support for System of Systems Engineering Applications.* John Wiley & Sons, Inc., Hoboken, NJ.

Roca, R., Pace, D. Robinson, S. Tolk, A., and Yilmaz, L. (2014) Paradigms for conceptual modeling. *In Proceedings of the 2014 Annual Simulation Symposium, Spring Simulation Multiconference, Society for Modeling and Simulation*, pp. 293–300, 2015.

Ruth, M., Pratt, A., Lunacek, M., Mittal, S., Wu, H., and Jones, W., (2015) Effects of home energy management systems on distribution utilities and feeders under various market structures, *23rd International Conference on Electricity Distribution, At Lyon.*

Sarjoughian, H. and Mayer, G.R. (2011) Heterogeneous Model Composability, in *Discrete-event Modeling and Simulation: Theory and Applications*, (eds G. Wainer, P. Mosterman), CRC Press, Boca Raton, FL.

Schmidt, D.C., (2006) *Model-driven Engineering*, IEEE Computer, Orlando, FL, 29 (2), 25–31.

Sommestad, T., Ekstedt, M., and Holm, H. (2013) The cyber security modeling language: a tool for assessing the vulnerability of enterprise system architectures. *IEEE Systems Journal*, 7 (3), 10.1109/JSYST.2012.2221853

Taylor, S.J.E., Fishwick, P.A., Fujimoto, R., Uhrmacher, A.M., Page, E.H., and Wainer, G. (2012) Challenges for Modeling and Simulation, *Proceedings of the 2012 Winter Simulation Conference*, (eds C. Laroque, J. Himmelspach, R. Pasupathy, O. Rose, and A. M. Uhrmacher,).

Tran, H.T., Domercant, J.C., and Mavris, D.N. (2015) Evaluating the agility of adaptive command and control networks from a cyber complex adaptive systems perspective. *Journal of Defense Modeling and Simulation*, 12 (4), 405–422.

US-Canada Power System Outage Task Force (2004) Final Report on the August 14, 2003 Blackout in the United States and Canada: Causes and Recommendations. Available at http://energy.gov/sites/prod/files/oeprod/DocumentsandMedia/BlackoutFinal-Web.pdf

Vangheluwe, H.L.M. (2000) DEVS as a common denominator for multi-formalism hybrid systems modeling. *Proceedings for the 2000 IEEE International Symposium on Computer-aided Control System Design, Alaska, USA*.

Vangheluwe, H., De Lara, J., and Mosterman, P.J. (2002) An Introduction to multi-paradigm modeling and simulation. In *Proceedings of AIS2002*. pp. 9–20.

Volter M., Stahl T., Bettin J., Haase A., and Helsen S. (2006) *Model-driven Software Development: Technology, Engineering, Management*. John Wiley & Sons, Ltd Chichester.

Wainer, G. and Mosterman, Pieter (eds), (2011) *Discrete-event Modeling and Simulation: Theory and Applications*, CRC Press, Boca Raton, FL.

Winsberg, E. (2010) *Science in the Age of Computer Simulation*. The University of Chicago Press.

Wymore, A.W. (1967) *A Mathematical Theory of Systems Engineering: the Elements*, John Wiley and Sons, Inc., New York.

You, Y.J. and Chi, S.D. (2009) SIMVA: simulation-based network vulnerability analysis system. *Proceedings of the 2009 Spring Simulation Multiconference, Society of Computer Simulation International*.

Zeigler, B.P. (2012) Activity and System Specification Levels, *ACTIMS'12, Cargese*.

Zeigler, B.P. (2013) Grand Challenges in Modeling and Simulation: What Can DEVS Theory Do To Meet Them? Parts 1 and 2, Seminar to School of Automation Science and Engineering, Beihang University, Beijing, China. Available at: http://duniptechnologies.com/jm/images/dunip/docs/behangseminarv2.pdf

Zeigler, BP. and Hammonds P.E. (2007) *Modeling and Simulation-based Data Engineering: Introducing Pragmatics into Ontologies for Netcentric Information Exchange*, Academic Press.

Zeigler, B.P. and Nutaro, J.J. (2015) Towards a framework for more robust validation and verification of simulation models for system of systems. *Journal of Defense Modeling and Simulation*. 1 (1), 3–16.

Zeigler, B.P. and Sarjoughian, H.S. (2013) *Guide to Modeling and Simulation of Systems of Systems*, Springer.

Zeigler, B.P. and Zhang, L. (2015) Service-oriented modeling and simulation for system of systems engineering, *Concept and Methodologies for Modeling and Simulation*, Springer-Verlag, Germany.

Zeigler, B.P., Praehofer, H., and Kim, T.G. (2000) *Theory of Modeling and Simulation: Integrating Discrete Event and Continuous Complex Dynamic Systems*. Academic Press.

Part V

Economics

13

Funding an Academic Simulation Project: The Economics of M&S

Saikou Y. Diallo,[1] Christopher J. Lynch,[1] and Navonil Mustafee[2]

[1]*Old Dominion University, Norfolk, VA, USA*
[2]*University of Exeter, Exeter, UK*

13.1 Introduction

In the United States, the discipline of modeling and simulation (M&S) is steadily growing and expanding. It is now taught in at least one major university at the undergraduate and graduate level. In the military, M&S has been recognized as a specialization in the United States Army (FA 57). The United States Department of Defense (DoD) is also a main funder of M&S research through their acquisition, analysis, and training processes (Balci et al., 2000, 2002; Hillegas et al., 2001).

However, outside of the DoD, researchers who focus exclusively on advancing the discipline of M&S and who work to bring about improvements in the design, verification, validation, and execution of models and simulations might find it difficult to find funding for their work. This is due mainly to the fact that M&S does not have its own directorate or line of funding within the two major federal agencies that fund scientific research: The National Sciences Foundation (NSF); and the National Institutes of Health (NIH). Instead, since it is considered an endeavor that cuts across disciplines, M&S funding is often lumped with the discipline from which the object of the study originated. As a result, while it is not directly funded, many projects that utilize M&S as the main method of investigation find support within those agencies.

In this chapter we take an empirical approach to examine to what extent federal research resources allocated to NSF and NIH are used to fund projects that use M&S as a method, technique, or tool. Our approach is twofold: (1) we perform content analysis on three sets of abstracts from funded NSF and NIH projects that use M&S as one of the methods of

The Profession of Modeling and Simulation: Discipline, Ethics, Education, Vocation, Societies, and Economics, First Edition. Andreas Tolk and Tuncer Ören.

investigation; and (2) we compare and contrast NSF funding across geographical and disciplinary dimensions. The goal of the content analysis is to identify concepts and themes that are commonly found in projects that use M&S as a method of investigation. The role of the third dataset is to provide researchers clues as to what disciplines and topics related to M&S tend to receive the most funding in NSF over the past 5 years.

With our twofold approach, we hope to give the reader insight into projects that use M&S and display not only the wide variety of disciplines and topics involved, but also the geographic variety in funding opportunities. The remainder of the chapter is organized as follows; we begin by briefly describing the datasets in terms of their content, size, and time span and present a brief background of content analysis and the tool we use for our analysis. We then discuss our findings from applying content analysis to the datasets. Following the analysis, we focus on the third dataset in order to derive a geographic and disciplinary breakdown of NSF funding before we conclude.

13.2 Background

In this section, we briefly describe the datasets and present the method and tool used to perform our analysis.

13.2.1 Evaluated Datasets

The first two datasets are obtained through the use of a machine learning classification model to classify and separate awarded NIH and NSF grants that use M&S as one of the main methods of investigation (Gore et al., 2016). Gore *et al.* (2016) examined NSF awards from 1990 to 2003 (Pazzani and Meyers, 2003) and NIH awards from 1990 to 2012 (NIH, 2003) and identified 10,097 NSF M&S-based awards and 9,688 NIH M&S-based awards during these time periods. Based on the M&S-classified awards, the total research value provided by NIH toward the discipline of M&S totaled 2.41 billion dollars over a 22-year period while the research value provided by NSF was 722.98 million dollars over the 13-year period.

The third dataset contains three thousand awarded NSF and NIH abstracts for the last 5 years (2011–2016). The limit of three thousand awards is imposed by the NSF download site. Table 13.1 shows a brief description of each dataset along with the name under which we will refer to each dataset in the balance of this chapter.

In Section 2.2, we describe content analysis and the tool we used to analyze our datasets.

Table 13.1 Overview of datasets.

Source	Time span	Content	Size (number of abstracts)	Short name in the chapter
NSF	1990–2003	Abstracts of NSF funded projects that use M&S as the method of investigation	10,097	Historical M&S-based NSF Awards
NIH	1990–2012	Abstracts of NSF funded projects that use M&S as the method of investigation	9,688	M&S-based NIH Awards
NSF	2011–2016	Abstract of NSF funded projects with the keyword "modeling and simulation"	3,000	Past five years M&S-based NSF Awards

13.2.2 An Overview of Content Analysis and Leximancer

Qualitative content analysis has been applied to various types of texts, including abstracts (Cretchley et al., 2010), newspapers (Tse et al., 1989), journals (Mustafee et al., 2012), and conferences (Diallo et al., 2015a). Studies have profiled disciplines (Lonner et al., 2010; Mustafee et al., 2014), identified research trends (Cretchley et al., 2010), or compared concepts across bodies of knowledge (Crofts and Bisman, 2010). Various M&S-specific content analyses have explored topics such as healthcare simulation (Brailsford et al., 2009; Katsaliaki and Mustafee, 2011), interoperability (Diallo and Lynch, 2016), and the domain of M&S as a whole (Diallo et al., 2015b). We refer the interested reader to Poser *et al.* (2012) and Smith and Humphreys (2006) for evaluations on Leximancer – a software tool designed for analyzing natural language text data using statistics-based algorithms, and extracting semantic and relational information and visualizing their output through concept map, network cloud, and concept thesaurus.

We use Leximancer to conduct automated content analysis of the collected abstracts in order to identify the main themes and concepts (Leximancer, 2011) occurring within NIH and NSF awards. Content analysis is a technique for creating content categories from the words contained within a body of text to identify the main themes, ideas, and topics in a systematic and objective manner (Berelson, 1952; Kassarjian, 1977; Stemler, 2001; U. S. G. A. Office, 1996). Themes help to identify trends and patterns by highlighting the main assertions, beliefs, and attitudes within the text (Kassarjian, 1977). Inferences and interpretations can then be drawn from the text as a whole (Weber, 1990).

Leximancer identifies concepts by counting the frequency of each word and calculating its relative co-occurrence to every other word within the selected texts. Relative frequency is calculated as the conditional probability that a concept (C_O) is present given a particular category (C_A) as provided in Equation 13.1, strength is measured as the conditional probability of a category (C_A) given that a particular concept (C_O) is present as provided by Equation 13.2, and the prominence of a concept is the joint probability of a concept (A) given a category (C) divided by the marginal probabilities of the appearance of A and C across the total number of context blocks within the entire body of text as shown in Equation 13.3 (Leximancer, 2011). We specify Leximancer to create context blocks using every two sentences or the end of each paragraph as boundaries.

$$P(C_O|C_A) = \frac{P(C_O \cap C_A)}{P(C_A)} \tag{13.1}$$

$$P(C_A|C_O) = \frac{P(C_A \cap C_O)}{P(C_O)} \tag{13.2}$$

Prominence (given A&C)

$$= \frac{\left(\dfrac{\text{co-occurrence count for A\&C}}{\text{total number of context blocks in data set}}\right)}{\left(\dfrac{\text{occurrence of A}}{\text{number of context blocks}}\right) \times \left(\dfrac{\text{occurrence of C}}{\text{number of context blocks}}\right)} \tag{13.3}$$

Themes help illuminate trends and specific foci within the awards that can help us draw out the main features within the abstracts that led to their selection for funding. Themes emerge based on the frequencies of co-occurrence and strengths of occurrences between concepts within the abstracts. In the next section, we present and discuss our results.

13.3 Concepts and Themes of NSF and NIH M&S Related Funding 1990–2012

13.3.1 Properties of M&S-Based NSF Awards Between 1990 and 2003

We examine the NSF dataset in order to identify the main themes and concepts within NSF funded projects that utilize M&S. Figure 13.1 shows the concept map for the NSF awards during the time span of 1990 to 2003. The bubbles convey the primary themes and the nodes represent the main concepts identified within the NSF texts. Each bubble is color coded by connectivity meaning themes with higher connectivity have higher intensity.

We examine the top five themes within Figure 13.1. For each theme, we provide (1) the theme's percent connectivity calculated as the co-occurrence count of each concept within the theme against the total of all concept

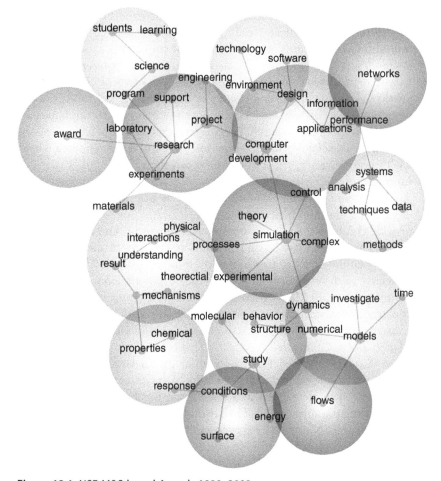

Figure 13.1 NSF M&S-based Awards 1990–2003.

occurrence counts, and (2) the main concepts that comprise the theme. The connectivity percentage provides an estimation of how well each theme is connected across all of the NSF awards. Interestingly, two themes, simulation and research, have 100% connectivity across the NSF abstracts which confirms that the dataset is representative of the use of M&S across the board. A more detailed discussion follows:

1) *Simulation (100% Connectivity), Processes, Control, Complex, Experimental, Theory*: Simulation is commonly employed as an artifact for conducting experimentation or testing theory. NSF funded simulation-assisted research utilizes simulation to learn about complex processes. Simulations are especially helpful when examining theoretical solutions as it can be much easier to update the parameters of an experiment and

alter the control variables and rules than when experimenting with the real system. This indicates that M&S researchers have to be part of multidisciplinary teams and fill the role of simulation experts.

2) *Research (100% Connectivity), Project, Experiments, Engineering, Support, Laboratory*: Funded research is driven by the goals of a project and generally involves experimentation and laboratory involvement. NSF funds research pertaining to science and engineering; however, only engineering appears as a main concept under research. This suggests that M&S is perceived more as a tool than a discipline during this period of time and that simulation lends itself toward engineering solutions or experimentation.

3) *Computer (93% Connectivity), Design, Applications, Development, Performance, Information*: NSF-awarded grants utilize computers for the design and development aspects of research. Computer applications provide conduits for human–computer interactions and computer performance concerns itself with the ability to conduct studies that require network connectivity or that deal with multiple systems. The computers also serve as the artifacts that provide the output information that allows for analysis within M&S-based studies. This is an indication that digital/constructive simulation (as opposed to live, virtual, or a mix of live virtual constructive simulation) is the dominant form of M&S in NSF.

4) *Models (82% Connectivity) Models, Dynamics, Investigate, Numerical, Time*: Models are used for investigation into a study's questions and commonly handle capturing the dynamics of the system being examined in the study. Dynamic aspects of a system can involve understanding and representing the structural and behavioral changes that the system undergoes over time. Numerical refers to numerical methods that are used in engineering to approximate a solution or model continuous behavior. This is further indication that engineering simulation (as opposed to social and behavioral M&S) is the dominant form of M&S in NSF.

5) *Systems (78% Connectivity), Data, Methods, Techniques, Analysis*: System thinking is strongly related to M&S. System science provides the methods and techniques needed to represent a referent in order to conduct analysis. It also means that M&S is used to represent and study a system but a simulation is also considered a system with all that it entails in terms of components, data, and methods of analysis. This theme is an indication that M&S is used in the context of system science and system thinking in NSF.

Table 13.2 provides the top 10 word-like concepts that commonly co-occur with the primary concept within each of the five themes listed above and their corresponding prevalence values within the NSF abstracts.

Table 13.2 confirms that past funded NSF M&S awards are in-line with the use of M&S as a tool and a method and not as a discipline. It also highlights the

Table 13.2 Prevalence (Prev) of the top 10 word-like concepts that commonly co-occur with the main concept from the top five NSF 1990–2003 themes.

Simulation		Research		Models		Computer		Systems	
Word-like concept	Prev	Word-like concept	Prev	Word-like concept	Prev	Word-like concept	Prev	Word-like concept	Prev
Numerical	1.96	Award	2.11	Data	1.58	Science	2.47	Software	2.02
Molecular	1.70	Program	1.78	Behavior	1.53	Software	1.99	Control	1.89
Computer	1.56	Support	1.74	Numerical	1.52	Engineering	1.79	Complex	1.85
Software	1.55	Science	1.58	Conditions	1.40	Complex	1.73	Performance	1.60
Experiments	1.50	Students	1.50	Simulation	1.40	Program	1.59	Dynamics	1.55
Experimental	1.49	Theoretical	1.48	Dynamics	1.36	Simulation	1.56	Design	1.52
Laboratory	1.48	Engineering	1.34	Flows	1.33	Theoretical	1.55	Environment	1.49
Dynamics	1.42	Development	1.31	Response	1.30	Methods	1.52	Applications	1.46
Performance	1.42	Applications	1.27	Analysis	1.29	Applications	1.47	Analysis	1.45
Models	1.40	Environment	1.27	Complex	1.28	Analysis	1.47	Networks	1.42

association of students with research that involve M&S. This is an important acknowledgement that projects involving M&S can have an impact on the training of future generations and aligns very well with the mission of NSF.

Next we examine the NIH dataset to see if we can isolate M&S related research in health sciences.

13.3.2   Properties of M&S-Based NIH Awards Between 1990 and 2012

Figure 13.2 displays the concept map for the NIH awards during the time span of 1990–2012. We describe the five main themes as follows:

1) *Research (100% Connectivity), Testing, Techniques, University, Tissue, Cancer, Treatment, Genetic:* Research, as expected within NIH awards, focuses primarily on health-related fields, such as tissue, cancer, treatments, and genetics. A primary concept within research focuses on training or testing. This theme is an indication that M&S plays a much smaller role in NIH than it does in NSF, which is not surprising.

2) *Protein(s) (92% Connectivity), Molecule(s), Receptor, Regulation, Inhibitors:* This theme illuminates that protein modeling is the primary application area for M&S that receives NIH funding. Modeling protein involves capturing the properties of protein molecules and examining receptors, inhibitors, and regulation throughout the body. The protein theme links to the theme of research through the theme of screening. The path from protein to simulation includes molecular structures and dynamics. This theme is an indication that M&S plays a very specialized role in NIH studies, namely, to help in the study of structured and computable objects as often found in computational biology for instance.

3) *Structures (91% Connectivity), Molecular, Interactions, Dynamics, Chemical, Mechanism:* This theme highlights aspects that for the structure of simulation design and construction. The concept of dynamics provides the bridging link from the structures theme to the simulation theme. Structural focus includes molecular interactions and mechanisms, chemical interactions, and the dynamic formation of molecule structures. This theme is an indication that M&S is used in support of training and testing studies where researchers are either teaching or know basic behaviors and are interested in complex emergent behaviors.

4) *Clinical (41% Connectivity), Patients, Program:* This theme connects to the theme of research through training, center, and grant. Clinical centers serve as locations that train physicians on treating patients. Additionally, clinical grants frequently involve blood and tissue research. This is an indication that research involving M&S has to be fully integrated with numerous aspects within the traditional way healthcare studies are conducted.

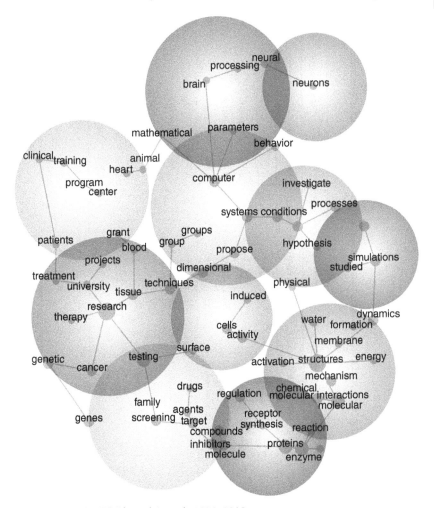

Figure 13.2 NIH M&S-based Awards 1990–2012.

5) *Computer (35% Connectivity), Systems, Propose*: Computer systems allow for the creation of mechanisms to explore dynamic behavior using simulation. This theme is similar to the system view of simulation found in the M&S NSF dataset, which is to say that one of the main focuses is on constructive simulations implemented using digital computer systems.

Table 13.3 provides the top 10 word-like concepts that commonly co-occur with the primary concept within each of the five themes listed above and their corresponding prevalence values within the NIH abstracts.

We observe that computer simulation is the fourth most prevalent grouping in the computer category, which is not surprising since healthcare

Table 13.3 Prevalence (Prev) of the top 10 word-like concepts that commonly co-occur with the main concept from the top five NIH 1990–2012 themes.

Research		Protein		Structures		Clinical		Computer	
Word-like concept	Prev	Word-like concept	Prev	Word-like concept	Prev	Word-like concept	Prev	Word-like concept	Prev
University	1.88	Proteins	2.32	Proteins	1.85	Patients	2.73	Mathematical	2.02
Training	1.81	Enzyme	2.21	Enzyme	1.84	Therapy	2.54	Program	1.79
Testing	1.79	Inhibitors	2.17	Protein	1.83	Treatment	2.52	Chemical	1.76
Screening	1.73	Reaction	2.15	Inhibitors	1.83	Blood	1.95	Simulations	1.72
Projects	1.64	Molecule	2.12	Reaction	1.81	Animal	1.84	Physical	1.62
Grant	1.55	Receptor	2.04	Molecule	1.81	Training	1.83	Behavior	1.56
Clinical	1.52	Chemical	1.99	Compounds	1.80	Cancer	1.81	Training	1.52
Center	1.51	Molecules	1.98	Molecules	1.77	Heart	1.78	Center	1.50
Molecule	1.50	Compounds	1.98	Synthesis	1.76	Program	1.69	Projects	1.48
Therapy	1.48	Interactions	1.97	Chemical	1.76	Grant	1.69	Parameters	1.47

simulations involve live people simulating patients, virtual simulations for training, and constructive simulations that solve complex methodical problems. The focus on training, testing, and screening within the concept of research contributes to the potential for empirical data collection that can be used to create domain-specific healthcare models and simulations.

With these datasets, we obtain the major characteristics of funded projects that involve M&S and have been funded historically by NIH and NSF. However, especially with NSF, we only have data up to 2003 that brings into question the incorporation of major technological and social changes since then. As a result, we examine the third dataset in order to determine what trends have emerged over the last 5 years.

13.3.3 Landscape of NSF Funding in the Past 5 Years

In this section, we seek to identify the current role of M&S within NSF projects over the past 5 years from September 2011 to September 2016. We use the NSF's online advanced award search feature to conduct a keyword search and obtain abstracts for M&S research (https://www.nsf.gov/awardsearch/). We configure our search using the keywords "modeling" and "simulation" with award dates starting on or after September 01, 2011. We extract the maximum allowable 3,000 awards in order to (1) apply content analysis to identify the primary types of M&S projects that are receiving NSF awards in the past 5 years using content analysis, (2) identify the distribution of NSF funding across the United States, and (3) examine the distribution of funding across various topics.

The information obtained from the NSF's award search included information such as the NSF Directorate, the type of award instrument, and names of the Principle Investigators in addition to the abstract. In order to gain insight into how the concepts within the NSF are associated with the various NSF Directorates we setup the content analysis within Leximancer to apply the NSF Directorate labels within the texts as tags within the analysis. Tagging the directorates results in mapping each concept against each of the directorates and produces the frequency, strength, and prominence of each concept with respect to the directorate. Figure 13.3 displays the concept map for the 3,000 NSF abstracts. The NSF Directorates are represented as nodes with names beginning with "NSFDirectorate_" within the figure.

The top five themes within the past 5 years of NSF awards that utilize M&S are research, models, develop, systems, and design. Not surprisingly, these themes suggest that from a general perspective M&S is utilized within NSF research for designing and developing models of systems. This observation is in-line with what we found within the historical dataset. In order to provide more granularity, we group the awards based on the awarding NSF

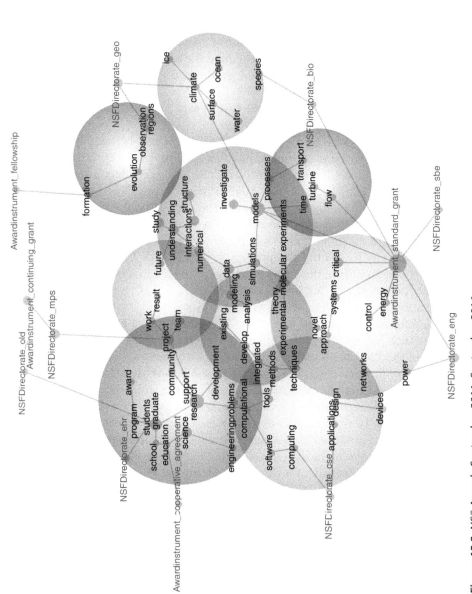

Figure 13.3 NSF Awards September 2011–September 2016.

Table 13.4 List of NSF Directorates and the Office of the Director.

Directorate	Focus	Short name
Directorate for Education and Human Resources	Science, technology, engineering and mathematics education at every level, pre-K to graduate	EHR
Directorate for Computer and Information Science Engineering	Fundamental computer science, computer and networking systems, and artificial intelligence	CSE
Directorate for Engineering	Bioengineering, environmental systems, civil and mechanical systems, chemical and transport systems, electrical and communications systems, and design and manufacturing	ENG
Directorate for Social, Behavioral, and Economic Sciences	Neuroscience, management, psychology, sociology, anthropology, linguistics, science of science policy and economics	SBE
Directorate for Biological Sciences	Molecular, cellular, and organismal biology, environmental science	BIO
Directorate for Geosciences	Geological, atmospheric and ocean sciences	GEO
Directorate for Mathematical and Physical Sciences	Mathematics, astronomy, physics, chemistry and materials science	MPS
Office of the Director	General Counsel; Integrative Activities; International Science and Engineering; Legislative and Public Affairs; and Diversity and Inclusion	O/D

directorate and investigate which group is most associated with a theme. Table 13.4 shows a summary of NSF directorates and their focus areas. The results of our analysis show the following associations:

1) *Research (100% Connectivity), Project, Graduate, Students, Computational, Education, Science, Development, Engineering, Work, School, Program, Problems, Community, Award, Support*: Graduate schools and graduate students play a pivotal role in NSF research grants in the past 5 years. Graduate students support research but can also be the object of research, that is, how to improve teaching through M&S. This theme is most closely linked to the Directorate for Education and Human Resources (EHR) and is most commonly associated with science, technology, engineering, and mathematics (STEM) fields.

2) *Models (64% Connectivity), Simulations, Data, Processes, Understanding, Study, Dynamics, Numerical, Experiments, Interactions, Time, Structure, Investigate*: Models are used for experimentation, study, investigation, and to gain understanding into the structure, dynamics, and processes of the systems involved with the research problem. The models theme connects to research through the themes of develop and results. Models are most commonly associated with the Directorate for Computer and Information Science Engineering (CSE). This is perhaps the closest directory to M&S as a discipline. The main NSF award instrument associated with models is continuing grants (1.696 prominence) followed closely by cooperative agreements (1.570 prominence).

3) *Develop (36% Connectivity), Modeling, Methods, Analysis, Experimental, Integrated, Molecular, Theory, Existing*: This theme involves developing methods for modeling and methods for analysis that allow for theory to be constructed in such a way that it can be simulated and efficiently analyzed. This theme serves as a bridge that connects the concepts of models and simulations to the theme of research. The theme is not strongly connected to any directorate.

4) *Systems (24% Connectivity), Energy, Approach, Control, Critical, Novel, Networks, Power*: This theme highlights the types of systems that are researched through the application of models and simulations, including energy, power, networks, and control systems. NSF looks for novel approaches toward studying these types of systems. Awards for systems research commonly come for the Directorate for Engineering (ENG) or the Directorate for Social, Behavioral, and Economic Sciences (SBE).

5) *Design (20% Connectivity), Tools, Applications, Software, Techniques, Computing, Devices*: This theme covers a range of topics including the design of tools, devices, and applications for modeling or analyzing the research problem as well as the development of the computing software used to model and simulate the research problem. The Directorate for Computer and Information Science Engineering (CSE) is commonly associated with awards that focus on design.

Next, we examine the distribution of awards and funding throughout the United States during these 5 years, including the District of Columbia. The total funding amount from these 3,000 awards is 1.35 billion dollars. California has the most awards (346 or 11.5%) and the most funding (163 million or 12.0%) followed by New York with 207 awards (6.9%) and 103 million dollars (7.6%) of funding. Examining the distribution of funding by region (Midwest – 11 states, Northeast – 10 states, South – 17 states, and West – 13 states) reveals a close split in funding across regions. Comprised of the most states, the South received $382 million or 28% of the M&S-based research funding while the Western region received $365 million or 27% of the funding. The

Northeast on the other hand received $339 million or 25% of funding, while the Midwest was awarded $267 million or 20% of the funding.

Table 13.5 provides the number of awards per state, the amount of total funding per state, the region that the state is classified under, and the total number of awards and funding.

Table 13.5 Distribution of the number of NSF awards and total funding by state and region.

State	Region[a]	Number of awards	Total funding	State	Region[a]	Number of awards	Total funding
AK	W	17	$4,089,807	NC	S	95	$56,765,230
AL	S	20	$5,368,347	ND	M	5	$1,584,779
AR	S	16	$16,794,998	NE	M	12	$5,053,723
AZ	W	68	$36,253,549	NH	N	25	$7,606,882
CA	W	346	$163,206,098	NJ	N	82	$30,133,866
CO	W	112	$50,674,650	NM	W	25	$16,384,659
CT	N	41	$15,585,325	NV	W	14	$6,679,091
DE	S	30	$11,216,627	NY	N	207	$103,816,473
FL	S	84	$30,637,090	OH	M	94	$29,823,964
GA	S	90	$37,163,663	OK	S	20	$7,992,032
HI	W	15	$5,395,804	OR	W	48	$19,804,364
IA	M	36	$11,514,868	PA	N	154	$60,907,192
ID	W	9	$6,364,881	RI	N	26	$9,926,759
IL	M	135	$75,334,392	SC	S	32	$9,744,839
IN	M	80	$46,687,198	SD	M	5	$1,440,461
KS	M	21	$8,564,867	TN	S	51	$22,870,378
KY	S	10	$4,813,078	TX	S	181	$76,174,476
LA	S	26	$23,356,789	UT	W	37	$18,353,525
MA	N	176	$85,131,262	VA		82	$37,319,963
MD	S	72	$31,504,830	VT	S	4	$2,502,311
ME	N	13	$4,852,483	WA	W	82	$32,435,911
MI	M	123	$53,365,267	WI	N	54	$18,722,693
MN	M	43	$25,108,440	WV	S	2	$533,448
MO	M	34	$9,313,502	WY	W	9	$2,853,785
MS	S	6	$1,280,868	DC	S	24	$8,921,404
MT	W	7	$2,637,485	Total		3,000	$1,354,568,376

a) Regions are Midwest (M), Northeast (N), South (S), and West (W).

Table 13.6 Distribution and amount of M&S-based NSF funding by category.

Category	NSF directorate abbreviation	Number of awards	Amount of funding	Percentage of funding
Biological Sciences	BIO	148	$78,407,080	5.79
Computer and Information Science and Engineering	CSE	507	$277,601,175	20.49
Education and Human Resources	EHR	69	$67,709,671	5.00
Engineering	ENG	804	$330,072,485	24.37
Geosciences	GEO	652	$271,382,960	20.03
Mathematical and Physical Sciences	MPS	725	$255,943,164	18.89
Social, Behavioral and Economic Sciences	SBE	70	$32,801,423	2.42
Office of the Director	O/D	25	$40,650,418	3.00

We then analyze the distribution of award funding by scientific category using the seven NSF Directorates and the NSF's Office of the Director. Table 13.6 provides the total number of awards, the cumulative amount of funding, and the percentage of funding for each category from the 3,000 awards.

We observe that M&S is used across all directorates that is encouraging for researchers because it means that M&S experts can participate in any multidisciplinary team. As we expect sciences, computing, engineering, and mathematics account for over 80% of the funded projects, which is not surprising. However, we hope to see a rise in funding in SBE and EHR to reflect to emergence of computational social science and complex adaptive system studies in Systems Engineering.

13.4 Conclusion

As the M&S community continues to promote M&S as its own discipline, evidence shows that the moment it is a versatile and ubiquitous method in NSF and NIH. Our focus on NSF shows that projects using M&S are funded within every directorate. However, M&S remains mainly favored by

Table 13.7 Checklist for what to do when applying for NSF and NIH grants while utilizing M&S.

Item	NSF	NIH
Form a multidisciplinary team	✓	✓
Frame the M&S endeavor as a tool for engineering a solution?	✓	
Utilize M&S to handle computational components of the study?		✓
Focus on constructive simulations built on digital computers	✓	
Focus on all types of simulation including Live Virtual and Constructive		✓
Describe simulations as systems	✓	✓
Utilize M&S to support experimentation?	✓	
Utilize M&S to support training and testing?		✓
Target the Directorate that is the closest match to the problem domain?	✓	✓
Utilize graduate programs and students within the study?	✓	✓

engineers and computational scientists. Table 13.7 shows a checklist derived from the content analysis performed on the datasets.

Most of these findings are not surprising to the experienced M&S practitioner but we see slight differences between the NSF and NIH funded projects. The main take away is that for M&S researchers, funding for their research exists as long as it is within an application domain. As a result, it is incumbent on M&S researchers to advance their field while grounding their work in practical areas.

References

Balci, O., Ormsby, W.F., Carr J.T. III, and Saadi, S.D. (2000) Planning for verification, validation, and accreditation of modeling and simulation applications. *Proceedings of the 32nd Conference on Winter Simulation*, pp. 829–839.

Balci, O., Nance, R.E., Arthur, J.D., and Ormsby, W.F. (2002) Improving the model development process: expanding our horizons in verification, validation, and accreditation research and practice. *Proceedings of the 34th Conference on Winter Simulation: Exploring New Frontiers*, pp. 653–663.

Berelson, B. (1952) *Content Analysis in Communication Research*, Free Press.

Brailsford, S., Harper, P., Patel, B., and Pitt, M. (2009) An analysis of the academic literature on simulation and modelling in health care. *Journal of Simulation*, 3, 130–140.

Cretchley, J., Rooney, D., and Gallois, C. (2010) Mapping a 40-year history with Leximancer: Themes and concepts in the Journal of Cross-Cultural Psychology. *Journal of Cross-Cultural Psychology*, 41, 318–328.

Crofts, K. and Bisman, J. (2010) Interrogating accountability: an illustration of the use of Leximancer software for qualitative data analysis. *Qualitative Research in Accounting & Management*, 7, 180–207.

Diallo, S.Y., Padilla, J.J., Papelis, Y., Gore, R., and Lynch, C.J. (2015a) Content analysis to classify and compare live, virtual, constructive simulations and system of systems. *The Journal of Defense Modeling and Simulation: Applications, Methodology, Technology*, 13, 1548512915621972.

Diallo, S.Y., Gore, R.J., Padilla, J.J., and Lynch, C.J. (2015b) An overview of modeling and simulation using content analysis. *Scientometrics*, 103, 977–1002.

Diallo, S.Y. and Lynch, C.J. (2016) A Roadmap for Building a Digital Patient System. *The Digital Patient: Advancing Healthcare, Research, and Education*, 207–224.

Gore, R., Diallo, S., and Padilla, J. (2016) Classifying modeling and simulation as a scientific discipline. *Scientometrics*, V 109, 615–628.

Hillegas, A., Backschies, J., Donley, M., Duncan, R.C., and Edgar, W. (2001) The Use of Modeling & Simulation (M&S) Tools in Acquisition Program Offices: Results of a Survey, DTIC Document 2001.

Kassarjian, H.H. (1977) Content analysis in consumer research. *Journal of consumer research*, 4, 8–18.

Katsaliaki, K. and Mustafee, N. (2011) Applications of simulation within the healthcare context. *Journal of the Operational Research Society*, 62, 1431–1451.

Leximancer (2011) Leximancer Manual: Version 4.

Lonner, W.J., Smith, P.B., van de Vijver, F.J., and Murdock, E. (2010) Entering our fifth decade: An analysis of the influence of the Journal of Cross-Cultural Psychology during its first forty years of publication. *Journal of Cross-Cultural Psychology*, 41, 301–317.

Mustafee, N., Katsaliaki, K., Fishwick, P., and Williams, M.D. (2012) SCS: 60 years and counting! A time to reflect on the Society's scholarly contribution to M&S from the turn of the millennium. *Simulation*, 88, 0037549712456448.

Mustafee, N., Katsaliaki, K., and Fishwick, P. (2014) Exploring the modelling and simulation knowledge base through journal co-citation analysis. *Scientometrics*, 98, 2145–2159.

NIH (2003) National Instititute of Health Research Awards 1990–2012 via Exporter, ed.

Pazzani, M. and Meyers, A. (2003) NSF Research Award Abstracts 1990–2003 Data Set, ed.

Poser, C., Guenther, E., and Orlitzky, M. (2012) Shades of green: using computer-aided qualitative data analysis to explore different aspects of corporate environmental performance. *Journal of Management Control*, 22, 413–450.

Smith, A.E. and Humphreys, M.S. (2006) Evaluation of unsupervised semantic mapping of natural language with Leximancer concept mapping. *Behavior Research Methods*, 38, 262–279.

Stemler, S. (2001) An overview of content analysis. *Practical Assessment, Research & Evaluation*, 7, 137–146.

Tse, D.K., Belk, R.W., and Zhou, N. (1989) Becoming a consumer society: a longitudinal and cross-cultural content analysis of print ads from Hong Kong, the People's Republic of China, and Taiwan. *Journal of Consumer Research*, 15, 457–472.

US GA Office (1996) *Content Analysis: A Methodology for Structuring and Analyzing Written Material: Program Evaluation Methodology Division*, BiblioGov.

Weber, R.P. (1990) *Basic Content Analysis*, Sage.

14

Why Spend One More Dollar for M&S? Observations on the Return of Investment

Steven Gordon,[1] Ivar Oswalt,[2] and Tim Cooley[3]

[1]*Georgia Tech Research Institute, Orlando, FL, USA*
[2]*Simulation U Analytics, LLC, Fredericksburg, VA, USA*
[3]*DynamX Consulting, Castle Rock, CO, USA*

14.1 Introduction and Overview

New or expanding businesses formulate a strategy that describes how their products or services will satisfy the needs of consumers, followed by development of a pertinent business plan that includes financial data, such as return on investment (ROI). The strategy and business plan form a basis from which to evaluate if the new product or service will be a good investment. Similarly, some initiatives in the government – especially in the Department of Defense (DoD) – like computer-based modeling and simulation (M&S), often require analyses and justifications much like ROI, or at least a comparison of costs and benefits. Models are physical or mathematical representations of an object or system, and simulations execute the models over time or through events. Simulations can portray future environments and can realistically immerse personnel in current or planned technologies or systems to evaluate how they are likely to perform. Simulations are also used in support of product development, acquisition, and manufacturing programs; to provide synthetic representations for training; and as an experimental environment for analysis and assessments. In the early 1990s, M&S became increasingly useful to the DoD, other government organizations, industry, and academia. Simulation could be used to help answer very complex combat or disaster-relief questions relative to current or future conditions and environments.

However, in the early 1990s, the art and science of simulation was not as mature as it is today. Simulation use at that time was helpful but expensive and problematic. Later that decade, new and improved M&S programs were

The Profession of Modeling and Simulation: Discipline, Ethics, Education, Vocation, Societies, and Economics, First Edition. Andreas Tolk and Tuncer Ören.
© 2017 John Wiley & Sons, Inc. Published 2017 by John Wiley & Sons, Inc.

being designed to answer questions more adequately, but DoD budgets were facing significant shortfalls. As a result, senior leaders in DoD and leaders in industry asked "Why spend one more dollar for M&S?" before any funding was approved. Military and business leaders in many other countries were also asking the same question in terms of their currency. When your boss is one of the leaders asking this question, finding a good answer is especially important.

This chapter discusses the journey through alternative approaches to answering that question. Those answers came in several forms, but initially none of them were quite right for our bosses. Then, through serendipity at an M&S conference, we met a gentleman, Mr. Bill Waite, who was wearing dollar bills clipped to his suit coat, handing the dollars out one by one, and asking us to attend a conference session on "Why Spend One More Dollar on M&S?" (Waite *et al.*, 2005). For the next 25 years, he helped the M&S community in formulating rigorous and yet creative ways to answer that question through cost-benefit assessments, ROI analysis, and cost avoidance justifications. At one point in those 2.5 decades, Bill spearheaded a ground-breaking 2008 technical report, "Metrics for Modeling and Simulation (M&S) Investments" for DoD's M&S Coordination Office, from which several insights and examples will be discussed in this chapter (AEgis, 2008). Along with these insights and examples, we will define key terms and concepts, describe historical examples, characterize dimensions of M&S value, provide example metrics, and discuss measuring the utility of M&S-associated areas, like standards.

14.2 The Need to Know M&S Costs and Benefits

Government officials, particularly DoD leaders, want to and should make informed decisions about current and future investments under their authority. There is significant motivation for them to plan future investments to achieve the best economic and practical advantages possible, both within their own operations and across the DoD. These decision-makers need metrics to enable them to understand the level and degree of success of any particular investment over time in order to be able to deploy the most cost-effective systems and infrastructure possible.

M&S in the 1990s was increasingly being used to analyze future force structures and deployments of forces around the world, to develop and acquire new weapon systems, and to train warfighters to use their weapon systems, often in large command and control (C2) exercises. These exercises trained leadership and staff at many military command levels, sometimes exercised subordinate elements, and prepared the commanders for the most likely conflict scenarios.

However, the overall quality of M&S at the time was fairly low. Sometimes the force employment simulations had flaws that appeared to favor one military service or one type of weapon system, so there were calls to develop new simulations that embodied fair play and fair fights. Too often the simulations used to portray new weapon systems did not adequately reflect the benefits of all weapons fairly, so there was considerable support for new simulations to better support analysis of alternatives. In the early days, even with first-generation computer-driven M&S, large C2 exercises often required the deployment of thousands of personnel and their weapons systems, plus deployment of a significant number of controllers and role players to manage the training scenario. In the 1990s, simulation to support these C2 exercises matured and reduced the need for deployment and redeployment, limited the wear and tear on people and weapon systems, and avoided damage to terrain and structures. However, these training simulations often did not adequately reflect or reward the use of correct warfighting procedures and normally did not include key features (like weather) that significantly affect warfighting, disaster relief, and similar operations.

Consequently, programs to build new simulations for force structure analyses, acquisition of new weapon systems, and C2 training were initiated. Because these areas of force structure planning, developing new weapon systems, and training warfighters were so different, at least three new simulations were needed to correct the deficiencies. These simulation development programs were very expensive, and in lean budget years, spending on M&S competed with funding for weapons systems, spare parts, and other immediate warfighting needs. Leadership was also skeptical about how good the new simulations really would be. So at that time, spending on M&S, and any additional dollar for M&S, had to be justified in terms of benefit(s) that outweighed the costs and/or outweighed the benefits of alternative uses for that funding – for every dollar. In all areas where simulation proved to be useful and grew in importance, the value of simulation relative to cost had to be justified.

In part to help meet this need, the "Metrics for Modeling and Simulation (M&S) Investments" study (AEgis, 2008) was sponsored by the DoD Modeling and Simulation Coordination Office in fiscal year 2007. When this study began, no effective processes had been established to generate and describe metrics like ROI or cost avoidance for DoD M&S. Metric categories and levels had not been defined, their relationships were not described, and methods to combine relevant qualitative metrics with those that are naturally numeric or quantitative had not been explored. However, there were strong indicators that M&S had value and that its contributions were significant and important.

14.3 A First Attempt to Answer Why We Should Fund M&S

It is difficult to prove that M&S is a good investment if you do not know how to begin the analysis. How can we prove M&S is a good investment? The first attempt to answer the question, "Why should we spend one more dollar for M&S?" was to divide the areas that M&S supports – where simulation has unique purposes and distinct value propositions – and concentrate first on the training segment. At the time, documentation of the value (or benefits) of training were plentiful and the training directives and doctrine from the DoD Joint Staff and the Military Services were great initial sources for authoritative statements on why training is vital. Justifying *how* M&S supports training would be the next step. Reports from Operation Desert Storm (a military operation that began January 17, 1991) highlighted the importance of training and helped connect M&S used for training to combat readiness.

14.3.1 The Value of Training to Readiness

In portraying the benefits of training, it is important to remember that history shows that preparing military forces for combat is clearly important:

> "To be prepared for war is one the most effective means of preserving peace."
>
> –President George Washington
> (Washington, 1790)

In the 1990s, DoD publications and doctrine stated that military forces fight in the same way as they have trained and that combat performance is determined by training performance. These are logical statements that are difficult to argue with, but how does M&S support training? Does M&S contribute to readiness? At that time, simulation was growing in use. Simulation could add realism by replacing exercise controllers that gave written or verbal prompts and responses to the training audience; M&S was being linked to or integrated with the C2 systems that the trainees would use to send directives and report intelligence in combat. Simulation was becoming good enough to provide more realistic training, allowing training audiences to be immersed in an environment closer to the reality of war.

Distributed simulation environments used for training in the early 1990s often allowed forces to train at or near home station vice deploying to other locations, and this ability to distribute M&S reduced personnel travel (improving their quality of life) and improved weapons' longevity (by reducing wear and tear).

14.3.2 Feedback and Results from Desert Storm

The Battle of 73 Easting (a map location in southern Iraq) began on February 26, 1991 (IDA, 1991). This was the first battle in the war where Iraqi forces did not withdraw after they were initially attacked. The Iraqi Republican Guard in this battle had significant prior combat experience, and they attacked from entrenched positions, fought gallantly, and counterattacked repeatedly (Gorman and McMaster, 1992). This battle was homeland defense for the Iraqi forces who were protecting the important port city of Basra and their fellow Iraqi forces retreating from Kuwait (Clancy, 1997). The United States forces had several disadvantages: no previous combat experience, fighting in a blinding sand storm, and no air cover. However, the United States forces also had advantages: M1A1 tanks with thermal sites, dependable manual loaders for their munitions, greater weapons ranges, and crews that had participated in intense readiness training before deployment into combat. Within minutes, 28 Iraqi tanks, 16 personnel carriers, and 39 trucks were destroyed, and 1300 Iraqi prisoners were taken with no American losses (Clancy, 1997). When commanders were asked how their forces had fought so well in their first combat engagement, they answered that *they had already been in combat* – on the training ranges and in the simulation training environments at home station. In some cases, the soldiers volunteered that the training they experienced prior to the battle was more challenging than the battle itself (however, it was clearly less dangerous). A Congressional Committee, hearing testimony from commanders and soldiers about the Battle of 73 Easting, was told that simulation-based training was one of the benefits that enabled our forces to prevail and that simulation is fundamental to readiness for war (Gorman and McMaster, 1992).

14.3.3 Narrative Descriptions of M&S Value

M&S has value and provides benefits to many areas of application (also known as communities, domains, or use-cases) (Gordon, 1995). Along with the value of simulation to training, M&S provides tailored synthetic environments for mission rehearsal and mission preview, experimentation, wargaming, assessment of forces, combat operation course-of-action analysis, acquisition, testing and evaluation, and logistics. Dividing up M&S use by its purpose is only one of the many ways of differentiating the DoD domains, yet M&S has a role within them all.

In training and rehearsal simulations, M&S advantages include mistakes are not deadly, crew errors can be corrected with repeated run-throughs of problematic events, we can simulate many possible environmental conditions, and the scenarios can be practiced covertly to provide information assurance. Experimentation puts future weapons and processes into realistic scenarios a decade or more in the future, and simulation is ideal to portray those futures in terms of several alternatives in a distributed, multisite environment. M&S can provide wargamers with tailorable injects that portray alternative futures – perhaps decades in the future, and simulation allows the wargame sponsors to simulate multiple scenarios and vignettes without the burden of retasking the historically large numbers of role players and control staffs. For assessment of forces and operational decision support, M&S allows the users to investigate many courses of action for potential enemy and coalition forces in a secure environment where our intentions are not openly advertised – as they would be on ranges or in live field exercises. Analysis-based collateral damage tools were improved with M&S technologies in the 1990s, and these tools helped warfighters in the second war in Iraq get faster approval for targeting while saving noncombatants from injury or death (Martin and Gordon, 2004; Yaukey, 2003). In acquisition, initial prototypes and follow-on versions of weapons systems are designed, developed, tested, and assessed in M&S. M&S also augments test and evaluation (T&E) events and provides an arena for analyzing the system in any combat zone around the world: at any time and in many environmental conditions. In the industries that provide products and services for consumers, M&S is used in product development, manufacturing design, and safety and effectiveness testing. In all these use areas, M&S and model-based systems engineering (MBSE) matures as the weapon system's concept matures. Any one of these benefits, in each of these areas, is significant. However, we discovered that no matter how well we described those benefits and other results of M&S as narratives, we were still asked by our bosses what the ROI was for any particular M&S investment.

14.4 Initial Numeric Statements of M&S Value/Return

Early data on the return (or utility) of M&S use were often numeric but anecdotal. Value estimates were generated from past experiences and judgments about M&S, relative to how systems were acquired or forces trained before computer-based M&S was used (Waite *et al.*, 2005). The reason for having to rely on these judgments was that neither rigorous nor long-term studies using comprehensive quantitative methods had been planned or funded to answer these questions. However, as M&S technology and capabilities improved, these numeric assessments helped define key dimensions of value and cost, and began to generate the data needed for a more complete answer to why spend one more dollar on M&S.

14.4.1 Benefits of M&S Use in Product Development, Acquisition, and Manufacturing

Here are a few of the anecdotal examples of the contribution of simulation to acquisition in 1998. Some analysts estimated that 3–15% of acquisition program budgets were being spent on M&S, and, with over 200 defense programs at the time between \$1 M and \$50B each in funding size, the potential for M&S impact was high (O'Bryon, 1999). It was estimated that, through the intelligent use of M&S, a reduction in total acquisition time of 50% on average could be realized (Joint Simulation Based Acquisition Task Force, 1998). By using simulation early in a program and throughout a program's life, M&S could potentially reduce system life cycle costs by 2% (easily a reduction in the \$Bs); and for small programs M&S has an estimated ROI of 25 (2500%) (National Academy of Sciences, 1997).

More impressive examples came from industry, where they showed the results of changes driven by M&S. For example, the Boeing Corporation invested heavily in M&S for the design and development of the 777 aircraft (Gordon, 2001). A comparison of the 777 with the earlier 747 aircraft design and development (when M&S was used minimally) showed that the 777 required 1% of the number of shims for ill-fitting panels needed for the 747, 70% of the scrapped material, 10% of the reworked parts, and much less factory floor space (Joint Simulation Based Acquisition Task Force, 1998). In the case of the F-15, M&S was used to plan the manufacturing process. Compared to earlier variants of the platform where M&S was not used for this purpose, the M&S-supported process led to a 33% reduction in design release time, a 27% reduction in design cost, a 19% reduction in manufacturing cycle time, a 20% reduction in factory floor space, a 24% reduction in parts count, and a 78% reduction in fasteners required for assembly (Joint Simulation Based Acquisition Task Force, 1998). These metrics reflect positive change, much of which may be attributable to M&S, but they do not link the improvements with the resources invested, nor do they allow for trade-offs between result categories (benefit, cost, risk, and time), and finally they do not associate direct results with mission impact. As we will see, this was true also for readiness changes due to M&S use in training.

14.4.2 Data Compilation for Readiness Training

In the early 1980s, a training event's control staff manually scripted the entire exercise and developed printed input cards to provide scenario inputs to the training audience, and the training audience generally deployed to common coalition locations. The scripting meetings took many weeks – spread out over many months of planning. The ratio of training audience (warfighters) to control staff was 1-to-1, and every engagement was decided based on dice rolls, completion of required training events, and manual input cards. (The real enemy did not plan on handing out input cards for our warfighters to read.) In 1980, an exercise called POSITIVE LEAP required 2800 controllers and trained 2800 warfighters. In 1988, REFORGER trained the warfighters in 35 maneuver headquarters supported by a sizable exercise control force, and the opposition forces (OPFOR) consisted of many combat aircraft, 97,000 soldiers, 7,000 tracked vehicles, and 1,080 tanks (Gordon, 2001). The OPFOR received minimal training and mostly played roles to make the scenario more realistic for the maneuver headquarters' staffs that were the primary training audience. The control staff was not being trained, and even worse, they often had to insert inputs manually – which decreased the training's realism. Clearly, costs could be avoided if the use of simulation could reduce the need for the control staff and replicate (replace the need for) the live OPFOR. The cost for REFORGER 1988 was $53.9 M in direct costs and $20 M in maneuver damage to terrain and structures (Gordon, 2001). Using more robust M&S support, REFORGER 1992 trained 35 maneuver headquarters and required 20,000 soldiers and 135 tracked vehicles as training aids, costing only $19.5 M with no maneuver damage. In this case, the use of distributed simulation to drive the events and reduce the need for the maneuver OPFOR provided a cost reduction for REFORGER 1992 versus 1988 of $54.4 M while training the same number of maneuver headquarters staffs (Gordon, 2001). If we knew the cost of the M&S laydown, we could calculate a form of the ROI; however, it was clear that the use of M&S avoided costs.

14.5 M&S Return on Investment Development

Businesses survive (or expand) because their products and services are sought by customers and the pricing structure for these offerings is sufficient (the ROI is positive) to permit the business to go on. Although DoD does not manage to a bottom-line, its leaders and key stakeholders are charged with spending taxpayer funds wisely. To do this, they need a methodology to evaluate investment opportunities (Brown *et al.*, 2000).

Leaders, particularly in the military services, wanted to see an indicator of positive return on the funding they were spending, and those voices were heard. Thus, the business construct of ROI was extended to DoD M&S.

The initiative, "Why should we spend one more dollar for M&S?" began to mature with working groups at several international conferences such as the Simulation Interoperability Standards Organization (SISO) Workshops, the Society for Computer Simulation (SCS) Conferences, and the Interservice/Industry Training, Simulation and Education Conference (I/ITSEC) (Waite *et al.*, 2005). We will briefly introduce the ROI equation, and then discuss efforts/projects that applied it to M&S, as well as insights and extensions, before concluding with a discussion of drawbacks, benefits, and possible future initiatives.

14.5.1 Introduction to the Calculation of ROI

ROI is a very simple and powerful metric for new, improved, and ongoing products, services, and businesses. The calculation can be stated in words as the ratio of the difference between return and cost with this difference being divided (normalized) by the cost. In most cases, the return is the added revenue realized (or estimated to be realized) by a change, while the cost is the cost of making the change:

$$\text{ROI} = \frac{\text{return-investment}}{\text{investement}} = \frac{\text{return-cost}}{\text{cost}} = \frac{\text{revenue-expenditure}}{\text{expenditure}}$$
$$= \frac{\text{benefit-cost}}{\text{cost}}$$

In the formula for ROI above, the difference between revenue and expenditure can be considered profit or loss for the particular area being investigated, and the difference term in the numerator is divided by the cost. So, if the ROI for a new product over a specific period of time or for a specific volume of sales is 0.20, then the revenue was 20% more than the cost of providing the product to customers. A business will not stay in business unless the overall ROI is positive enough over the long haul. The calculation of ROI is the same whether the situation is one of starting a new business, selling a new product, adding a barber and chair to a barbershop, adding patio service to a restaurant, or generating more business by training the staff in order to deliver different or better service. It is critical that the return and the cost be directly related (attributed) to the change that is being considered; for instance, one would not put the total cost of running a restaurant into the same ROI calculation with just the new revenue from the

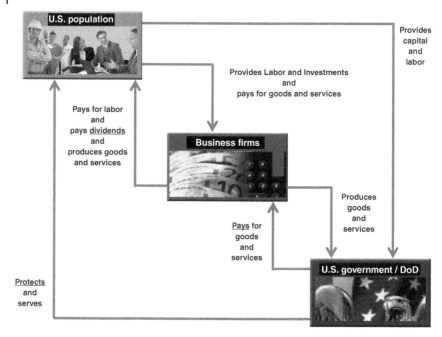

Figure 14.1 Macroeconomics view of population, business, and government.

remodeled patio dining area. Only the costs attributed to the remodeled area should be used in this specific calculation.

While ROI is widely used in business, some modifications to the calculation and meaning of ROI need to be made when looking at government, particularly Department of Defense (DoD), expenditures. In government, profit and, to a lesser extent, investment may need to be renamed – or at least thought of differently. Government does not follow the same practices as commercial business; in government there is a lack of profit incentive, government is not in a competition for a distribution of customers that purchase services or products and realize value in return, and the success of government is not measured primarily in terms of dollars (i.e., revenue stream). The depiction in Figure 14.1 seeks to illustrate key transactions and differences between how the government, business, and the population, in general, interact (adapted from AEgis (2008), p. 22).

It is important to point out that the tangible benefits for providing/ producing from a business or population point of view (e.g., pay and dividends) can be measured using standard business ROI techniques; however, the returns from government activities (e.g., protect) are not exclusively (nor often primarily) monetary and many times are not naturally quantitative (e.g., the value of life, liberty, and the pursuit of happiness). So, in the case of DoD or government, ROI calculations become multi-attribute in nature.

14.5.2 Alternative Definitions for ROI in Government/DoD

In government, and particularly in DoD, expenditures are not investments that return a profit. In fact, proclaiming a return – or savings – from a decision would likely be noticed by a comptroller that has many extremely underfunded programs and who would gladly redirect the proclaimer's saved funding! In addition, it is highly unlikely that any office is fully funded, period. Often, all of the critical needs are not resourced completely, and judgment is needed to spend the available dollars in the most beneficial way to meet the mission. Finally, and perhaps most importantly, 'return' in ROI for the government, and particularly in DOD, is not solely monetary or naturally quantitative. It includes force readiness, mission accomplishment, system acquisition within an acceptable risk, and that platforms are delivered in time (given the threat), for instance. Thus, the form of the equation for ROI presented in Section 14.5.1 is the one on the second line, which casts ROI in terms of benefit and cost, where both can include results, expenditures, risk, and time.

14.5.3 Advances in Calculating ROI for M&S

There have been a number of projects that have advanced the state of the art in the calculation of the ROI of M&S. Some of them were very system specific, some took a more comprehensive and methodological point of view, while others concentrated on one particular value dimension (results, cost, risk, or time), and some focused on M&S within a particular domain or application area (training, acquisition, etc.). What follows is a selection of these efforts that we hope will provide a satisfactory summary of many of the results to date.

14.5.3.1 Development of a More Comprehensive ROI Methodology

One of the initial attempts to calculate a comprehensive set of M&S ROI metrics was sponsored by the DoD M&S Coordination Office (MSCO) and conducted by a team led by AEgis Technology Group. The resulting 314-page report was generated in 2008 (*Metrics for Modeling and Simulation Investments*), and a summary was subsequently composed for the *Acquisition Research Journal* in 2011 (*Calculating Return on Investment for U.S. Department of Defense Modeling and Simulation*). Overall, this effort developed an approach for comparing different M&S

investment opportunities using an ROI-like measure. It describes methods to evaluate "benefit" (i.e., increased readiness, more effective training, etc.) received from an investment by using metrics in a decision analysis framework to provide a consistent input to the ROI-like calculation. Finally, it concludes by discussing the importance of viewing M&S investments from a DoD Enterprise point of view, evaluating investment over multiple years, measuring well-structured metrics, and using those metrics in a systematic way to produce an ROI-like result that can be used to evaluate and prioritize M&S investments.

14.5.3.2 Branching Out to Other Fields: Medical Simulation ROI

Along with the development of the ROI methodology, the approach has been tailored to specific user communities. For instance, a small business's innovative research (SBIR) effort (OSD11-H19) managed by the Office of Naval Research (ONR) in support of the Department of Health Services (DHS) produced Medical M&S-Based Training ROI Decision Models. One project, MedRISCalc, focused on using the methodology developed previously and applying/modifying it for the medical community (Severinghaus *et al.*, 2012; AEgis, 2008). The report detailed how to construct a systematic process for assessing costs and benefits of medical simulation systems and then formulated a decision-making process to support medical M&S investment decisions. The report refines the 2008 methodology and succinctly states some of the requirements as shown in Table 14.1. Additionally, the team developed a case study that described two different M&S cardiopulmonary resuscitation (CPR) training options and the analysis process to determine the ROI of each in order to make the investment decision.

Table 14.1 *M&S ROI key assessment considerations.*

M&S is critical, ROI assessment is key, and it can be calculated if one:
• Considers the stakeholder's point of view
• Considers every factor within a rigorous overall, system-of-systems methodology
• Scopes the analysis in terms of enterprise, application, program, system, or other
• Designs and includes metrics to fit the analysis at hand
• Scales the approach to the data that is or can be made available
• Clearly distinguishes between terms like metric, scale, range, assigned value, and so on
• Includes results, cost, time, and risk measures
• Takes account of non-cost, qualitative values in a quantitative way
• Includes life cycle cost (perhaps projected) in a net present value (NPV) calculation
• Can include an M&S system's direct, derived, and deduced impact
• Deducts or subtracts the influence or contribution of coincident activities
• Accounts for correlation and avoids double counting
• Employs a robust computational approach and includes sensitivity analysis

Another output from the SBIR effort was the development of an ROI-based decision support tool for medical M&S training designers focused on helping them to make decisions based on trade-offs between individual component fidelity, performance, and cost. The overall goal was to improve the effectiveness and efficiency of health service delivery (Cohn et al., 2013).

The medical application of ROI may have some differences from the previously described usage within DoD. For instance, in this case, there is a potential for a revenue stream (i.e., some hospital systems are for profit). However, to understand the complexities of training systems that ultimately impact patient safety and quality of care, ROI calculation is necessary. This medical case study shows that each situation is unique and requires a tailored analysis depending on the evaluation of what is required. The MedRISCalc process and methodology were presented at the 2013 International Meeting of Simulation in Healthcare, the presentation was well-received, although medical institutions had not yet become widely concerned about calculating ROI.

14.5.3.3 Additional ROI Efforts for DoD

Between 2006 and 2016. there were other efforts that attempted to answer the question of the value of M&S. In particular, a tutorial titled Return on Investment (ROI) for Modeling and Simulation was developed and presented a number of times at I/ITSEC (Degnan, 2010). It noted that ROI is a performance measure used to evaluate the efficacy of an investment and/or to compare the efficacy of several competing investments. In DoD, knowing the level of positive ROI facilitates the decisions made by leadership, provides life cycle justification for maintenance and upgrades, and provides a baseline from which to measure alternatives and follow-ons. This study stated that the impact of M&S is difficult to quantify because acceptable effectiveness metrics are lacking, the supporting data from simulation use are difficult to discover and gather, and, in some cases, it is not obvious from what baseline the simulation benefits can be measured. There is a danger in tracking the current training system with periodic ROI calculations because it could obscure a more important perspective of what is the best of all possible courses of action (for this type of training) that could be put in place. Strict ROI may look at numbers for costs and benefit, but there is another part of the benefits story. For instance, there are measures that represent social costs and benefits or operational effectiveness improvement – These are beyond strict economic or accounting numerical frameworks.

Relative to M&S in particular, this tutorial states that the benefits of M&S include reuse (computer code, models, and simulations), readiness (training, rehearsal, and mission previews), efficiency (distributed to the user and allows rewinds and re-dos), risk reduction (mistakes are not fatal and information is secure), effectiveness (can target specific performance goals), money (simulation is cheaper than live), environmental protection (damage to the environment is avoided), lives saved (limits danger of injuries or death during training and trains the most dangerous situations so that the trainees are mission ready), and time saved (limits deployments, sets up in minutes or hours vice days, and enables reruns). It recommends that an undertaking to calculate ROI on any system, process, or procedure first establish the programmatic or business goals (purpose, reason, or focus) of the proposed investment. This discussion and process of agreeing to that statement of goal(s) establishes the entire team and ROI process on the right path, focused on the clearly stated goal(s) of the ROI undertaking. This tutorial also states that, no matter how mathematically solid or technically sophisticated an ROI analysis may be, it (ROI) will not likely be the sole determinant of an investment decision. Nor is there one correct way to conduct or present an ROI analysis. It is likely that ROI findings are only part of the decision-making process combined with internal organizational factors (employee morale, and social and political relationships), customer or client satisfaction and faithfulness (retention, feedback, and word of mouth recommendations), production (output and quality metrics), and other external relationships or agreements.

Another effort that focused on Joint Staff training was captured in a Ph.D. thesis titled "Developing a repeatable and reliable methodology to determine return-on-investment." This thesis notes that the U.S. military is faced with rising costs for weapon systems, fuel, and personnel while political and budget reduction pressures increase (Nesselrode, 2008). It further states that, while training of individual warfighters and small to large teams in live and simulation-based environments is a key strength, training of operational staffs using simulation is not widely used. Coincidentally, operational staff training is ideally conducted in a synthetic (simulation) environment because all the supporting and reporting forces can be portrayed by constructive simulation entities or role players vice requiring the use of many live forces. The research hypothesis was that, due to the complexity of the required scenarios, training at the operational level for staffs in a synthetic environment is more cost-effective and provides a greater ROI. In order to test this hypothesis, it is necessary to establish operational-level costs of training and common standards to measure accomplishment of training objectives; and neither of these types of data were readily available.

The thesis included a set of common measures and a model to assess the performance of operational-level staffs after training, a model of costs for live training versus synthetic (simulation-based) training, and a tool to measure ROI for these types of training events for operational staffs. It tested the proposed system in operational-level events to evaluate the measures, model, cost data, ROI results, and weighting schemes. Once the framework is tested in the field, modified, and confirmed, the intent is that it could be useful to DoD, Services, and also nongovernmental agencies and large corporations that respond to crises or operational emergencies.

Another I/ITSEC Tutorial on the topic, *Establishing the Value of Simulation (VoS): Methodology and Case Study*, states that ROI is a means of comparing investments, and it is used to get the most out of limited resources and determine if new capabilities are measurably better than the current solution (Numrich, 2013). It states that the returns for using M&S for training are often nonmaterial and not readily converted to a numerical ROI – specifically that the most important return for use of M&S for training is enhanced operational effectiveness. Converting operational effectiveness to a monetary amount may be possible; however, the conversion and getting buy-in could be problematic. At the time of this research, DoD's strategic guidance was changing, and as a result, the requirements for training were also changing. At the same time, resources were shrinking, and there was a continuing discussion of whether simulation for training was a better buy than healthcare, improved facilities, or new weapons and systems. So, to capture the breadth of the benefits of simulation, Dr. Numrich's team decided to look at the Value of Simulation vice ROI. Her method allowed a more flexible approach, recognized that all benefits were not fiscal, and combined the benefits with cost data.

In a Value of Simulation study, the team assessed the value of simulation in command and staff collective training with emphasis on the military decision-making process and the staff execution of the operational concept (Riecken *et al.*, 2013). In this study, the team used results from three surveys and general rule of thumb costs for various types of deployments to training locations. The results appeared to be sufficient to test the key hypothesis: Simulation environments differ among home stations. Survey results were mixed for comparing field exercises, classroom instruction, and simulation-based training in terms of skill improvement. In terms of cost-effectiveness, the simulation-based environment showed some advantage over the classroom instruction, and very significant improvement over live training. The study concluded that cost data were complicated and difficult to acquire, requiring organizational relationships (and cooperation and time), and an explicit definition of what data to capture was necessary. Also, it appeared that commanders and senior staff provided reasonable estimates of costs in order to evaluate live, schoolhouse, or simulation-based training. Differential performance evaluations were difficult to conduct; however, surveys at key milestones during training events did provide some information on training effectiveness.

14.5.4 International M&S ROI Assessments

Under the theme "Why spend one more dollar for M&S?" government, military, and business leaders in many other countries asked this question in terms of their currency. The professional organizations that chartered working groups to explore this theme at SISO Workshops, SCS conferences, and I/ITSEC were international. It was inevitable that like-minded researchers in the North Atlantic Treaty Organization (NATO) and in other countries organized similar efforts. A common theme was collaboration and sharing of information and M&S assets.

14.5.4.1 NATO

The NATO M&S Task Study Group (MSG) 031 studied the exploitation of M&S by the member defense communities with the goal of saving time and money and increasing performance under the theme, "The Cost-Effectiveness of Modeling and Simulation (M&S)" (Waite *et al.*, 2005). MSG-031 reviewed the Simulation Based Acquisition (SBA) efforts in the United States and the Synthetic Environment-Based Acquisition (SeBA) in Canada and the United Kingdom. A similar effort in Sweden included collaborative web sites and studies on best practices. The fundamental strategy for MSG-031 was to compile and leverage national information into a single NATO analytical context that could establish findings on the management of M&S and cost-benefit of M&S.

14.5.4.2 Australia

With significant M&S expenditures, in the range of $2B–$3B (Australian), over the last decade, an initiative took a collaborative approach by the government (with industry partners and international simulation organizations) to ensure every dollar was well spent (McFarlane and Kruzins, 2006). The Australian Defence Simulation Office produced a reference guide to identify where, why, and how simulation could be applied to support defense outcomes at each phase of system development. Through the intelligent use of simulation, the benefits were found to be capability enhancement, resource saving, and risk reduction; however, it was thought to be essential to identify the benefits and the limitations associated with the selected M&S prior to progressing to the detailed assessment of M&S used for acquisition (Kruzins, 2006). In order to find the simulation resources that were best value, the Australian Defence Simulation Office developed a process of conducting studies to look at previous uses and risks, conducting a market survey of available in-country resources, and then looking at third-party sources offshore. The reference guide described the cumulative synergistic benefits of simulations as more and more simulations were used and understood, and the guide also highlighted that simulations tend to be reused for additional benefits. (*Note:* This benefit of reuse across programs and military services is what could be called an "enterprise-level" realization of benefits, and we discuss this later.)

14.5.4.3 The United Kingdom

Under the SeBA initiative, there was an emphasis on representations of objective systems via simulation and databases, collaborative decision support systems for access by distributed leadership and agents, and evolution of the representations in simulation and data. This vision included collaboration of industry and government within that synthetic environment across the acquisition phases and stages. The goal was to share information concerning best practices, lessons learned, and current activities; promote integrated use of systems engineering and M&S throughout the acquisition process, and identify and adopt common metrics for cost-effectiveness descriptions (Waite *et al.*, 2005).

14.5.4.4 Canada

Here, the Department of National Defense (DND) initiated the Canadian Advanced Synthetic Environment (CASE) project. CASE prototyped a collaborative environment and standard business practices that could be grown and improved by every subsequent user of CASE. In recognition of the power of M&S and the convenience of having a collaborative environment to share best-of-breed models, simulations, and databases, Canada provided a standing offer to vendors to have their products listed in the collaborative space so that government users could select sole-source products conveniently. This M&S market environment was extraordinarily unique, convenient, and efficient.

14.5.4.5 Sweden

The Swedish Material Administration (FMV), as a component of the Swedish Armed Forces, initiated a study of M&S to help identify the evidence necessary to make decisions on M&S investments. FMV also produced a best practices guide to assist leaders in deciding on the optimal investments in M&S; the goal was to inform leaders so that they could make the best choices in deciding on investments in M&S for specific purposes. FMV acknowledged that for training, it is obvious that M&S is cheaper than training with real equipment against a complete live opposing force; however, intelligent M&S use in every phase of the product life cycle must be planned. This purpose for M&S support was thought to have been the most promising for wisely avoiding unnecessary spending of considerable amounts of funding. That decision-making and planning was the purpose of the Swedish Armed Forces' Economical Benefits of M&S study (Waite *et al.*, 2005). This effort would require a renewal of methods, material, and doctrines; and the Swedish Armed Forces recognized that they needed to be compliant with their allies. The Economic Benefits of the M&S study was limited to the early phases of the acquisition process because the effects of decisions in this early stage are most significant to the cost and outcomes of the program, and the study goal was to provide project managers in acquisition with knowledge and confidence in using M&S for their program. So, for Sweden (like Australia, the United Kingdom, and Canada) and NATO MSG-31, M&S was looked at as a tool that could avoid costs in acquisition programs, and they shared a desire to provide information and M&S assets in accessible collaborative environments.

14.5.5 Select ROI Insights and Artifacts

Over the years, as the approaches described above have been developed and applied, the authors have created a number of M&S ROI frameworks and techniques. Some of them describe processes; some define and associate metric types; while others are methods that we have found to be useful in measuring or aggregating measured values.

14.5.5.1 Methodology Process Description

When calculating the value or ROI of an M&S system, it is useful to consider going through a systematic process as depicted in Figure 14.2 (Oswalt *et al.*, 2011). By doing so, it helps to structure and bound the problem and increases the likelihood that the results will match the question(s) being asked by the sponsor.

First, it is important to define the context of the analysis. That is, what type of value assessment is needed and from what viewpoints will the

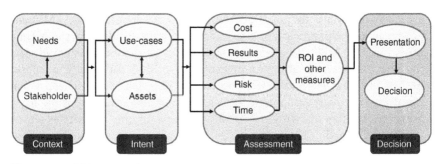

Figure 14.2 ROI process description.

assessment be conducted. Will the results inform a choice within a program office between alternative acquisition strategies that incorporate M&S, will it provide insights to a trainer of recruits on how best to provide weapons instruction, or will it inform Program Objective Memorandum (POM)-level investment decisions being made by a service? In terms of viewpoint, there are significant differences between what metrics are important and how they are characterized depending upon that perspective. Some decision-makers are concerned about features of the M&S system itself (e.g., the "ilities" such as usability or reliability), others on the relative strengths and weaknesses when compared to non-M&S alternatives to meeting mission goals, while still others may be concerned with enterprise management of M&S within a technology portfolio.

Next, it is critical to characterize the intent of the analysis: both the use-case or scenario under consideration and the assets that are in and out of the box. The use-case defines the activity under examination in enough detail to identify critical concerns and drivers (e.g., to test an aircraft's ability to search, track, and engage multiple supersonic off-axis threats). An asset list provides a parallel set of items potentially used in the activity (see Table 14.2). Both help to specify the value proposition being calculated.

The third step in this ROI process is the definition and assessment/measurement of the metrics. As we will discuss in the next section, these normally include entries in results, cost, risk, and time categories. Although these metrics can be examined individually, in order to calculate a

Table 14.2 Sample ROI cost categories.

Hardware	• Computers • Hardware-in-the-loop • Mock-ups • Electronic hardware • Spares	Software	• Models • Simulations • Tools (CAD/CAM) • Data/datbases • Repository	People	• Expertise • Experience • Skills/education • Operational knowledge
Networks	• Lines • Architecture • Transaction protocols	Facilities	• Buildings • Labs • Ranges • Physical models	Products and procedures	• Plans • Policies • Analysis results • Conceptual models • Management processes • Standards

comprehensive ROI value, it is necessary to also weight and combine them into a single/composite value. Finally, the decision-maker is presented the ROI data, calculations, and output. These should be presented relative to the sensitivity of the parameters, the data's pedigree and legitimacy, and the window of time within which the outputs are valid. Considering all of these, the decision-maker makes a choice.

14.5.5.2 Metric Categories and Samples

In the many years of developing an M&S ROI approach, the need for a comprehensive set of metric categories became apparent. Such categories allow a fair fight assessment between M&S options that often score very differently within specific metric categories: for example, a comparison of one choice that is predicted to have high results but also high risk versus an alternative that has high cost but is likely to significantly reduce elapse time. For this reason, four metrics categories have proved to be useful. The first metric category is *results*. We use the term results since it more naturally allows for both positive and negative outcomes, unlike the words value and utility that tend to imply a positive improvement. The term results is intended to convey the impact/outcome of an M&S application; it can include both positive and negative outcomes; it encompasses various expressions of value, utility, contribution, benefit, impact, return, and similar; and it allows for both naturally quantitative and qualitative inputs.

The metric category of *cost* includes the expenditure of funds, consumption of resources, employment of personnel, or other item of monetary value spent on the development and use of M&S. It can include cost savings, cost avoidance, and cost reduction (i.e., cost-offsets). It should be calculated for the life cycle of the system and adjusted as appropriate (e.g., net present value, the current year value of costs realized over time). Some example categories and cost types are listed in Table 14.2 (AEgis, 2008), although there are more complete lists (Knapp and Orlansky, 1983).

Third is the category of *time*. In terms of the impact of M&S on mission success, time is often measured in terms of overall training time, duration/retention of lessons learned, time needed for decision-making or mission accomplishment, or the incremental learning that occurs over time (Champney *et al.*, 2006). One analysis examined the cost to develop, produce, and deliver 320 hours of training to specialized skills students using technology-based instruction. The results were a savings of about 136 hours for each student and an overall ROI of 3.36 (Cohn and Fletcher, 2010).

The fourth and final metric category is *risk*. In general, risk is the likelihood and magnitude of a negative outcome. In terms of M&S application, this manifests itself in metrics like the risk that M&S will, for instance, not work as advertised, not be accurate enough for the intended use, incur

Table 14.3 Sample types of M&S risk.

Risk type	Description
Inaccuracy	The degree that an M&S system is not a 'faithful representation of the relevant features of the original'
Error	Mistakes. Incorrect assumptions or implementations
Uncertainty	Suspicion, doubt, or lack of faith in the truth or correctness of the outputs

upfront investment that is not recouped downstream, or that it will cause negative impacts on cost, schedule, or performance. More formal terms that characterize M&S risk, especially as it pertains to analytic application, are provided in Table 14.3.

Another potential risk is that the use of M&S may lead a decision-maker to place too much confidence in the accuracy of the M&S results. Over specificity or spurious specificity should not be attributed to the outputs of M&S tools. Outputs must be appropriately characterized or contain necessary caveats, in both display systems and numeric values. Risk also includes the changes in decision-making risk as the result of the use of M&S, for example, the early use of M&S to identify errors and omissions in the acquisition of a complex system, thereby reducing programmatic risk.

14.5.5.3 Metric Levels/Relationships

In conducting an ROI analysis (like in other types of analysis), it is useful to decompose the area of investigation into parts, if for no other reason than for manageability. But in this case it is particularly critical, since division enables the explicit association of benefits and costs at different levels. For M&S value to be accurately assessed, its life cycle/downstream impacts must be included. For instance, if an organization develops an M&S standard designed to save time or funding, then the ROI calculation of it needs to include the cost to develop it, its impact on developmental risk, the associated changes in M&S interoperability it produces, how those changes influence training for instance, and how those changes are reflected in personnel proficiency and ultimately in combat/operational outcomes. This is much easier said than done, although to discount or ignore downstream effects leads to an underassessment of the ROI of an M&S asset. (*Note:* This underassessment may include both positive and negative impacts.)

With this in mind, it is helpful to mimic the military operations research community and their sequence of metric types: dimensional parameters (DP)/measures of activity (MOA); measures of performance (MOP); measure of effectiveness (MOE); and measures of force effectiveness (MOFE), measures of political effectiveness (MOPE), and measures of overall outcome (MOOO). In the case of M&S ROI, we call these levels foundational, direct, derived, and deduced. Foundational metrics reflect the value of essential activities, but they do not generate direct/near-term payback. Things like standards, area-specific high-performance teams, repositories, and similar improvements are foundational. The next level is direct measures, which reflect the accomplishment of a mission area in a way that is "better, faster, or cheaper" and includes metrics on the M&S's "ilities" (availability, usability, extensibility, etc.), its impact on time (relative to a system acquisition or training delivery), and the change in cost (often cheaper, but after significant upfront investment). Next are the derived and deduced categories. These reflect the impact of M&S, in a more proximate or near-term (direct/derived) way and in the long-run (deduced). A summary of these levels is provided in Table 14.4.

Considering and including all of these levels could significantly increase the complexity of an ROI analysis. However, the unwanted consequences of limiting the areas covered in an analysis could be the possibility of rejecting an M&S alternative based on high cost at one level even though ultimately that M&S alternative could significantly improve the odds of winning the war. An example is provided in Table 14.5. We hope that the inclusion of this approach underscores the value of including the multidimensional nature of both M&S ROI calculations and also their potential impact.

Table 14.4 ROI metric levels.

	Foundational	Direct	Derived	Deduced
Example	Planning, working groups, standards catalogs/ repositories	Better Faster Cheaper	Proficiency Readiness Performance Sustainability	Warfighting effectiveness
Definition	Activities that foster, facilitate, or coordinate	First-order impact Relative to a community or mission	Second-order impact of the resulting product. Also domain specific	Consequence to the overall outcome
Measurement level	Measure of activity/ dimensional parameter	Measure of performance	Measure of effectiveness	Measure of force effectiveness

Table 14.5 M&S ROI impact across metric levels.

MOA/foundational (unit/system)	MOP/foundational (task)	MOE/direct (mission)	MOFE/derived (campaign)	MOOE/deduced (strategic)
From working group collaboration, meetings with analytic users, and so on. The data provider better understands military environmental representation requirements	Data provider delivers improved environmental representation products to military analysts for their inclusion in applicable M&S systems	Improved environmental representation allows the inclusion of cloud-cover in M&S systems used for assessment. This results in Blue's aircraft having limited visual acuity. Blue Concept of Operations (CONOPS) limits identification of friend or foe (IFF) transmission due to emissions control (EMCON). Result is significant Blue on Blue fratricide. So, Blue mission-level ROE is changed to prevent friendly fire (positive IFF before firing). This significantly exposes Blue assets and results in improved Red anti-aircraft results	Inclusion of cloud-cover in scenario at the mission level changes force-level warfighting CONOPS to minimize Red gains. *Decision:* Attack Red's rear-area logistics using surface-to-surface missiles (SSMs). Red is repelled	This generates the need to evaluate adequacy of the SSM inventory in theater. Also the role of mobile assets for use in theater. All of these inputs are made available to decision-making regarding aircraft EMCON and number and type of SSMs deployed. *Bottom-line:* inclusion of M&S environmental representation – beginning at the foundational-level can lead to significant deduced impacts

14.5.5.4 Accounting for Perspective

In reporting the ROI of M&S in DoD, perspective matters. Stakeholders within DoD have different motivations to use M&S, and they have different perspectives on the benefits realized. This is similar to different observers having different perspectives on a moving aircraft – left to right or higher or lower than a landmark depending on the observer's position. In the DoD M&S marketplace, each type of stakeholder has a different perspective, and the most important overall benefit of M&S will probably be realized at the DoD enterprise level. The M&S stakeholders could be described in terms of this role-dependent hierarchy (Oswalt *et al.*, 2011):

- *Project or program:* Concentrating on the benefits to their task and their system, perhaps as part of a system of systems
- *Community:* Acquisition, testing, experimentation, wargaming, training, force structure analysis, operational decision support, and so on
- *Enterprise:* Military services, combatant commands, DoD agencies, or DoD in general
- *Federal:* Across departments and agencies
- *Society:* Across government, industry, academia, and consumers

It is assumed that as the level of the stakeholder increases to above the program or project level, the probability of multiple uses and reuse of the M&S assets increases, and the stakeholders control more funding accounts that, as combined, are more likely to have funded the M&S and realized the benefits of the M&S. The first three listed stakeholder groups (program, community, and enterprise) could be considered to be internal to DoD, and the stakeholders that deal in M&S for DoD may be consumers/users, buyers, sellers/producers, investors (financial programmers and budgeters), approvers, reviewers, and promoters/advocates. Much like discounting future costs and results to the current year, translating (aggregating) the costs, benefits, returns, cost avoidance, and ROI to the enterprise level is a worthy goal to show the value of M&S at the DoD enterprise level (Oswalt *et al.*, 2011).

14.5.5.5 Measuring ROI Metrics

The metrics that are measured in an ROI calculation are only one item within an overall assessment structure that applies to the M&S under examination (see Table 14.6). The examination might focus on a simulator, a Live, Virtual, Constructive (LVC) event, an M&S management activity, a faster-than-real-time analytic tool, and so forth. Arrangements of live simulation (warfighters operating their wartime equipment), virtual simulation (warfighters in simulators or mock-ups of their weapon systems), and

Table 14.6 An M&S ROI assessment structure.

Component	Example
Analytic classes/use cases that divide the application domain space	Training, analysis
Hierarchical structure of metric types from activity to effectiveness	Foundations, direct, derived, deduced
The *referent* or metric category under examination	Result, cost, risk, time
Referent *elements* that reflect specific features or functions of the referent	*Result:* Maintainability
Element *properties* that describe their attributes, quality, or characteristics	Mean time between failure
Metrics, which are standards of measurement, like variables	1–10 h
Metrics values are relative to a *scale* (a specified graduated reference) and may be nominal, ordinal, interval, or ratio in type	0-1-2-3-4-5-6-7-8-9-10
They may *range* from 0/nothing to X, where X represents all of the quantity possible	0-10
Metrics are *assigned values*, based on the presence of the properties (the act of measurement)	9
Values can be *combined* into aggregate measures of merit	9^* Cost to Repair = $X

constructive simulation (weapons, systems, and warfighters represented in computer programs) are becoming very popular, and ROI calculation of various mixes or blends of different levels of each will be necessary. However, for any analysis, at the top of the structure are analytic classes or categories that define the area of M&S application. Then, there is the hierarchical structure of the metric types; that also has a parallel set of levels and associations (as were discussed earlier and presented in Table 14.4). In some ways, this structural component is orthogonal to the others listed, but it is important to include.

The next component of the structure is the referent – or the metric category under examination. Said another way, it is the dimension of value being referred to. Here, they are the results, cost, time, and risk associated with application of a simulator or simulation system. For each referent, there are a set of elements, verbs describing what the item under examination does (e.g., SISO develops standards) or nouns that describe important characteristics (e.g., the maintainability of Naval Simulation System (NSS)). It is critical to include the dominate elements/factors within an ROI assessment. In some cases, these factors will be readily apparent (e.g., software development costs of a simulation system), but for others, the determination may be more difficult. In all cases, it is useful to begin with as complete a list of class elements as possible and then consciously and explicitly reduce the list. Next are properties that reflect key features, characteristics, or qualities of the element.

For each property, metrics are defined that reflect the degree to which the property is found in the item under examination. It is important to emphasize that metrics can be positive or negative. Metric values are relative to a scale (e.g., tons, pounds, or ounces – for weight), there are assigned numeric values of varying specificity, and these values fall within a range – that can be combined into composite terms. An example that follows this general pattern is provided in Table 14.7 (Severinghaus *et al.*, 2012).

Measuring ROI metrics requires an encompassing set of instrumentation tools. Some M&S ROI metrics are naturally quantitative and fall nicely into neat categories (e.g., the cost, in dollars, of a 6-degree-of-freedom aircraft simulator), but many/most do not. In overall terms, metrics may be naturally quantitative (e.g., number, percent change, cost, time); qualitative but can be converted to quantitative (e.g., boredom converted into elapse time); qualitative but very hard to convert to quantitative (e.g., degree of misunderstanding); not measurable (e.g., perfection); or unknown. As a result, the measurement of many metrics must use techniques from measurement sciences, subjective assessment, and multiattribute utility theory (Hubbard, 2014). Relative to data availability, data may exist and are collected (e.g., number of rounds expended in a simulator); exist but are not collected/easily available; not currently exist but could be collected; not exist; or be unknown. Again, innovative use of collection techniques is required.

14.5.5.6 Combining Disparate Measures

Decision-makers prefer to use metrics for proposed M&S projects in order to objectively and quantitatively describe the net value relative to the costs (AEgis, 2008). The authors of the 2008 study selected the multiattribute decision method (MADM) as a very appropriate method to combine a countable number of weighted metrics to derive a score for each alternative in the decision process. The team settled on the most straightforward linear version of MADM, which at its simplest is a weighted sum. MADM techniques were adopted and tailored to various M&S decisions for case studies in the 2008 study. The described process is tailorable to any multi-attribute scenario, where the attributes could be called factors, inputs,

Table 14.7 A results element structure for medical training M&S systems

Category	Type	Element	Property	Standards	Scale	Range	Example
Result	Direct	System characteristics – accredited	Level degree	3 Cat 0–100%	H/M/L 1%/ratio	X 0–100	M 33%
Result	Direct	Technical features – maintainability, modifiability, adaptable	Level degree	5 SEI/CMM 0–100%	1–5 1%/ratio	X 0–100	4 50%
Result	Direct	Technical features – reonfigurable, expandable, flexibility	Level degree	3 Cat 0–100%	H/M/L 1%/ratio	X 0–100	L 30%
Result	Direct	Training tool context – gains attention and motivates	Level Degree	2 Cat 0–100%	H/L 1%/Ratio	X 0–100	H 75%
Result	Direct	Training tool context – represents needed situations	Degree	0–100%	1%/ratio	0–100	80%
Result	Direct	Training tool context – accommodates learning styles	Degree number	0–100%X	1%/ratio $1–\sum$ #	0–100 1–N	50% 100
Result	Derived	Training transfer – degree lessons are learned	Degree number	0–100% lessons	1%/ratio 1/integer	0–100 1–N	20%5
Result	Derived	Training transfer – number of new skills acquired	Degree number	0–100% New skills	1%/ratio 1/integer	0–100 1–N	20% 50
Result	Derived	Knowledge longevity – duration learning is remembered	Time degree	Hours 0–100%	1.0 1%/ratio	1–N 0–100	5 22%

(continued)

Table 14.7 (*Continued*)

Category	Type	Element	Property	Standards	Scale	Range	Example
Result	Derived	Knowledge longevity – decay rates of information learned	Time degree	Days 0–100%	1.0 1%/ratio	1–N 0–100	20 45%
Result	Derived	Specificity of knowledge learned – time taken to identify, treat, and so on	Time degree	Hours 0–100%	1.0 1%/ratio	1–N 0–100	5 80%
Result	Derived	Specificity of knowledge learned – by procedure, injury type, and so on	Number degree	Number of categories 0–100%	1/integer 1%/ratio	1–5 0–100	2 40%

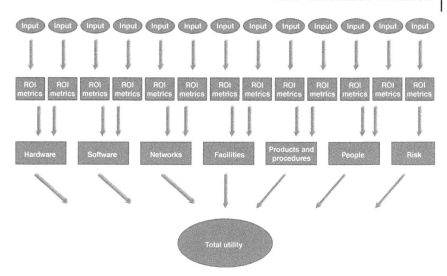

Figure 14.3 Diagram of MADM process for DoD M&S investment by asset categories and risk.

variables, or metrics without any change in the process. While ROI or cost avoidance (the numerator of the ROI ratio) are ways to make a decision, in those cases where one wishes to consider other factors in addition to or instead of ROI, MADM works exceptionally well. In general, MADM blends all the metrics into one quantitative score (total utility) that can be maximized or minimized for choosing the "best" selection. In Figure 14.3, as published in the 2008 study (AEgis, 2008), the depiction illustrates a generic tree of metrics blended to provide a total utility score (or this could be called the MADM score, response, or output) for the blended quantitative score of all the various alternatives.

MADM is robust, relatively explainable, objective, and consistent. MADM initially requires the setup of weighting criteria, and once that criterion is established, MADM can be executed fairly simply to help make complex decisions involving the best mix of several metrics. In MADM, the user decides on the weights of the various metrics. Across all the metrics, the weights are between 0 and 1, and they must total to 1.0 for that level. All possible metrics must be weighted, so if a metric is added or deleted from the MADM analysis, the remaining weights will change. For each metric, multiple measures are possible, and the user defines the highest possible measure and normalizes that as 1.0. All measures of each metric are distributed between 0 and 1. In Chapter 15, a case study will illustrate the use of MADM to provide utility scores for various alternatives.

14.6 ROI of M&S Extensions and Relations

The development and application of M&S is enabled by a significant infrastructure of organizations, activities, and products. In some of these areas, their value or contribution to the development and use of M&S has been examined.

14.6.1 Management and Standards

In the area of M&S enterprise-level management, one effort endeavored to develop metrics that show the worth or value of an enterprise leadership function for M&S (Cooley *et al.*, 2013). This requires evaluating the "benefit" of the enterprise function. While a benefit can be an outcome that includes obviously measureable results, in a case like this, it often takes on its more abstract meaning of effects: not immediately quantifiably measurable, but rather those that add to and increase the effectiveness of an organization. An example of the set of metrics developed by this effort, arrayed by role and asset type, is provided in Table 14.8.

Table 14.8 Enterprise management: sample metrics.

Role	Asset	Metric
Guide – lead, direct, communicate	Policy	Number of major command (or equivalent) organizations that have implemented new DoD-wide policies issued by the M&S Enterprise Organization
Guide – lead, direct, communicate	Policy	Percentage of subordinate commands that accurately communicate HHQ policy in their respective guidance documents
Guide – lead, direct, communicate	Policy	Costs avoided by adherence to an endorsed standard
Guide – develop, build, implement, execute	Processes	Average length of time from the requirement definition of a standard to registration
Guide – develop, build, implement, execute	Processes	Ratio of the organizations faithfully implementing HHQ sponsored processes
Support – sponsor, fund, invest	Technology	Percentage of available resources invested in studies on how to apply emerging technology to M&S

Table 14.8 *(Continued)*

Role	Asset	Metric
Support – sponsor, fund, invest	Technology	Number of enterprise sponsored M&S system components that are employed across communities
Manage – govern, oversee, administer	Processes	Number of collaborative M&S events facilitated that are consistently scheduled and held within which goals are established, minutes are accurately recorded, action items assigned, and open items brought to closure
Manage – govern, oversee, administer	Processes	Average number of distinct roles employed by enterprise management in striving to achieve each DoD-wide M&S goal
Manage – organize, team, coordinate	Technology	Number of technical informational exchange events that are HHQ sponsored/facilitated/attended where multiple M&S communities are represented
Manage – organize, team, coordinate	Technology	The ratio of repository entries that have been updated and accessed in the past 6 months
Manage – organize, team, coordinate	Technology	The ratio of repository entries that have not been updated in the past 6 months (measure of risk)
Acknowledge – monitor, aware	Knowledge/ studies	Percentage of resources expended gathering information on basic and applied research from institutions, government agencies, organizations, and universities

Regarding standards, an effort was undertaken and the results reported to assess their utility (Collins *et al.*, 2012). The authors point out that the use of M&S standards allows reuse, simplifies interoperability, and provides cost-savings. They also state that "standards are not only useful, but essential for the survival of our modern day society." However, they also conclude that it is difficult to determine the ROI of M&S standards.

14.6.2 Learning, Training, and LVC

In the area of learning, a recent article looked at the benefit of technology-based education and training, and the authors acknowledged that any significant expenditure in industry will be evaluated in terms of business success and other measures of output quantity, quality, cost, or time. The authors selected ROI as their metric, and presented a case study of a mobile learning program for software sales staff of a large business. They used an experimental group of 25 and a control group of 22 sales associates, and the study showed the mobile learning environment increased sales for an ROI of 3.11 (increasing sales over 300% more than costs of the sales program), and produced several significant intangible benefits (Elkeles *et al.*, 2015). In the area of serious games, an approach has been postulated to "calculate predicted future or past financial benefit that a group or organization obtains from designing, developing, implementing, and using/reusing a serious game for training" (Kirkley, 2007). Finally, relative to the ROI for Live, Virtual, Constructive (Smart) M&S environments, a "checklist" for return on investment (ROI) consideration in four classes of simulation – "constructive," "virtual," "Live," and "Smart" – has been outlined (Carter, 2001).

14.7 Cons and Pros of M&S ROI Calculation

As we have reviewed the development of M&S ROI calculation, from its humble beginnings in narrative descriptions of value, the level of complexity has increased dramatically. From numerical anecdotes that characterize specific dimensions of improvement to a much more complete set of ways to look at ROI calculations across four metric categories, four metric levels, at least three possible points of view, many application areas and use cases, and many possible alternative solutions with at least one being M&S, the complexity of analyzing ROI has increased. Although these improvements in characterization of the elements will yield significantly improved insights, they are not without cost or risk. Research and development in areas like metric cross-correlation (when metrics are not mutually exclusive/independent) and causal dependencies (quantitatively linking metrics at different levels) needs to be conducted. These more methodological considerations are often trumped by the very real challenges of data: its collection, proper formatting, extraction, and retention. The practical problem of gathering the right data to make decisions is often the weak link.

On the other hand, we "measure what we treasure" and without measurement there cannot be optimization – or even efficient application. ROI allows us to evaluate alternatives for efficient application and eventual optimization. It can be formulated to include the key dimensions of value (results, cost, time, and risk), stakeholder perspective, and application relevance in a balanced equation that can determine true M&S impact.

It provides a way to evaluate the differences between live and traditional simulation approaches within an LVC mix and also the consideration of the role of gaming technologies in enhancing access, user comfort, and interactivity. ROI coupled with or as one component of a MADM assessment is a powerful decision support tool. Every other approach, while persuasive, is comparatively *incomplete*. However, in many contexts, persuasive is good enough. Thus, the constructs of cost avoidance/savings are discussed in the next chapter as alternatives to ROI.

14.8 Future of M&S ROI Assessment

There are trends in the M&S community that (i) argue for the increased need for ROI calculation and that (ii) recommend use of ROI be tailored in its application relative to current business realities. In terms of the increasing need for ROI, it is especially true in areas that seek to efficiently and effectively employ M&S to meet dynamic multivariate requirements. This is the case in persistent (or persistently available) M&S (or LVC) environments and also in dynamically embeddable simulations within complex weapons systems – primarily for training. In each case, the challenge is to select from an array of possible M&S solutions and effectively pair them with requirements. In both cases, results/cost/risk/time all are important concerns, and databases are increasingly being developed that describe potentially applicable M&S assets (at least in terms of their capabilities). Assessing such an M × N array of capabilities and requirements to provide the efficient and effective composition of M&S assets relative to changing needs is a perfect application opportunity for ROI.

It is also important to acknowledge that DoD/Government use of M&S will increasingly be leveraging commercial developments. The Microsoft Connect® device was developed for the civilian/commercial marketplace, although it will be extended and used within DoD/Government. So, calculating the "marginal or incremental" ROI of M&S relative to DoD – as simulation increasingly becomes an embedded vice appended capability – needs to be formulated and calculated.

Finally, ROI will increasingly be needed to calculate the overall value of application-specific M&S capabilities (Cooley and Oswalt, 2012). No longer will it be sufficient to understand the ROI for a specific event, federation, or application. It will be critical to understand M&S ROI over time and across potential applications. This is especially true as M&S, LVC, virtual and synthetic environments provide blended solutions to multiple problems. In such an environment, calculating ROI to account for the combinatoric and long-run impact of a particular M&S investment will be critical.

14.9 Conclusion

In this chapter, we discussed the motivation and the methods used in the early days of computer-based M&S to prove the worth of M&S. In those days, M&S was growing in use but had some flaws. And, the expense to build or purchase was high enough to cause leadership in DoD to want to understand the cost-benefit relationship for M&S versus buying weapons or supporting the systems and military forces already in the field. The first attempts to show the worth of M&S were qualitative declarations of the value of M&S to training and the value of training to combat readiness. This relationship was validated through feedback from the war in Iraq in 1991. Later that decade, initiatives like SBA and SeBA began internationally to look at the value of M&S to acquisition of weapons systems – typically a very expensive process. In military departments in various countries, initiatives related to collaborative web-based sites that served as an M&S marketplace to exchange information on M&S systems, best practices, sources for help, and the M&S tools themselves were established. These sites promoted the most effective employment of M&S at the most advantageous timing in the acquisition process. Flaws discovered early in the acquisition cycle are much easier and cheaper to change; so, if M&S is used to put the system in a computer-based combat environment of the future, weaknesses can be found earlier. M&S had the ability to portray combat zones and environmental conditions that are impossible to enter other than in combat or that are decades in the future; therefore, M&S is increasingly important in force structure analysis, experimentation, wargaming, and in decision support in combat.

When leadership sees this growing use, they ask "Why spend one more dollar for M&S?" and more specifically they ask for a metric that would indicate the value of investment in M&S. The metric most often asked for was ROI, and the related metric of cost avoidance. ROI is a key metric for business globally, a key component of business plans of new endeavors, and an excellent way to capture the value of a future investment based on future sales or benefits. We covered the calculation of ROI within DoD and discussed one way to combine metrics in a quantitative analysis technique called MADM. We also covered ROI insights; metric categories (cost, risk, time, and results); a structure for metric levels, perspectives, and assessment of ROI components; ROI extensions and the results of related studies; and, in conclusion, an assessment of future trends in M&S ROI. In Chapter 15, we provide examples of the MADM calculations to show how multiple attributes can be combined for multiple alternatives in order to derive the optimal choice of one alternative in terms of a utility score. In the next chapter, we will also take this discussion further to cover how cost avoidance is combined with proficiency evaluations to show the specific impact of

M&S in training use cases and for predicting the best spending plans for each marginal dollar.

Acknowledgment

The authors wish to thank Ms. Margaret Callahan of Simulation U Analytics LLC for her outstanding assistance with graphics, formatting, and locating correct references.

References

AEgis Technology Group, Inc . (2008) Metrics for Modeling and Simulation Investments. Report No. TJ-042608-RP013. Available at http://www.ndia.org/ Divisions/Divisions/SystemsEngineering/Documents/Committees/M_S% 20Committee/2009/April%202009/Metrics%20for%20MS%20Investments_ Final%20Report_07_SBCI_009_20090319.pdf

Brown, D., Grant, G., Kotchman, D., Reyenga, R., and Szanto, T. (2000) Building a business case for modeling and simulation. *Acquisition Review Quarterly*, 24 (4), 312–315.

Carter, J. (2001) A Business Case for Modeling and Simulation. Special Report-RD-AS-01-02, Aviation and Missile Research, Development and Engineering Center.

Champney, R., Milham, L., Carroll, M., Stanney, K., and Cohn, J. (2006) A method to determine optimal simulator training time: examining performance improvement across the learning curve. *Proceedings of the Human Factors and Ergonomics Society*, 50 (25), 2654–2658.

Clancy, T. (1997) *Into the Storm*. Berkley Books, New York.

Cohn, J., and Fletcher, J. (2010) What is a pound of training worth? Frameworks and practical examples for assessing return on investment in training. *Proceedings of the Interservice/Industry Training, Simulation, and Education Conference*, Orlando, FL.

Cohn, J., Combs, D., Anglero, A., Johnson, B., Rozovski, D., Eggan, S., and O'Neill, E. (2013) Medical modeling and simulation based training return on investment decision model. *Communications in Computer and Information Science*, 374, 144–147.

Collins, A.J., Meyr, D., Sherfey, S., Tolk, A., and Petty, M. (2012) The value of modeling and simulation standards. *Proceedings of the 2012 Autumn Simulation Multi-Conference, San Diego, CA*.

Cooley, T. and Oswalt, I. (2012) Current trends in ROI calculation. *M&S Journal*, 7 (2), 16–18.

Cooley, T., Oswalt, I., and Callahan, M. (2013) Modeling and Simulation (M&S) Enterprise Metrics. Report No. N61340-12-0010.

Degnan, E. (2010) Return on investment (ROI) for modeling and simulation. Tutorial. *Proceedings of the Interservice/Industry Training, Simulation, and Education Conference*, Orlando, FL.

Elkeles, T., Phillips, P., and Phillips, J. (2015) *ROI Calculations for Technology-Based Learning*. Association for Talent Development.

Gordon, S. (1995) Determining the value of simulation. *Society for Computer Simulation Proceedings* (969–973).

Gordon, S. (2001) Economics of M&S task force: enabling technology for simulation science. *Proceedings of the SPIE*, 4307. Available at http://proceedings. spiedigitallibrary.org/proceeding.aspx?articleid=912863, pages 281–287.

Gorman, P.F. and McMaster, H.R. (1992) The Future of the Armed Services: Training for the 21st Century. Statement before the Senate Armed Services Committee.

Hubbard, D. (2014) *How to Measure Anything: Finding the Value of "Intangibles" in Business*, 3rd edn, John Wiley & Sons, Inc., New York.

IDA (Institute for Defense Analyses) (1991) 73 Easting: Lessons Learned from Desert Storm via Advanced Distributed Simulation Technology. IDA Document 1110 (AD-A253 991). Available at http://www.dtic.mil/dtic/tr/ fulltext/u2/a253991.pdf

Joint Simulation Based Acquisition Task Force (1998) A Road Map for Simulation-Based Acquisition. Report of the Joint Simulation Based Acquisition Task Force, Acquisition Council Draft for Coordination, Department of Defense, Washington, DC.

Kirkley, S. (2007) Navy Serious Games for Training (N-SGT) Methodology and Toolkit. N06-T006: Conventional Training versus Game-Based Training. Phase 1 STTR Report.

Knapp, M. and Orlansky, J. (1983) *A cost element structure for defense training*, Institute for Defense Analyses, Paper P-1709.

Kruzins, E. (2006) Defense Simulation Investment Reference Guide. Version 1.0. Australian Defence Simulation Office. Department of Defence, Canberra.

Martin, D. and Gordon, S. (2004) Collateral damage estimation: transforming time-sensitive command and control. *Proceedings of the Interservice/Industry Training, Simulation, and Education Conference*, Orlando, FL.

McFarlane, D. and Kruzins, E. (2006) *Australian Defence Simulation – Status*. NATO Publication Reference Number TRO-MP-MSG-045. Available at http:// ftp.rta.nato.int/public//PubFullText/RTO/MP/RTO-MP-MSG-045///MP-MSG-045-02.pdf

National Academy of Sciences (1997) *Technology for the United States Navy and Marine Corps, 2000–2035*. Available at http://www.nap.edu/catalog/5863/ technology-for-the-united-states-navy-and-marine-corps-2000-2035-becoming-a-21st-century-force

Nesselrode, M. (2008) Developing a repeatable and reliable methodology to determine return-on-investment. Dissertation, Old Dominion University, VA.

Numrich, S. (2013) Establishing the value of simulation (VoS): methodology and case study. Tutorial. *Proceedings of the Interservice/Industry Training, Simulation, and Education Conference*, Orlando, FL.

O'Bryon, J. (1999) Meet MASTER-modeling and simulation test and evaluation reform. *Program Manager*, March–April, 8–14. Available at http://www.dau .mil/pubscats/PubsCats/PM/articles99/obryonma.pdf

Oswalt, I., Cooley, T., Waite, W., Waite, E., Gordon, S., Severinghaus, R., Feinberg, J., and Lightner, G. (2011) Calculating return on investment for U.S. Department of Defense modeling and simulation. *Defense Acquisition Research Journal*, 18, 122–143.

Riecken, M., Powers, J., Janisz, C., Numrich, S., Picucci, P., and Kierzewski, M. (2013) The value of simulation in army training. *Proceedings of the Interservice/ Industry Training, Simulation, and Education Conference*, Orlando, FL.

Severinghaus, R., Cooley, T., and Oswalt, I. (2012) Medical M&S Based Training ROI Decision Model. SBIR Phase I Final Report No. N00014-12-M-0135.

Waite, W., Gordon, S., Öhlund, G., and Björk, Å. (2005) Review and update of findings from economics of simulation study groups. NATO OTAN, NATO Publication Reference Number NMSG-035/RSY005, 21-1. Available at http:// www.academia.edu/17833044/Review_and_Update_of_Findings_from_ Economics_of_Simulation_Study_Groups

Washington, G. (1790) Address of President George Washington to the United States Senate and House of Representatives. Available at http://founders. archives.gov/documents/Washington/05-04-02-0361.

Yaukey, J. (2003, March 21). *Shock and awe combines destruction, protection.* Gannett News Service. Available at http://www.bibliotecapleyades.net/ exopolitica/esp_exopolitics_R_1C.htm.

15

Does M&S Help? Operationalizing Cost Avoidance and Proficiency Evaluations

Steven Gordon,[1] Tim Cooley,[2] and Ivar Oswalt[3]

[1]*Georgia Tech Research Institute, Orlando, FL, USA*
[2]*DynamX Consulting, Castle Rock, CO, USA*
[3]*Simulation U Analytics, LLC, Fredericksburg, VA, USA*

15.1 Introduction and Overview

Chapter 14 showed that a great deal of effort was devoted to answering the question, "Why spend one more dollar for M&S?" Much of this effort was championed by Mr. Bill Waite, and his ideas and leadership in the 2008 DoD report *Metrics for Modeling and Simulation (M&S) Investment* that laid significant groundwork for applying the efforts described in Chapter 14 (AEgis, 2008). However, commissioning a full return on investment (ROI) study can be very costly and time-consuming if one considers all the factors described in that report (Cooley and Gordon, 2012). Thus, a follow-on study described in this chapter is primarily focused on gleaning information from the most important component of ROI (e.g., cost avoidance). Additionally, during the same time frame, a new and related question arose: How do I justify the monies expended on my simulation-based training systems? In an era of shrinking budgets, military leaders were reviewing every line item, attempting to find areas for reducing costs. It was within this context that the question of justifying spending on simulation-based training systems was raised. One effort sponsored by the United States Marine Corps (USMC) Program Manager for Training Systems (PM TRASYS) sought to estimate a value for expenditures on USMC simulation-based training systems. In this effort, Dr. Steve "Flash" Gordon and Dr. Tim Cooley culled a key subset of the metrics delineated in the 2008 report (AEgis, 2008) to help PM TRASYS evaluate the derived benefit of budget expenditures for their simulation-

The Profession of Modeling and Simulation: Discipline, Ethics, Education, Vocation, Societies, and Economics, First Edition. Andreas Tolk and Tuncer Ören.
© 2017 John Wiley & Sons, Inc. Published 2017 by John Wiley & Sons, Inc.

based training systems from a cost avoidance standpoint. That is the framework where this chapter begins. This first effort led to further refinement and exploration and laid the research foundation to help answer the fundamental question posed in the 2013 Government Accountability Office's (GAO) report, *Better Performance and Cost Data Needed to More Fully Assess Simulation-Based Efforts* (U.S. GAO, 2013). In particular, the GAO stated that avoiding costs through the use of simulation-based training systems is good news, but they also asked if the systems actually increased the proficiency of the trainees in the readiness areas desired. In researching the answer to this fundamental question from GAO, PM TRASYS authorized a proficiency study led by Dr. Rob Dunne. This proficiency study was designed to take a first step in answering the questions asked in the referenced GAO report. In essence, "By looking at a subset of the *benefits* side of M&S ROI calculations discussed in the 2008 report, we attempted to answer the question posed by the GAO with a proof-of-concept proficiency study" (Dunne *et al.*, 2014).

In parallel with the proficiency study, the USMC wanted to evaluate the expansion of Live, Virtual, Constructive (LVC) simulation for the Large Scale Exercise (LSE) 2014. This LVC mix of simulation was thought to promote effective training with some level of cost avoidance over purely live events. LVC may be defined as, Live (warfighters operating weapon systems), Virtual (warfighters in simulators), and Constructive (computer simulations of entities and interactions). Historically, LSEs were conducted using live and constructive simulation. The use of virtual simulators to replace live aircraft dropping live munitions would likely have a significant cost avoidance and may provide effective training. The results of this LVC effort were reported in two different articles: "Calculating Simulation-Based Training Value: Cost Avoidance and Proficiency" and "Measuring Virtual Simulation's Value in Training Exercises: USMC Use Case," both presented at the Interservice/Industry Training, Education, and Simulation Conference (I/ITSEC) in 2015 (Cooley *et al.*, 2015; Jones *et al.*, 2015). The analysis showed considerable cost avoidance. Additionally, in "Calculating Simulation-Based Training Value: Cost Avoidance and Proficiency," the 4-year longitudinal results of the PM TRASYS cost avoidance studies were reported.

While these efforts were ongoing, others such as Dr. Syed Mohammad with University of Central Florida, Khary Bates with the Naval Postgraduate School, and Dr. Maloney and Dr. Haines of Australia also made advancements in the cost avoidance/cost benefit of M&S. Their work is discussed throughout this chapter. Chapters 14 and 15 also present a discussion of where we predict future research is needed in determining the value of simulation for different application areas.

15.2 Application of Cost Avoidance: A Component of ROI

As we discussed in Chapter 14, ROI is a term used widely throughout the business community to determine if a specific investment of funds now or over time will provide a positive return – a return greater than the investment. The investment and the return should be defined in terms of the same currency and a standard base year to account for the value of money that varies over time. In business, the investment would only be undertaken if the risk and return combination was preferable compared to more traditional investments like savings accounts, real estate investments, or the stock market (Loper, 2015). Most businesses routinely look at investment, future expenses, and anticipated sales for new entrepreneurial ventures, new/improved products, or other changes to business facilities or processes for factors that influence their bottom-line productivity. Businesses use a profit basis. Improved sales, increased profits, or lower expenses in business are often rewarded with bonuses or promotions.

So, how can we begin to answer "What do I get in return for spending one more dollar for M&S?" in DoD? In the DoD modeling and simulation market, as we discussed in Chapter 14, proclamations of savings or returns on investment may be welcomed by comptrollers eager to withdraw one program's savings in order to cover another program's funding deficit. Claims of cost savings in government would likely be rewarded with immediate budget cuts because funding is seldom considered sufficient in government, and if one office claims savings, then the savings could go elsewhere to another program. The Federal Government, including DoD, does not invest in the same way that a business may. Government looks at cost, effectiveness relative to mission, benefit, and cost avoidance basis not in terms of a revenue stream or standard ROI. If costs can be avoided, programs can cover some of their already unfunded needs.

15.2.1 Practical Cost Avoidance Calculations: Cost Avoided by Using M&S Vice Live

It is well known that M&S has a "use cost," that is, it costs something to provide electricity and other utilities, secure the building(s), staff the simulators, and maintain the M&S. When we look at cost avoidance, we calculate the use cost of the simulation or simulator with as many of these costs (that we can identify) included (Loper, 2015; Cooley and Gordon, 2012). Then we subtract this simulator use cost from the cost the stakeholder would have paid for use of the ground, naval, and air vehicles and crews and for firing or dropping the munitions that would otherwise be funded if the training was live. These costs are use costs for use of the simulator and use costs for the live assets should they have been needed to conduct the training. The difference in the two costs is net cost avoidance (if the live use costs are largest, which they normally are). And, to take the process one step further, the net cost avoidance divided by the cost of the use of the simulation environment is the ROI for using the simulator/simulation vice using live training assets.

Discussions of these cost avoidance methodologies occurred first during the research leading up to the 2008 report (AEgis, 2008). At that time, another member of the Waite family, Mr. William "Elliot" Waite, Mr. Bill Waite's son, was part of the research team, and he explained to us how cost avoidance could be simply calculated in a straightforward way, where use costs of live assets and simulation were fairly compared. The idea is to make these calculations reasonable and functional, by simplifying the calculations to not include the costs to design and acquire the simulator buildings, the simulator, nor the communication networks. The building and networks have other uses and have ongoing value for other missions. Likewise, this approach does not use the costs for development, acquisition, or logistics support for the live weapons systems and their support infrastructure. If the acquisition, logistics, and support costs for both live and simulation assets are used, these calculations would be bogged down with unnecessary details (and live weapons systems' acquisition costs would likely be much higher than simulator use costs). After considerable trial and error, only use costs were used in the cost avoidance calculation (Cooley and Gordon, 2012; Cooley *et al.*, 2015). In this way, the difference between the costs of performing the same task without M&S and the actual use costs of the M&S system can be calculated to demonstrate and validate the worth of the M&S system. (That is, as the GAO report highlighted, cost avoidance is important to demonstrate the worth of using the M&S if the simulation-based system actually provides training that improves readiness to perform the combat/operational task it was designed to train.)

15.2.2 The First Study: USMC Simulation-Based Training Systems

In the fall of 2011, Dr. Gordon and Dr. Cooley were commissioned to assist PM TRASYS in measuring the value of PM TRASYS' simulation-based training portfolio. At first, the plan for this task was to use the ROI methodology in the AEgis 2008 study (AEgis, 2008). However, using the entire methodology was quickly determined to be extremely difficult in this case. In addition, some data that should be considered, such as the electricity to operate the simulators, were not readily available. Other types of data could be located but were almost impossible to apportion accurately to the correct simulation system. For example, the Indoor Simulated Marksmanship Trainer (ISMT) is a trainer used to simulate firing small arms at targets. Every round fired by the ISMT is one real round that is not used by the USMC. This part of the cost avoidance calculation is straightforward. However, every round not fired is less wear and tear on the live range where that training would otherwise have taken place. The cost of range maintenance is known, and the ranges where small arms are fired can be identified. Clearly, those range maintenance costs can be considered as costs avoided by using the simulator vice using live fire on the range. However, further research showed that it is much more complicated to apportion the range maintenance costs. Most ranges are used for multiple types of training. Therefore, the entire range of maintenance cost should not be used in the ISMT cost avoidance, but how much should be attributed to small arms training? One method would be to apportion the costs by hours used. However, there are times when small arms training is performed concurrently with artillery or other training and, therefore, the apportionment would not be clear. Even if this were not the case, one needs to consider if the range maintenance for small arms fire is at the same level of complexity and cost as that of other uses (such as artillery shell fire), or do other uses require more maintenance per hour of use?

The benefits side of the ROI calculation causes an equivalent quandary. How do you measure the benefit returned from a simulation-based training system? While there are ways to do this with proficiency testing (a study that is discussed later in this chapter), readiness testing, and training scores, none of these data are captured or readily available. This, coupled with the inability to accurately collect all the costs, makes a full ROI study seem impossible.

15.2.3 Modified Cost Avoidance: The Significant Data

Since some of the costs avoided are unavailable and the benefit side of the ROI calculation is equally unobtainable, a modified total cost avoidance was developed. Only those data points significant to the calculation are considered. However, it was difficult to know what data were available for the most significant costs of an M&S system. This required a small amount of upfront investigation to determine if our plan for a modified cost avoidance calculation was manageable and logical. We first considered ISMT and conducted the initial investigation into data that would be most beneficial.

In performing such studies, the first step is to determine the configuration of the system and what data are available. The system was fielded at multiple sites, and the number of simulated rounds fired by weapon was available as well as the Contractor Logistics and Support (CLS) costs for the system itself. Armed with that knowledge, we gathered the data for one site, selected because it seemed to have high usage, thereby giving us insight into a high percentage of total system usage. Additionally, since this was a quick order-of-magnitude analysis, we chose only a small sample of weapons – one that is very common, such as the M-16 (high rate of use) and one that might be fairly expensive to fire live, such as a shoulder-mounted rocket (high cost per unit). As an example of the calculations, let us assume the data show that for 1 month at this one site, the two weapons "fired" 125,000 simulated M-16 rounds and 500 shoulder-mounted rockets. Assume that is a typical month. The cost of per live round fired is $0.41 for the M-16 and $2000 for the rocket, and the O&M costs for that *entire* system at that site is roughly $100,000/month. A quick calculation shows that $125,000 * \$0.41 + 500 * \$2,000 = \$1,051,250$/month would have been spent if those munitions had been fired live. Subtract the O&M costs from the cost of the two munitions fired (remember the O&M costs are for the entire system at that site, not apportioned to just these two weapons) and the rough net cost avoidance is $951,250.00/month. It is relatively easy to see that the costs avoided by using simulated fire and not live fire are probably going to be the most significant costs, and even though there are some operational costs not included, it is most likely that a cost avoidance study is beneficial (Cooley and Gordon, 2012).

Indeed, as further research was completed, it became obvious that the costs avoided by not firing live ammunition were the main drivers in the cost avoidance calculations. After subtracting maintenance costs, the entire ISMT system avoided approximately $500,000,000 in annual costs by not firing live ammunition (Gordon and Cooley, 2013). To put this in perspective, suppose each ISMT system uses 5 kW of electricity per hour (a number much higher than actual usage) and 100 ISMT systems are turned on 24 h per day for 365 days. With an average cost per kilowatt of 10.5 cents (at that time), the electricity usage for these 100 ISMT systems would be approximately $460,000 or less than 0.1% of the costs avoided by not firing live ammunition. For another data point, the total range maintenance budget across all of the USMC was on the order of $7,000,000. If even 20% of that were apportioned to ISMT (approximately $1.4M), that would only be approximately 0.3% of the costs avoided by not firing live ammunition. From methods used in this analysis, the idea of calculating and using a modified cost avoidance that only considered the data points that were major drivers in the calculations was born (Cooley *et al.*, 2015). Table 15.1 provides a

Table 15.1 Cost avoidance methodology for USMC training systems.

Cost Avoidance Methodology for USMC Training Systems

- *Overall Principle*: Net costs avoided are costs avoided by not performing training live minus cost to operate and maintain simulator/training system. Cost avoidance is reported by responding site and system. Management metrics are reported as Marines trained, hours used, and percent availability
- *Costs avoided* are only those costs that would be expended if the training was performed live:
  - Ammunition costs are calculated by taking the number of rounds fired multiplied by the cost of round
  - Vehicle costs are ideally the number of miles driven multiplied by the cost per mile to drive vehicle (this is the most accurate calculation); when cost per mile is not available, we use the number of miles driven multiplied by the cost of fuel per gallon divided by miles per gallon expected for that vehicle (in this alternative calculation, other vehicle costs for maintenance and support are not included)
  - Aircraft sortie costs are calculated by taking number of sorties multiplied by the length of the sortie multiplied by the cost per hour to fly the specific type of aircraft
- *Simulator costs expended* are Contractor Logistic Support (CLS) costs only.
  - Costs of acquisition of trainers/simulators are not included in the calculations because the calculations concentrate on use costs avoided, not acquisition costs of weapons or training systems
  - Temporary Duty (TDY) costs for Marines to travel to training simulators (and TDY costs for live training) are not calculated in this phase of the study
  - CLS costs include standard operations and maintenance costs in addition to program management/overhead costs, cost of spares, and CLS workforce travel costs
- *Total costs avoided* on training systems are calculated based on best data available.
  - Some systems have incomplete usage data, and as a result, the true cost avoidance is higher since the simulator operational costs are included
  - Some live training costs such as range support costs, lead remediation, differences in TDY costs, and naval fire support costs are not included in this phase of the calculations; avoidance of these costs would increase the cost avoidance findings
  - Simulator costs such as electricity, infrastructure operation and maintenance, and security are not included for most systems; the expenses are estimated to be insignificant, but would reduce cost avoidance
- *Training system benefits* are not considered in this phase of the project. Follow-on phases would consider that training decreases the risk of future accidents, simulation avoids damage to vehicles or injuries to personnel in live training, training can simulate a wide variety of weather and terrain conditions, simulation avoids the cost of live range cleanup, training systems placed convenient to USMC locations reduce travel of large groups of trainees and supporting forces, and similar benefits
- *Cost Data Sources*
  - Ammunition costs from Standard Unit Price list provided by the Marine Corps Systems Command Program Manager for Ammunition
  - Vehicle use costs per mile, miles per gallon, and fuel costs from the Navy Visibility and Management of Operating and Support Costs data and the Army Operating and Support Management Information System data
  - Air Force ordnance and hourly flight cost from AFI 65-503

synopsis of the methodology that includes the reporting of use data, system availability, and cost avoidance.

15.2.4 The Results of the Data Gathering Efforts for Cost Avoidance

The initial purpose was to investigate the use of simulation-based training, the costs avoided by using simulation, and the monies expended to operate and maintain the simulation systems. Indeed, one effect of calculating modified cost avoidance values was that simulation-based systems were being viewed in a different light. In one instance, a system was "revived" after being eliminated from the budget; the restoration of the program was due to presentation of use data, cost avoidance information, and other collateral beneficial information. Examples of such information that can be gleaned from use data and cost avoidance data are as follows:

- Sites use different training methods or scenarios and fire different simulated munitions.
- Sites change training methods or scenarios, as indicated by simulated munitions fired.
- Trends in training throughput and system utilization rates for sites and systems.

System utilization rates by site (number of Marines trained and number of hours trained) are key management indicators that may show low usage of a simulator at one site and high usage of that same system at another site. These data can assist in optimizing simulator placement and, therefore, ensure that the USMC gets the most value for their simulator dollars. Some sites uniquely simulate firing a specific, sometimes unique, subset of the available munitions. This begs the question: "What training techniques or scenarios are they training that no one else does?" If there is a significant increase during a particular month in the simulated use of a certain type of weapon system or munition, this can lead to investigation of training practices or other procedures that precipitated the change. In one instance, a significant increase in the use of a particular weapon in the simulator looked unusual; this munition was seldom used there previously. An investigation found this increase in use was caused by the Army beginning to use that simulator at that USMC site for their training. Analyzing the number of Marines trained and the hours expended in training at each site (training throughput) can lead to discussion and sharing of best practices and the sharing of training techniques used by the sites with higher training throughput. These best practices could promote more efficient training across the USMC and, again, better utilization of scarce training dollars. As a result, all of these management indicators were consolidated and briefed to senior leaders using charts developed as a spin-off from the cost avoidance calculations (Cooley *et al.*, 2015).

15.2.5 Fighting Off the Challenges

Challenges to assumptions, data gathering, and conclusions are natural and beneficial to promoting rigorous analyses. For any analysis, there are always those who will argue that the assumptions are unrealistic, the data are skewed, the data collection method is flawed, or there is some flaw in an analysis. The efforts of computing cost avoidance are no exception. There are those who note the fact that the live ammunition, ordnance, and aircraft hours would never be used because there is not enough budget to support purchasing or expending those assets. For example, in FY14, there were greater than 8.6 million simulated 5.56 mm rounds fired in training systems. Were all of these rounds expended in legitimate training, and would they be fired if the simulator did not exist? How would the training be accomplished if not in a simulation environment? Without simulation, some of the live training would definitely need to be increased, but how much? Would it be 10, 30, or maybe 50% of the rounds shot in simulation? We often prepared tables that illustrated the cost avoidance realized if 10, 20, 30, and up to 100% of the simulation-based cost avoidance were counted. However, there must be some way to partition the simulation use into some portion being essential for training. Certainly, there are some instructions in Training & Readiness (T&R) manuals that require simulator use, and, in that case, the simulated rounds fired can be directly tracked to T&R requirements.

The following analysis was performed for M16/M4 recertification requirements. Using the number of Marines that must be recertified each year in M16/M4, the firing tables that must be completed for recertification, and the T&R manual that states what can be performed in a simulator, it was shown that approximately 35% of the 5.56 mm rounds fired in the simulator could be directly tied to the recertification requirement alone. This percentage is low because it does not consider familiarization training for basic training, initial certification training (which requires considerably more rounds to be fired), operational scenario training prior to deployment, and other required uses of the simulator. The tank gunnery proficiency study described later in this chapter (Dunne *et al.*, 2014) tied the simulated tank ammunition expended directly to semiannual certification training and calculated an annual net cost avoidance of $1M/crew just for semiannual certification. This use case accounts for virtually all the cost avoidance for the tank gunnery system. Certainly, there are some simulated shots fired in system maintenance and testing and perhaps more shots fired in a simulated training scenario than would be fired for the same scenario on a live range. However, even if only 70% of the total net cost avoidance (suspected to be a low number) were directly tied to training requirements, the total cost avoidance across the PM TRASYS portfolio would have been greater than $2B per year for FY12 and greater than $4B for FY14, with the increase in cost avoidance by FY14 due to increased usage and more accurate data collection (Cooley *et al.*, 2015). Therefore, the net cost avoidance methodology, as described, serves well to assist in justifying simulation-based training systems designed to prepare Marines for combat.

15.3 Beyond Cost Avoidance: Proficiency Proof-of-Concept Study

In 2013, the GAO did a study on simulation-based training across the DoD. Of note was a statement in the Navy and USMC report that said:

> "Without a means to assess the impact of using simulators on performance and to compare the costs associated with live training and the use of simulation-based training devices, decision makers in the Army and Marine Corps lack information to make fully informed decisions in the future regarding the optimal mix of training and related investment decisions." (U.S. GAO, 2013)

In layman's terms, this statement means "Cost avoidance is nice, but if the training is not increasing the skills of the trainees then cost avoidance is meaningless." This led to PM TRASYS commissioning a cross-disciplinary team to

- Determine if proficiency be measured and what are the metrics, and
- Perform a limited proof-of-concept study to show how this can be done if the answer to the first question is yes.

This team, led by Dr. Robert Dunne and U.S. Marine Captain Sam Oliver, supported by Dr. Cooley and Dr. Gordon, assessed the impact of simulator-based training on the proficiency of crews in the Abrams Main Battle Tank (M1A1) Advanced Gunnery Training System (AGTS) simulator (Dunne *et al.*, 2014). In looking for prior research on the topic of increased proficiency that could be realized from simulation-based training, we noted that Canadian researchers utilizing Virtual Battlespace 2 (VBS 2) and One Semiautomated Forces (OneSAF) had shown the mostly anecdotal increase

Table 15.2 Average AGTS simulator and live-fire scores for all tasks by crew.

Crew	Average beginning AGTS scores[a]	Average ending AGTS scores[a]	Δ = ending – beginning scores	Average live-fire qualification score[a]
1	63.6	93.0	29.4 (46% increase)	90.7 (result: qualified)
2	55.2	81.9	26.7 (48% increase)	85.0 (result: qualified)
3	53.6	87.0	33.4 (62% increase)	78.0 (result: qualified)

a) In conjunction with other evaluation requirements, a passing score is 70.

in training effectiveness with the use of Serious Games (Roman and Brown, 2008). But, there had not been a rigorous study of proficiency gained or not gained from the use of simulation-based training programs. The first attempt at a rigorous study involved following the only three tank crews that remained intact in training through several hundred firing exercises. Utilizing these data from the three crews through simulation-based training and live-fire qualification for the proof of concept, the team from PM TRASYS determined that the scores on individual gunnery tasks completed in live-fire qualification could be used as metrics to demonstrate proficiency in each task. They further noted that performance in the AGTS appeared to translate to performance on the live range. The increase in scores and the subsequent live range qualifying scores are summarized in Table 15.2.

As important as the proficiency study was, the connection of cost avoidance to real-world training requirements was also shown as significant. The usage data for the rounds fired by type were collected, and the simulator operations and maintenance costs were prorated to give a net modified cost avoidance just for the training progression from new gunner to qualified combat-ready gunner. The results in Table 15.3 show a net cost avoidance of

Table 15.3 Number of simulated rounds fired by crew and cost avoided.

Crew	Munition 1	Munition 2	Munition 3	Munition 4	Munition 5	Munition 6	Total
1	729	748	7	17	21	110	1,632
2	460	1,445	4	8	34	75	2,026
3	999	877	29	36	48	176	2,165
Subtotal	2,188	3,070	40	61	103	361	5,823
Cost	$2,079	$10,684	$107,037	$163,232	$275,621	$966,011	$1,524,663

approximately $500,000 per crew for the qualification training. Each entry is the number of simulated rounds fired of that munition by that crew to complete the qualification training. The total cost for all three crews for that specific munition is listed in the last row. If this training can be assumed to be representative of the required training leading up to qualification, and there are two qualification cycles per year, then the costs avoided by using the AGTS for training for these three crews per year is $3,034,910 or approximately $1M/year/crew to maintain combat qualification.

The 2013 GAO report made a logical point that calculating the economic benefit – or cost avoidance if that is how benefit is measured – of a simulation-based training system may be necessary, but this is not sufficient. The training system must also be shown to be an effective trainer, allowing warfighters to improve their operational readiness skills. So, for the AGTS, the study team was tasked to find a way to determine if the simulation-based training improved the effectiveness of the crews to perform combat operations in live-fire qualification. In this training proficiency study, the results showed trends of improved proficiency and considerable cost avoidance ($1M/year/crew) for using the simulation-based AGTS training system.

It should also be noted that, as in many simulation-based training systems, one purpose of the trainer is to make the warfighters and warfighting crews knowledgeable in the safe operation of the system. (This "safety" purpose is also present in sidearm and rifle-fire trainers and in emergency procedures trainers for combat vehicles.) In this case, the AGTS has fundamental purpose of safety training for tasks such as loading tank ammunition, moving to engage, and firing on and stopping the enemy. Therefore, the study team could not safely evaluate how the crew performed in live fire without having the benefit of AGTS training. This would have created a "control" group for the study, but would have caused safety concerns – at the least.

Finally, ideally, training effectiveness evaluations should be part of the systems engineering of new or modified training systems. And, where possible, the data collected from the simulation-based training systems (using an automated process) should be designed upfront to provide useful data for management decisions and cost avoidance/ROI calculations. This includes proficiency data for effectiveness studies and upgrade decisions for the fielded system.

15.4 Additional Contributions for Cost Avoidance

As noted in the overview, there are many who have contributed to this body of work in the past several years. Dr Mohammad's thesis from work at the University of Central Florida in 2012 entitled, "Application of Modeling and

Simulation to Reduce Costs of Acquisition Within Triple Constraints," provided a simple but useful data collection form for cost data from simulation use (Mohammad, 2012). His form was also used to track the cost of doing tests in a live environment (more traditional testing). The comparison of M&S that could be used in some acquisition case studies showed that M&S provides a useful venue for evaluating programs in design, analysis, and testing in the acquisition process while delivering considerable cost savings and other benefits. The benefits appear to be reductions in future failures and repairs. The author states that the benefits are too complex to be quantified, hence detracting from the capability to do a true cost-benefit analysis. In looking at various case studies where M&S could be leveraged, the author also concludes that not all situations may require or be appropriate for M&S use, and the author found numerous situations where M&S was a detractor to cost, schedule, or performance constraints (the triple constraints).

From the Naval Postgraduate School, Bates (2011), Lieutenant Commander, U.S. Navy, in his MBA thesis, "Cost Analysis and Effectiveness of Using the Indoor Simulated Marksmanship Trainer (ISMT) for United States Marine Corps (USMC) Marksmanship Training" built on the findings of a previous study (Yates, 2004) that determined there was no statistical difference in the scores of recruits whether they were trained in the simulation-based ISMT plus live-fire qualification versus a control group that trained only on a conventional firing range followed by qualification. In that earlier 2004 study, differences in weather conditions may have influenced the firing data and the outcome. For the current 2011 study, the author conducted a cost analysis to determine the financial advisability of using the ISMT in basic marksmanship training versus using live-fire training alone. This study focused on the M16A2 service rifle and the simulated and live-fire equipment, supplies, and logistical support for training. The author reasonably assumes that, as in the previous study (Yates, 2004), there will not be a significant reduction in training standards or marksmanship scores by using the ISMT before live-fire range training. "Therefore, an appropriate mix of ISMT with live-fire range training should not reduce the overall effectiveness of the Marine Corps' basic marksmanship training standards" (Bates, 2011). The author concludes that the cost analysis shows that ISMT yields the highest total net present value of savings compared to live-fire range operations (without ISMT). (*Note:* The term "net present value" indicates that savings realized over time are quantified (discounted) to the equivalent current year dollars because a dollar in your hand today is worth more than a dollar, for instance, promised in 3 years.) In other words, in terms of current year funding (2011 at that time), the use of ISMT in marksmanship training over its life cycle maximized the cost avoidance versus using live-fire only for training. The author also demonstrates that there is no significant difference in qualification scores between the ISMT plus live-fire test subjects and the live-fire control group.

As discussed in Chapter 14, the medical community is interested in cost-benefit and cost-effectiveness as discussed by Dr. Maloney and Dr. Haines in their article "Issues of Cost-Benefit and Cost-Effectiveness for Simulation in Health Professions Education" (Maloney and Haines, 2016). Interestingly, in the health professional education field, the authors recognize that the use of simulation in training must be justified, the metrics used may be judged in financial terms and in less tangible terms, and the perspective of the stakeholders must be considered. They state that there is a need for universities to identify efficient educational methodologies so that their courses can remain competitive in a global marketplace. The authors hypothesize that the appropriate use of quality simulation practices will more deeply engage learners, improve their motivation to learn and their clinical performance, and, over the long term, change their behavior. These changes could have cascading effects to promote positive changes in clinical practice, which may, in turn, lead to better health outcomes as postulated in "A financial benefit in simulation education may be the cost averted through students making fewer errors in real patient situations" (Maloney and Haines, 2016).

As also noted in the 2008 M&S ROI study (AEgis, 2008), whether the benefits are quantitative or less tangible, the costs and benefits may be distributed to different stakeholders. Someone pays, others benefit, depending on what level of stakeholder perspective is used. It is important to consider all the costs and benefits relevant to a particular stakeholder or viewpoint chosen for the analysis, and there should be an attempt to increase the level of the analysis to a stakeholder level that "owns" all the costs and benefits. The ability to conduct the analysis at the right level – in order to consider all the costs and benefits – is key to an accurate evaluation of the total cost avoidance.

In an article describing a system developed by Calytrix Technologies for a customer in Australia, the LVC Cost Counter, first fielded in 2012, was described in "Counting the Costs of Simulation" (Calytrix, 2012). The LVC Cost Counter monitors a computer network using typical architecture formats used for simulation environments (such as Distributed Interactive Simulation and High Level Architecture standards) and accumulates and displays the individual, category, and total costs associated with simulation events. It logs all events during an exercise that would incur cost in the real world and automatically calculates an approximate cost that is based on the resources consumed. The Cost Counter allows costing data to be collected and analyzed, and the system enables users to create a number of different costing tables, each containing data such as, but not limited to, the cost of munitions, operational cost per hour, operational cost per kilometer traveled, daily maintenance rates, the costs of simulation, average salary costs, and platform damage costs (combat damage if live).

15.5 Case Study of Multi-Attribute Assessment

We felt a case study would help clarify many of the thoughts we have presented in Chapters 14 and 15 and tie together several concepts and cost-benefit research. This example is derived from a larger case study presented in 2008 (AEgis, 2008), and it discusses the first experiment in Alaska that considered and actually used a Semiautomated Forces (SAF) simulation. SAF simulations have built-in and selectable human behaviors for all forces, and they only need a small number of control staff to run the simulation and scenario. The experiment was designed to look at various systems, processes, and assets that could be used to best rescue a downed airman. Traditionally, simulations are distributed throughout the continental United States (CONUS) and Alaska for these types of events. Sometimes, the events are conducted entirely with live forces. In this instance, the program manager (PM) was tasked to find the best simulation solution and conduct the experiment perhaps many times with various systems being evaluated to save downed airmen.

15.5.1 Problem Statement

The PM would like to conduct a small 4-day experiment in Alaska to test the combat benefit of a new system for position determination of friendly ground forces. The scenario will be the rescue of a downed aircrew member wearing the new position location system. The PM will need to evaluate alternative simulations for use in this experiment. The cadre of simulation operators, even for the most commonly used simulations, is limited in Alaska, so the PM must compare various simulations and evaluate the need to distribute the simulation environment from another location. Friendly forces could be brought into the experiment live, represented through a constructive simulation, inserted in the scenario by a simulator where the warfighters operate mock-ups of real systems (virtual simulation), or inserted as a combination of all these. The position determination system may need to be simulated, assumed, or worn, depending on the readiness of the position reporting technology. Simulation databases for Alaska are limited, particularly for SAF simulations, so databases for geography and other environmental factors may also need to be purchased with sufficient lead time. Connectivity and simulation architecture costs must be evaluated. A comparison of the costs and benefits of the different simulation alternatives, estimate of the cost and benefit of conducting the experiment using all live forces, and calculation of cost avoidance estimates must be considered. The experiment will use a weighting decision support system to combine various benefits with the cost avoidance metrics to evaluate the "best" solutions.

15.5.2 Scenario Description

The scenario selected for an event will affect the costs and results from the analysis of alternatives (AOAs). In this case, the PM develops an event scenario and force lay down that is the minimum necessity to evaluate the experiment's objectives. Here are the basic elements:

- Three helicopter crew members acting as downed crew members evading hostile forces in rough terrain. In some trials, crew members will use current radio and identification systems, and in other trials, the new position determination system will be used. These crew members could be represented in a simulation, yet for this event, in all cases, they will be represented live.
- Hostile, mobile ground forces searching for evading crew members can be portrayed entirely by live forces or, alternatively, by a mix of live and simulation. A limited number of live forces will be necessary to preserve the reality of the chase.
- Friendly ground command and control forces, in this case a Combat Search and Rescue (CSAR) team linked to an Air Operations Center, augmented with the necessary radar and radio feeds, will be used. This team and facility could be represented in a constructive simulation, but they will be live to provide feedback on the experiment.
- Friendly airborne surveillance forces will be used either in the scenario as live, virtual, or in computer simulations (SAF or distributed to Alaska), as available, to complete this role.
- Friendly CSAR rescue team will conduct the live rescue and reinsert the three evaders for subsequent trials, and the CSAR rescue team will need to be live.
- Friendly combat air patrol (CAP) forces will be represented, in this case, virtually in the simulation environment by three fighter cockpit simulators.
- Other necessary command and control, surveillance, and friendly and enemy forces will be assumed to be functioning as usual but will be depicted via simulation in the event.

15.5.3 Stakeholders and Options

The stakeholders are the Commander of Alaska Command (ALCOM/CC), PM for the position determination system, PM planning the experiment, T&E community, and warfighting commands that require the system. While costs are shared across these stakeholders, an overall cost to the entire enterprise is used to evaluate the alternatives. One decision criterion is based on selecting the simulation alternative for this event that is the lowest cost only if the predicted *cost avoidance* as compared to the traditional simulation support is at least 20% more. The simulation assets considered in this case study are in these three alternatives: traditionally used simulations distributed from CONUS; SAF simulations run in Alaska; and live forces.

- Alternative A is the traditional M&S solution used for other events in Alaska; it can be distributed to the Alaskan theater from another simulation center. This alternative uses 100 personnel at one or more locations for the simulation operators, control force (controlling the simulation), white cell (monitoring experiment or training progress), and role players (representing unmodeled forces). Twenty-five personnel will need to attend two planning conferences in Alaska. No databases will need to be developed, no training of simulation personnel in Alaska will be required, and there will be negligible residual benefit to the Alaskan simulation center. Floor space in two facilities will be required, but no hardware or software will need to be purchased. Some distribution lines to and within Alaska will have to be reserved and leased. Reuse during subsequent years will be as if the support were a new start.
- Alternative B uses SAF simulation to reduce personnel required to run the simulation. For this SAF simulation, two people plus a backup team member will be required in Alaska. Three operators will double as white cell, control force, and role players; they will need a 2-week SAF simulation course in Virginia. A SAF simulation database will be developed at a cost of $50K with a lead time of 60 days. The leave-behind for this option is a trained cadre of three SAF operators to conduct similar events using the same database on subsequent occasions. This alternative requires two planning conferences for 10 personnel on travel to Alaska. One new laptop with additional memory will have to be purchased, but no software will have to be purchased. Software to link the SAF simulation to the Common Operational Picture (COP) is government-owned. Distribution lines within Alaska will be reserved and leased. This option will allow reuse during the current year and for the following years with minimal additional cost and advanced notice.
- Alternative C uses all live forces for the transport, rescue, humanitarian relief, and disaster recovery roles. All other secondary roles will be

represented in simulation due to the realities of costs and operational tempo. The use of the locator device will be simulated by downed aircrew in the same way for each alternative. Live forces will require planning conferences similar to Alternative A, and the operating cost for the live assets will be computed by using fictional cost values, but for actual analyses, the current year military budget documents and munitions cost lists can be used for cost per flying hour and munitions costs, if used. Additional communications lines do not need to be leased. This alternative would be difficult to repeat without long lead times and an equivalent expense for each event.

15.5.4 Multiple Attributes and Weighting

For this case study, the PM and stakeholders (mainly ALCOM/CC) want to use the simulation system for short notice experiments and training events in Alaska. So, the weight of this "reuse" attribute will be assigned as medium. Another attribute that is important to the stakeholders is the "comfort level" with the simulation or ease of use. This will be rated as a low. A third attribute is the number of training and readiness "tasks" that the simulation environment can exercise for training the force. This is a fairly important attribute, and the weighting for the attribute will be high. Finally, the "cost avoidance" attribute will be rated as medium. In a multiattribute decision-making (MADM) system, the weighting of all attributes must total to 1.0; so the weights will be Reuse (.2), Comfort Level (.15), Tasks (.35), and Cost Avoidance (.3).

15.5.5 Costing Data and Cost Avoidance Results

Since costs for this event will be expended in the current year (even though some planning must take place many months before the event), no discounting of costs is necessary. Avoided costs in terms of personnel and systems will be expressed in current year costs for this comparison (to avoid appreciating and/or discounting). Some of the data that the PM needs would typically come from or through management. So, whether the PM is directed to do the cost element analysis or decides independently to do the analysis, the PM will need some information on labor and per day costs for personnel, the number of personnel needed for specific duties, systems' cost data from the DoD Budget processes or other cost tables, and infrastructure costs. Some of these data can be found or calculated by the PM, but some, or all, of the data should be reviewed/approved by management so that the analysis starts with approved data. This prior review/approval reduces the likelihood that decisions will stall on the disagreement of where the data originated. This is a sample of the estimated costs used for this comparison, and these costs are used to populate Table 15.4:

- Travel costs are $1500 per trip per person, and $200 per day per person.
- Labor costs per person are 8 h per day, 40 h per trip, and $100 per hour.
- Operational assets are estimated at 4 h of use per day at $20K/h for the helicopter, $25K/h for the ground C2, $100K/h for the AWACS, $150K/h total for the three-ship of air-to-ground assets, and $200/day for the truck.
- Ground support for the rescue helicopter is 6 personnel; 48 ground personnel is deployed to support the live assets, if used, such as the AWACS aircraft and 3 air-to-ground fighters.
- Planning conferences are expensed at 40 labor hours, 6 days, and 6 nights in lodging.
- Network use (8 h daily) is $1K/h in Alaska; $2K/h to and from Alaska.
- Computer purchase is $2000, with the cost for use in an event estimated at $200/event.
- Facility charges are $5K per day per structure used in Alaska or elsewhere.

As depicted in Table 15.4, the bottom-line total costs indicate that use of the SAF simulation is the lowest cost alternative; it also provides a trained resident cadre of SAF operators who can run follow-on events, a leave-behind SAF database, added realism on the COP, and a smaller support staff requirement. Since SAF simulation is a proven asset that runs in the simulation center in Alaska, it will reduce risk because of

Table 15.4 Costs for elements of training event alternatives.

Cost element	Alternative A = traditional distributed M&S	Alternative B = semiautomated M&S	Alternative C = all live forces
Hardware systems			
Computer for COP ($200.00/event)		$200.00	
Truck for hostile forces (same for each)	$800.00	$800.00	$800.00
Helicopter for evading force insert	$320,000.00	$320,000.00	$320,000.00
AWACS aircraft			$1,600,000.00
3 Fighter aircraft			$2,400,000.00

(continued)

Table 15.4 *(Continued)*

Cost element	Alternative A = traditional distributed M&S	Alternative B = semiautomated M&S	Alternative C = all live forces
1 Control and Reporting Center (CRC)			$400,000.00
Networks			
In Alaska ($1K per hour)	$1,600.00	$1,600.00	
To/from Alaska ($2K/hour/center)	$8,000.00		
Facilities ($5K per day/structure)			
Alaska Simulation Center	$5,000.00	$5,000.00	$5,000.00
Simulation Building #1	$5,000.00		
Simulation Building #2	$5,000.00		
Aircrew & Ground Support Building #1	$5,000.00	$5,000.00	$5,000.00
Aircrew & Ground Support Building #2			$5,000.00
People (40 h and 6 days and nights/week)			
3 Evading crewmembers	$20,100.00	$20,100.00	$20,100.00
6 Hostile forces	$40,200.00	$40,200.00	$40,200.00
SAF simulation training		$40,200.00	
CSAR team of 20	$134,000.00	$134,000.00	$134,000.00
AWACS team of 10	$67,000.00	$67,000.00	$67,000.00
One pilot each in three Fighter Sims	$20,100.00	$20,100.00	$20,100.00
Simulation staff	$670,000.00	$20,100.00	
Planning Conference #1	$335,000.00	$67,000.00	$335,000.00
Planning Conference #2	$335,000.00	$67,000.00	$335,000.00
5 Rescue helicopter crewmembers	$33,500.00	$33,500.00	$33,500.00
Ground support for live assets	$40,200.00	$40,200.00	$321,600.00
Products			
SAF database		$50,000.00	
Total cost (current year)	$2,045,500.00	$932,000.00	$6,042,300.00

dependability and lack of dependence on long-haul distribution from distance simulation centers.

The SAF database and cadre of operators provides an accelerated reaction time for future needs. The SAF simulation is more than 20% cheaper than the traditional distributed simulation solution, so the SAF solution surpasses that stated hurdle. Since cost avoidance is calculated as Live-versus-SAF or Live-versus-Traditional, the cost avoidances are positive at $5,110,300.00 and $3,996,800.00, respectively.

15.5.6 MADM Calculations

The attributes and weights are reuse (0.2), comfort level (0.15), tasks (0.35), and cost avoidance (0.3). For each attribute, the maximum value that gets a score of 1.0 is as follows:

- Reuse maximum 1.0 = 1.0
- Comfort level maximum of 1.0 = 1.0
- Tasks completed maximum of 1.0 = 1.0
- Cost avoidance maximum of 1.0 = $6,000.000.00

In Table 15.5, the overall MADM utility score (highest is best) for the SAF alternative is a winning 0.65. The traditional alternative is a close second with a utility score of 0.58. Please note that the score would change if the weights or the ratings of the various alternatives relative to the metrics change. The weights and ratings are commonly generated by consensus from a group of stakeholder representatives and approved by senior management prior to the computations of the MADM utility scores.

Table 15.5 MADM utility score calculations.

Metric (input, factor)	Best for input	Traditional	SAF	Live	Weight for input
Ease of use	1	0.4	0.8	0.2	0.2
Comfort level (stakeholders)	1	0.6	0.4	1	0.15
Task completed	1	0.6	0.5	0.8	0.35
Cost avoidance	$6,000.000.00	0.66613	0.85172	0	0.3
	Sum = MADM score =	0.579839	0.650516	0.47	

15.6 Future Efforts and Parting Thoughts for Chapters 14 and 15

It is difficult, at best, to see into the future and predict what funding might be available for future studies. However, even the most pessimistic view must concede that cost avoidance studies can be persuasive and were, for some systems, a resounding success in their preservation. The results and conclusions assisted in protecting funding for simulation-based training by showing net costs avoided in the billions of dollars. Additionally, the proficiency research was key to tying cost avoidance to actual required training and readiness requirements. Tracking and matching the levels of simulated munitions fired to the actual weapons firing requirements in T&R manuals further solidified the validity of some cost avoidance studies. In the DoD environment, additional proficiency studies of simulation-based training systems with ties to cost avoidance should be completed. This further supports the value of this methodology and provides decision-makers better data to efficiently optimize their training systems. In the MADM utility score calculations, the relative proficiency improvements of various options could also be included in the calculations; proficiency would likely deserve a high or medium weighting.

In other communities, particularly the medical community, it is likely that training budgets will shrink to a level where they do not meet requirements. When that time arrives, the techniques and approaches outlined in both Chapters 14 and 15 will help determine and manage alternative, competing M&S investments. Additionally, those same techniques could assist in managing and optimizing live, virtual, and constructive types of medical training: determining the best mix of each of these modalities to produce the best trained medical professionals – at the lowest cost and in the shortest period of time.

There is another way the methods presented in this chapter may be useful. These data are used for management information (use data), cost avoidance, and proficiency evaluations; the data can also be very useful in systems engineering and training effectiveness evaluations of modified or improved training systems. For instance, a customer may have different modifications that they can afford to finance for an existing training system, and bidders for development and fielding of a new training system may offer alternative solutions to the training needs. These alternative choices may be evaluated based on cost and technical adequacy. Also, evaluating the alternatives based on predicted use data, cost avoidance (or ROI), and improved proficiency in existing or new tasks could provide key decision metrics for better buying power also. This topic deserves full development in future papers and technical reports. It is important here to briefly explain how the work described in Chapters 14 and 15 can help us buy better training systems in the future. There are ways that proposals may differ beyond cost and technical adequacy to allow differential evaluations of these types of benefits: predicted use data, predicted cost avoidance, and predicted proficiency improvement. If the proposed solutions differ in terms of the predicted number or rate of personnel trained (throughput), operational availability, ease of learning and use, training duration, maintenance costs, or other categories of use data, the differences can be evaluated. If the proposed solutions differ in terms of training and readiness (T&R) tasks or the realism of T&R tasks that can be trained, then some potential solutions may allow different training scenarios, use of different weapons and munitions, and, thus, differences in predicted cost avoidance for equivalent throughput. Finally, because of differences in the proposed solutions, some potential solutions may improve proficiency in distinct combat or operational readiness areas. These solutions may be more effective than the training effectiveness designed into the required solution. These three areas of use data, cost avoidance, and proficiency improvement can be combined with acquisition cost data, life cycle cost data, and technical compliance in a weighted score derived from a methodology like MADM to find the winning proposal for better buying power.

We think the techniques proposed in the paragraph above will, in some form, be incorporated in evaluations of alternatives for future training systems. Simulation-based training systems will continue to be fielded because live training using more complicated systems, tactics, and interdependencies will be too expensive and disclosure sensitive. Constructive and virtual M&S, where it can be made an effective alternative, will be the training environment of choice. M&S takes on many forms and has been around for many years; however, it only came to prominence with the advent of the computer age. In some cases, such as modeling landings on distant planets or modeling nuclear explosions, the only rational choice to conduct an analysis or solve a problem is to employ M&S. In all other use cases for M&S where classroom lectures or live training could be used, the question of the value of M&S will continue to be asked. Over the past 20+ years, techniques have been refined to allow better answers to those value questions. As society moves more and more into wearable technologies, driverless vehicles, and increasingly faster data speeds, more opportunities for M&S will arise – creating even more questions about value, proficiency, and ROI. With proper foresight, the data will be there to answer these types of questions; techniques will be refined, and new techniques will be developed just as they have been over the past 20 years. One thing is certain: The quest started in the 1990s to calculate the value of M&S, championed chiefly by Mr. Bill Waite – the consummate marketer, wearing a U.S. dollar on his lapel and handing out dollars with invitations to attend his first sessions on "Why Spend One More Dollar on M&S?" (We still visualize this in our minds!) – was timely and wise, and it will continue, carried on through those mentored and influenced by Bill, in this ever-changing technological world.

Acknowledgment

The authors wish to thank Ms. Margaret Callahan of Simulation U Analytics LLC for her outstanding assistance with graphics, formatting, and locating correct references.

References

AEgis Technology Group, Inc . (2008) Metrics for modeling and simulation investments. Report No. TJ-042608-RP013. Available at http://www.ndia.org/Divisions/Divisions/SystemsEngineering/Documents/Committees/M_S%20Committee/2009/April%202009/Metrics%20for%20MS%20Investments_Final%20Report_07_SBCI_009_20090319.pdf

Bates, K. (2011) *Cost analysis and effectiveness of using indoor simulated marksmanship trainer (ISMT) for United States Marine Corps (USMC) marksmanship training.* Naval Postgraduate School, California.

Calytrix. (2012) *Counting the costs of simulation, simulation savings: LVC cost calculator.* Available at http://www.calytrix.com/products/costcounter/

Cooley, T. and Gordon, S. (2012) Cost avoidance for M&S training systems: a subset of return on investment. *M&S Journal*, 7 (2), 35–40.

Cooley, T., Seavers, G., Gordon, S., Roth, J., and Rodriguez, J. (2015) Calculating simulation-based training value: cost avoidance and proficiency. *Proceedings of the Interservice/Industry Training, Simulation, and Education Conference,* Orlando, FL.

Dunne, R., Cooley, T., and Gordon, S. (2014) Proficiency evaluation and cost-avoidance proof of concept: M1A1 study results. *Proceedings of the Interservice/Industry Training, Simulation, and Education Conference,* Orlando, FL.

Gordon, S. and Cooley, T. (2013) Phase One Final Report: Cost Avoidance Study of USMC Simulation Training Systems.

Jones, N., Seavers, G., and Capriglione, C. (2015) Measuring virtual simulation's value in training exercises: USMC use case. *Proceedings of the Interservice/Industry Training, Simulation, and Education Conference,* Orlando, FL.

Loper, M. (2015) *Modeling and Simulation in the Systems Engineering Life Cycle,* Springer, London, UK.

Maloney, S. and Haines, T. (2016) Issues of cost-benefit and cost-effectiveness for simulation in health professions education. *Advances in Simulation*, 1, 13.

Mohammad, S. (2012) Application of modeling and simulation to reduce costs of acquisition within triple constraints. Doctoral dissertation, University of Central Florida, LD1772.F96T45 2012 No. 207.

Roman, P. and Brown, D. (2008) Games: just how serious are they? *Proceedings of the Interservice/Industry Training, Simulation, and Education Conference*, Orlando, FL.

U.S. GAO (2013) Army and Marine Corps Training: Better Performance and Cost Data Needed to More Fully Assess Simulation-Based Efforts. GAO-13-698.

Yates, W. (2004) *A training transfer study of the indoor simulated marksmanship trainer*. Naval Postgraduate School, California.

Part VI

Policy

16

Building a National Modeling and Simulation (M&S) Coalition

Randall B. Garrett,[1] James A. Robb,[2] Richard J. Severinghaus,[3] and Richard Fujimoto[4]

[1]*National M&S Coalition, Washington, DC, USA*
[2]*National Training & Simulation Association, Arlington, VA, USA*
[3]*CRTN Solutions, LLC, Washington, DC, USA*
[4]*Georgia Institute of Technology, Washington, DC, USA*

16.1 Chapter Summary

From its early beginnings of mechanical calculators through the development of the first true algorithm-based computing machines in the late 1940s, the emergence of "computing science" and expansion of computer simulation and modeling techniques have caused a revolution in how industry, academia, and government employ the technologies we now term "M&S." Computer science, first recognized in the early 1950s and 1960s as a distinct academic discipline, has expanded and transformed into a much broader technological discipline and practice. Initially embraced by the military and research communities, M&S has grown to impact nearly every aspect of national commerce, research, business, national defense, and economic analysis (Goldsman *et al.*, 2010). More specifically, over the past seven decades, M&S has contributed immensely to advancements in myriad of disciplines, prominent among them nuclear testing and disarmament, space exploration, medicine, homeland security, weather and climatological forecasting, national defense (including military strategy and operations), manufacturing, urban planning, and national-level policy analysis, and economic planning and forecasting, just to mention a few. These contributions have driven substantial improvements in national productivity and the United States economic growth.

Over this period of growth, the M&S communities of interest came to recognize a distinctive discipline, workforce, industry, and market for its products and services. One example of this is the evolution of M&S as a

The Profession of Modeling and Simulation: Discipline, Ethics, Education, Vocation, Societies, and Economics, First Edition. Andreas Tolk and Tuncer Ören.
© 2017 John Wiley & Sons, Inc. Published 2017 by John Wiley & Sons, Inc.

profession, resulting in 2002 with the creation of a Certified M&S Professional (CMSP) certification program. The idea is to leverage the body of knowledge and provide industry with its own professional certification, much like a professional engineer (PE). Today the M&S community of practice has expanded and recognizes an identifiable body-of-knowledge, supported by academic degree programs and professional certifications. Accordingly, the discipline has been embraced by professional societies and organizations among varied domains that include manufacturing, engineering, medicine and healthcare, automotive industry, aeronautics, cybersecurity, human behavior, and their associated academic forums, to name a few. Five years later, Congress recognized M&S in House Resolution (H. Res.) 487 on July 16, 2007, as a "National Critical Technology" (House Resolution 487, 2007).

Today, M&S is recognized as a distinct discipline, embodying a professional workforce, industry, and market for its products and services. The United States leads the world in the application of M&S, particularly in defense, where these technologies have been used to support analyses of strategic operations, systems acquisition, and training of military forces. Typical estimates for annual direct expenditures made in the United States for M&S products and services are approximately $50 billion. In recognition of this transformation of computing science into a broad-based approach to addressing and solving complex problems, a small group of dedicated "simulationists" initiated an effort in 2012 to create the National Modeling and Simulation Coalition (NM&SC)[1]. This was the direct result of the growing influence and impact of M&S in many areas of the national economy.

Current efforts continue to "expand the coalition" with a full understanding that M&S supports a foundational science for mathematics, analysis, engineering, testing, training, and operations of systems critical to industry, defense, homeland security, emergency management, healthcare, electronics, energy management, transportation, education, banking and finance, and work force creation. To this end, M&S has been endorsed to the President's Council of Scientific Advisors (PCAST) as a key to increasing the security and economic well-being of the State and the Nation.

16.2 Early Years

Some of the earliest work using models and simulations for computational (electronic/computation-based) products dates to the late 1920s, when Edwin Link built the first flight simulator (Bergeron, 2006). The field

---

1 National Modeling and Simulation Coalition. http://thenmsc.org/.

received a big boost in the years of WWII and beyond, and another in national response to the Cold War years. The 1950s saw the first use of computer-based military simulation games, and during the 1960s, researchers in the field of medicine began work in the area of computer-based medical instruction. The introduction of the personal computer (PC) in the late 1970s (the first Apple computer was introduced in 1977), IBMs entry into the PC market in 1981, and the creation of the U.S. Army's Simulation Network (SIMNET) and its validation as an effective training system during desert storm, created yet another huge surge of development. In parallel, the commercial gaming industry developed *Pong*, widely regarded to have started with the 1972 debut of Atari's arcade game.

M&S as a formal discipline that affected actions and outcomes in real life, along with the acknowledged trajectory of events described above, lead to creation of the NM&SC, arguably conceived at the first meetings of a U.S. Army group in Orlando, FL in 1989. These meetings eventually led to creation of the Distributed Interactive Simulation and High-Level Architecture computer simulation protocols that are today's mainstays of military M&S. Concurrent meetings led to the stand-up of the U.S. Defense Modeling and Simulation Office (DMSO). From its beginnings, DMSO strove to improve M&S interoperability and reuse as well as to drive growth of M&S for both military and commercial benefit. The year 2002 saw the establishment of an all-volunteer, internationally inclusive, nonincorporated organization, called the "SimSummit" Roundtable. The Roundtable was organized to create cross-disciplinary interaction among three organizations closely identified with M&S; the Simulation Interoperability Standards Organization (SISO) established in 1996, the National Training and Simulation Association (NTSA), and the Society for Modeling and Simulation International (SCS).

Recognizing and responding to this multidecade trajectory of M&S contributions to industry, the Department of Defense (DoD), the scientific community and the U.S. House of Representatives established the Congressional M&S Caucus in 2005. The Caucus was envisioned to showcase M&S initiatives, promote the industry, and to provide a forum to understand the policy challenges facing this growing and versatile technology. Adopting the SimSummit topical agenda[2] as a framework for discussion during its first four annual meetings, the Caucus attracted a growing membership of individuals and institutions having either an economic interest in M&S activities or concerns about the advancement of technology

---

2 Four topics, Technology, Professional Development, Industrial Development, and Business Practice, each address to four domains: Commercial, Education, Government, and Societies.

in the United States. Today, the Caucus has grown to over 30 members of Congress actively promoting the growth of M&S activities throughout the United States via advocacy and legislation. Actions taken by the Caucus include passage of House Resolution 487 (described below) and in 2008, the Higher Education Opportunity Act (20 USC 1161v) renewing the 1965 Act of the same name. Included in this Act are incentives for colleges and universities across the country to establish programs dedicated to the study of M&S.

16.2.1 M&S Leadership Group

The concept of an M&S leadership group and Congressional Caucus began with a series of annual meetings under the title "M&S Leadership Summit." In 2006, this annual summit brought together key members of the M&S community of interest from industry, government, and academia to discuss how Congress could address issues facing the community at large. Previous events under SimSummit focused on a contemporary agenda designed to discuss and debate progress in areas such as M&S workforce development, business practices, and technology. For the 2006 summit, meeting organizers working with the NTSA brought together leaders from many domains within the M&S community to address new levels of M&S utilization and sophistication within their specific areas.

From the initial meeting, the Caucus agenda expanded to address important economic and technical sectors within our community, and to develop actionable recommendations to the Congress for M&S educational support legislation. In essence, the community realized the national-level impact on both the nation's security and economic well-being, and in July 2007, the U.S. Congress officially recognized M&S as a National Critical Technology (H. Res. 487). The Leadership Group and M&S community of interest, working with members of the Congressional M&S Caucus, have been able to achieve the passage of the following significant legislation:

H. Res. 487 – A Congressional resolution, adopted July 16, 2007, that recognizes the contributions of M&S to the nation and its economy, and recognizes M&S as a *National Critical Technology*. (Note: Historically, a National Critical Technology is one that has particular value to the national security of the United States and/or significant economic impact on the country).

H. Res. 4165 – Provides grants to encourage and enhance the study of M&S at colleges and universities. Grants are made available to schools that have an established M&S program and also to schools that wish to establish a new program.

H. Res. 4321 – Enhances the SIMULATION (Safety In Medicine Utilizing Leading Advanced Simulation Technologies to Improve Outcomes Now) Act of 2007.

The SIMULATION Act of 2007:

• Creates medical M&S Centers of Excellence across America to provide leadership and research into advancing the field;
• Establishes medical M&S grants for academic and professional organizations;
• Promote innovation in medical M&S within the Department of Health & Human Services; and
• Establishes a coordinating council for federal government collaboration on medical M&S efforts.

H. Res. 5658 – A resolution to develop M&S standards for the Department of Defense.

16.2.2   M&S as a National Critical Technology

The United States leads the world in the application of M&S in many industries and across federal programs, particularly defense, where these technologies have been used to support analyses of strategic operations, systems acquisition, and training of military forces. Estimates range from $4 billion to $7.5 billion spent each year by the U.S. Department of Defense (DoD) on M&S tools, processes, services, and products.

Formally acknowledging this impact, the U.S. House of Representatives recognized M&S as national critical technology, one that has provided the following list significant contributions community to government, industry, and academia:

1) Commends those who have contributed to the M&S efforts that have developed essential characteristics of our Nation.
2) Urges that, consistent with previous legislation passed by the current and previous Congresses, science, technology, engineering, and mathematics remain key disciplines for primary and secondary education.
3) Encourages the expansion of M&S as a tool and subject within higher education.
4) Recognizes M&S as a National Critical Technology.
5) Affirms the need to study the national economic impact of M&S.
6) Supports the development and implementation of governmental industrial classification codes that include separate classification for M&S occupations.
7) Encourages the development and implementation of ways to protect intellectual property of M&S enterprises.

16.3 The National M&S Coalition (NM&SC)

The NM&SC is a cross-disciplinary organization of stakeholders in the M&S industry. It is the only national organization representing M&S across all domains, disciplines, and organizational affiliations. Formed by a committed group of individuals, corporations, national and international associations, academic and research institutions, and government organizations, NM&SC is committed to leveraging M&S technologies to further advance these domains and disciplines.

Advocates for forming the NM&SC consisted of a community of interest "Who's Who" that included M&S pioneers such as Frederick Lewis, RADM USN (ret.), then president of the NTSA, and Mr. Bill Waite, Chairman of AEgis Technologies Group and a lifelong promoter for the M&S profession and workforce. Mr. William (Bill) F. Waite, to whom this book is dedicated, remained a consistent and determined advocate to formally establish and recognize M&S as a discipline within the North American Industry Classification System (NAICS). NAICS codes are the standard for classifying business establishments for the purpose of collecting, analyzing, and publishing statistical data related to the U.S. business economy.

16.3.1 National M&S Coalition (NM&SC) Inaugural Congress

Supported by the M&S Congressional Caucus, the First Inaugural NM&SC Congress, hosted by the NTSA, convened on February 6, 2012, in Washington DC. Presenters and facilitators chosen from a broad cross-section of M&S professionals were invited to participate in the first nationwide NSWC Congress of this community of interest. The event was the culmination of numerous efforts to bring all the facets of the discipline into a defined national organization. Knowledge and experience of each segment of the discipline and professional practice were leveraged to establish a permanent body dedicated to fully realizing the national potential of M&S. The overarching concept was to strongly promote efforts to cooperate, collaborate, and coordinate M&S activities across all domains, and to move forward with the growth and expansion of M&S contributions to addressing issues of national importance and consequence.

16.3.2 Building the Coalition

Following the inaugural meeting and election of the NM&SC Policy Committee (PC) members, with Dr. Richard Fujimoto serving as Chair, standing committees were created to help "build the coalition." Five standing committees were established, chartered to define a national action plan focused on issues identified by each committee in accordance with the national goals of the Coalition. Initial interaction with NM&SC members included the formation of "framing questions" later to be used as inputs for the overall PC strategy and guidance. The initial five standing committees are defined as described below, along with the original framing questions.

Technology, Research and Development (R&D): This committee focuses on promoting research in, and development of, the technical applications of M&S and the definition, instantiation, and implementation of a strong, collaborative research agenda. Questions this committee addresses include the following:

1) What part of the DoD methodology for simulation interoperability is reusable across other domains?
   a) What is available now that could be reused?
   b) What new technologies enable this effort?
   c) What are the challenges and road blocks that need to be overcome?
2) What should the M&S Body of Knowledge contain?
   a) Why is M&S a discipline?
   b) What is the best mechanism for developing, maintaining, and distributing the body of knowledge?
   c) What key technologies need further research in an M&S Body of Knowledge?

Education and Professional Development (EPD): This committee is focused on establishing common educational standards, curricula, professional certifications, guidelines for the profession, and an M&S body of knowledge. EPD addresses national-level needs to ensure the education, development, and training of an M&S workforce that can meet the science, technology, engineering and mathematics requirements of our nation. Questions this committee addresses include the following:

1) What are the near and long-term challenges facing M&S workforce creation and development?
   a) Is the existing M&S workforce well understood? Are the needs for workforce development understood? Are there adequate data?
   b) What are the impediments to increasing the quantity and quality of the M&S workforce?
   c) Is the M&S workforce sufficiently diverse? If not, why not?
2) What actions or initiatives should NM&SC develop to address these challenges?

a) How can NM&SC ensure education and workforce development efforts are integrated with national efforts in areas such as manufacturing, energy, sustainable growth, healthcare, etc.?

b) How can NM&SC help to improve effectiveness and/or reduce costs of education and professional development?

Business Practice (BP): This committee is focused on the policies and business processes that are most relevant to the integration and use of M&S technologies. This committee addresses issues regarding establishment of codes and other standards within the North American Industry Classification System NAICS that drive and support best business practices. Questions this committee addresses include the following:

1) The significance of M&S technology, and particularly its economic implications worldwide, will continue to mount, and so on (Why is this important?).

2) The "economics of M&S" is relevant to the whole M&S community, and so on (Why does M&S relevance to economics matter?).

3) Systematic, inclusive, collegial exploration can produce a body of useful findings that can be pursued to concrete effect in improving M&S business practice, and so on (Can we make M&S business better?).

Industrial Development (ID): This committee is focused on ways to include the existing organizations using and integrating M&S with their roles, relationships, and inter-relationships across content domains. This committee fosters collaboration and participation by industry, government, academia, and professional societies to explore ways in which M&S can address future industrial needs for the nation, and support the achievement of national goals for the common good of our industrial base. Questions this committee addresses include the following:

1) Should we collectively promote the use of M&S in new application areas and business sectors? Assuming yes,
   a) Can we (industry) do that collectively through NM&SC?
   b) Do we need help from the government? Who? How?
   c) Is some other type of trade organization needed to represent M&S business interests?
   d) How do we accomplish this and maintain competitive separation/ intellectual property/customer relations?

2) What resources or support does industry need in order for us to "develop" (i.e., expand customer base, create and provide innovative products)?
   a) Workforce development support
   b) Standards

c) Federal government recognition of the "industry" – NAICS code approval
3) Is there a universal value proposition to justify use of M&S, if not, strong success stories, endorsements, or methods to predict return on investment?

Public Awareness (PA): This committee is focused on communicating a consistent, well-articulated message across all M&S disciplines and to provide outreach for the priorities established by the NM&SC. Questions addressed by this committee include the following:

1) How does the committee focus on communicating a consistent, well-articulated message across all M&S disciplines and to provide outreach for the priorities established by the NM&SC Charter?
2) Primary focus of the subcommittee:
   a) How to publicize NM&SC events & promote value to a national constituency as well as to NM&SC sponsors and members?
   b) How do we communicate the value of NM&SC and M&S to national level policy makers and leadership?
   c) How does the committee establish strategic action and implementation plans with estimates of resources needed?
   d) How do we identify volunteer members for first year activities?
3) How does the committee perform baseline outreach, communications and identify marketing resources?

Since the formation of the NM&SC, the Industrial Development and Business Practice committees have been combined into one body, titled Industrial Development/Business Practice (ID/BP), and the Public Awareness committee renamed to be Communications Outreach & Public Awareness (COPA), reducing the number of subcommittees to four. Also, supporting the NM&SC Organization is an informal "Advisory Committee" made up of acknowledged national experts in M&S who serve as elected members. Figure 16.1 outlines the NM&SC relationships.

16.4 Current State of Affairs

M&S remains one of the most innovative and exciting technologies to emerge over the past decades. From relatively restricted applications only a few short decades ago, it now pervades the entire U.S. economy. Today, it is nearly impossible to identify anything that is designed, developed, deployed, or disposed of that has not been touched by M&S at some stage. But this has been more a dependence on the technology rather than solely its

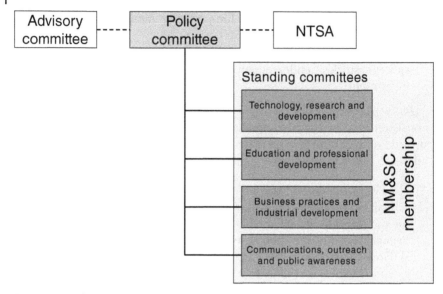

Figure 16.1 The NM&SC Organization.

dissemination into new and unanticipated areas such as 3D printing, geospatial interaction, or even Live, Virtual, and Constructive (LVC) simulations. As one successful manufacturing colleague put it, "The M&S community should declare victory; without M&S, there would be no such thing as advanced manufacturing."

As an example, there was a time when M&S was an adjunct to other ways of doing things. We could choose between live training and a relatively straightforward version of virtual training. Each had its strengths and weaknesses, and both could contribute to a better outcome, but we had an option to substitute one for the other. Over time, live training options are increasingly limited because of weapon's ranges (that now often exceed the geographic boundaries of our ranges), system-of-system employment concepts of operations, limited operational forces, and so on. In the meantime, simulation continued to evolve, incorporating features that could not be duplicated in real world environments. As a result, M&S emerged as a preferred method for many training, acquisition, and research applications. As the technology matures, this trend will continue to grow – significantly and perhaps exponentially.

M&S achieved preeminence in the field where it originated: that of training and research to perform complex activities or observations and often offering an alternative to performing dangerous tasks. While emerging into preeminence in these areas, M&S is also doing something revolutionary. It is infusing itself into new areas of observation or development that have never been accessible via existing technology.

16.4.1 Growth and Geographic Recognition

Early in the working relationships between NM&SC and Congressional M&S Caucus, it became evident that beyond large established M&S "Centers of Excellence" in locations such as Washington DC, Florida, Virginia, Alabama, Texas, and California, that many in industry and government including members of Congress were unaware of M&S activities and the economic impact within their districts. The NM&SC decided to attack this problem on two fronts. First, NM&SC would submit a proposal recommending a new NAICS code or adaptation of existing codes specific to M&S. Achieving this provides both an industry classification and a way and means of tracking economic impact. The second approach is to provide a "mapping" of all M&S-related regional activity within the United States. A successful mapping of these activities would allow members of the Caucus to show their colleagues where M&S is having direct impact within their districts.

One of many examples of growth of M&S in a specific geographic location is found within the state of MI. As the NM&SC is growing, a new active M&S Center of Excellence is emerging within MI. With Automation Alley (a MIbased 501(C)) as an active member of the NM&SC team, this technology innovator and business accelerator is dedicated to growing the economy of Southeast MI and enhancing the region's M&S reputation around the world. Automation Alley remains a strong supporter and leader in the organization of Modeling, Simulation, and Visualization activities in MI.

16.4.2 North American Industry Classification System (NAICS) and Industry Codes

NM&SC continues to recommend changes and updates to the current classification of M&S within NAICS. Much of this work is undertaken by the Industrial Development and Business Practice Standing Committee. The importance of M&S-specific codes within the NAICS is that it makes it explicit that M&S is a distinct industry classification for the purposes of collecting, analyzing, and publishing statistical data related to the U.S. economy. The NAICS industry codes define establishments based on the activities in which they are primarily engaged.

Previous attempts to have M&S as a distinctive NAICS code resulted in encouraging guidance from the Economic Classification Policy Committee

(ECPC), a collective body that periodically reviews and updates the code according to the appropriate industry sector. The ECPC acknowledges the support for the creation of a new M&S industry, and formally recognizes that "the occupation is growing, and specialized training and education are clearly present." They also note that "Professional standards have been developed and are being more widely applied."

The ECPC recommendation supports adding the phrase "modeling and simulation" as a part of the definition of NAICS Industry Group 5417, Scientific R&D Services, due to its growth in prominence in academia and in both the private and public sectors of the economy. The Committee also recommends that statistical programs consider adding inquiries for relevant industries to better quantify the production of M&S services across the range of producers. As a result of these findings, the NM&SC has developed the following long-term and short-term goals for including M&S within the NAICS industry group:

- Develop a strategy with regards to the ECPC's recommendation that "statistical programs consider relevant industries to better quantify the production of M&S."
- Research and monitor the development of the North American Product Classification System (NAPCS) and take appropriate actions to make sure that M&S products and services will be properly represented.
- Build a matrix of organizations (with an interest in M&S) beyond the defense industry. Forge and strengthen relationships and build a stronger coalition.
- Recruit community of practice representatives to participate in the NAICS & NACPS M&S codes effort.
- Provide sustained effort, via NM&SC, to examine all aspects of the issue, and to build a clear and compelling case.

16.4.3 M&S Industry Mapping

The NTSA has undertaken a plan to identify and locate organizations involved in the design and development of simulation products. This project, called the "NTSA Mapping Initiative" has completed its first phase. Organizations are listed and "placed" on a national map. For those engaged in M&S, it helps to identify areas of the country where there are higher concentrations in the development of this technology. It also assists government, university, and industry organizations to identify potential partners for collaboration. The list is updated as new organizations are identified. To date, there are nearly 1400 organizations across all 50 states included in this listing.

16.5 Toward a National M&S Research Agenda

Computer-based models and simulations are vital technologies needed in advanced economies to guide the design of complex systems and to support myriad needs for complex analysis, production, management, and to accelerate innovation (Markovits *et al.*, 2005). For example, M&S is essential in areas such as the creation of smart cities and sustainable urban growth, advanced R&D, security, and defense, among others. Over the past three decades, however, there has been little attention paid to analyzing and identifying those aspects of M&S technologies, tools, and processes that are needed to serve the needs of commerce, academia, and government in addressing the complex issues facing the nation. Complex engineered systems are but one example of the continually increasing complexity and scale of solutions being developed to address the issues just listed, and others.

What has become increasingly clear is that advances in M&S are essential to keep up with the ever-growing demand for technology-based solutions and to maximize the effectiveness of computer-based tools in the engineering. This includes the implementation of the increasingly complex products, services, and systems that will be needed to serve the economies of the future. Cloud computing, big data analytics, natural language processing, machine learning, new computer architectures, and the Internet of Things all represent aspects of technology advances that can benefit greatly from new developments in M&S. Indeed, advances in M&S technologies are increasingly essential to facilitate exploration of/ fully exploit complex problems and issues – whether engineering, societal, financial, healthcare, economic, or defense and homeland security related.

The NM&SC Research Agenda initiative, led by NM&SC, addresses both short- and long-term needs. The near-term plan focuses on "training the workforce" strategies and on identifying industry/academic contacts interested in creation and pursuit of a National M&S Research Agenda. The Research Agenda also includes efforts to reach out to members of National Congressional Subcommittees regarding critical M&S technical and R&D issues. Subcommittee include Armed Services, Education and the Workforce, Energy and Commerce, Transportation and Infrastructure, and the newly formed Science, Space, and Technology. Short-term efforts include panel discussions with recognized professionals, such as published by Tolk *et al.* (2015). Long-term plans include identifying technological and R&D areas of convergence across differing domains, continuing the identification of emergent technologies impacted by M&S.

16.5.1 A M&S Research Agenda Workshop

In direct support of creating a National M&S Research Agenda, a select group of top, nationally recognized scientists, government officials, and industry leaders convened on January 13 and 14, 2016, for a workshop at the National Science Foundation (NSF) in Arlington, VA. The primary objective of the two-day meeting was development of a foundation document and articulation of a "way forward" for guiding a national M&S research program. Workshop sponsors included NSF, National Aeronautics and Space Administration (NASA), Air Force Office of Scientific Research, and both NM&SC and NTSA. Participants included a number of major university and laboratory R&D organizations and select members from NM&SC and NTSA.

Leading researchers, representing multiple disparate disciplines, identified and articulated a number of the most important, intractable research problems, for example, "Grand Challenges" and questions facing the country, and addressed at length how to maximize the impact of new M&S technologies in addressing them (Taylor *et al.*, 2015). This included future research investment recommendations for M&S.

Critical M&S research challenges in the design of engineered complex systems were defined and articulated. The workshop focused on the following four key areas:

1) Conceptual modeling, for example, to determine how teams of individuals from different disciplines can best create sophisticated, reliable models of complex systems.
2) Advanced computational methods including topics such as exploitation of emerging computing capabilities and technologies for simulation, model checking, and inference.
3) Approaches to manage uncertainty and to address model fidelity concerns.
4) Approaches to enable and facilitate model reuse in order to accelerate and reduce the cost of creation of effective computational models.

The workshop also discussed the impact of advances in M&S on critical application areas. These included urban growth and sustainability, healthcare, aerospace, manufacturing, and defense and homeland security.

A guiding principal for the workshop included identifying and developing a consensus around critical research challenges in the M&S field (challenges whose solution will significantly impact and accelerate the solution of major problems facing society today). Although M&S has been an active area of study for some time, new developments such as the need to model systems of unprecedented scale and complexity, the well-documented deluge in data arising today, and revolutionary changes in underlying computing platforms are creating major new opportunities and challenges. The workshop focused on the following five areas to generate challenges for M&S research:

1) Conceptual modeling
2) Computational methods and algorithms for simulation, model checking, and other types of inference
3) Fidelity issues and uncertainty in M&S
4) Model reuse, composition, and adaption

The workshop is one of a series of events organized by the NM&SC R&D committee aimed at developing and building consensus around a common M&S research agenda. Activities leading to the workshop included presentations and panel sessions at the Winter Simulation Conference, MODSIM World Conference, Simultech, and the Simulation Innovation Workshop ("Interoperability Workshop" until a name change in early 2016) in addition to the NM&SC National Meeting. Subsequent to this workshop, follow-up included presentations and panel discussions at the M&S Congressional Caucus Leadership summit and related Training, Simulation, and Education forums.

16.5.2 Conceptual Modeling

Although one of the first steps in the development of a model is the development of a conceptual model, such conceptual models have too often been informal and document-based products. As the complexity of simulation models increases and the number of domain experts contributing to a single model grows, there is an increasing need to create formal, descriptive, and computer-based, models of the system under investigation and its environment. This is particularly important for the engineering of complex systems for which multiple system alternatives are explored, compared, and gradually refined over time. The descriptive model of each system alternative – describing the system of interest, its environment, and interactions between them – can serve as a conceptual model for a corresponding analysis or simulation model. Formal modeling of these descriptive, conceptual models poses the following significant research challenges:

- How can models expressed by different experts in different modeling languages be combined in a consistent fashion?
- What level of formality is suitable for efficient and effective communication?

- What characteristics should a modeling environment have to support conceptual modeling in an organizational context – a distributed cognitive system?
- What transformations of conceptual models to other representations are possible, and useful? What are the major impediments to realizing such transformations?

16.5.3 Computational Methods: Algorithms for Simulation, Model Checking, and Other Types of Inference

The main reason for modeling is to extend human cognition and scientific understanding. By expressing our knowledge in a mathematical formalism, the rules of mathematical inference implemented in computer algorithms can be used to draw systematic conclusions that are well beyond the natural cognitive ability of humans. For instance, for complex systems with millions of state variables and relationships, simulation provides capability, far beyond that of the human mind, to project how the state of such systems will change over time. Advancing the algorithms for such inference so that ever-larger models can be processed more quickly is likely to remain a crucial capability for engineering and science. Besides simulation, there is an increasing role for model checking, especially for engineered systems that are affected by high-impact low-probability events.

This leads to the following questions for discussion:

- What are current trends in computing affecting M&S, and how can they best be exploited?
- How will these trends change the nature of simulation and reasoning algorithms?
- What are the major gaps in computational methods for M&S, and what are the most important research problems?
- How can one best exploit the vast amounts of data now becoming available to synergistically advance M&S for engineering complex systems?

16.5.4 Uncertainty in M&S

The goal of M&S often is to make predictions, either to support decisions in an engineering, business, or policy-making context, or, in a scientific context, to gain understanding and/or test hypotheses. However, too often, it is impossible to prove that a model is well-suited to its intended use (they are never "correct"/complete representations of reality) – the predictions are always more uncertain than desired. Despite this, many models and simulations have been proven to be useful, and their results are routinely used for many purposes. To further improve the usefulness of models, it is important that we develop a rigorous theoretical foundation for characterizing the accuracy of the predictions. Within the M&S community, there is still a lack of agreement on how best to characterize this uncertainty. A variety of frameworks have been proposed around concepts of validation and verification, and a variety of uncertainty representations have been proposed, many based on questionable mathematical foundations (e.g., fuzzy numbers). A related question concerns determining the appropriate level of detail that a model should contain, and there are issues combining models that may operate at vastly different temporal or spatial scales. This leads to the following questions for discussion:

• What is the most appropriate approach to representing and reasoning about uncertainty in complex systems consistently?
• What is the best approach to characterizing the uncertainty associated with a simulation model in order to enable and facilitate reuse?
• How should one aggregate knowledge, expertise, and beliefs of multiple experts across different domains?
• What is the best approach to take advantage of the large and diverse datasets for characterizing uncertainty and for improving model accuracy?
• What are the most promising approaches to accelerate the validation of models for specific application contexts?
• What are the key challenges in multiresolution modeling and the most promising approaches to addressing them in the targeted application domains?

16.5.5 Model Reuse, Composition, and Adaptation

Although modeling is now indispensable in engineering and science, the cost of creating a good model can be considerable. This raises the question of how these costs can be reduced. One approach is to encode domain knowledge into modular, reusable libraries of models that can then be specialized and composed into larger models. Such a modular approach allows the cost of model development, testing, and verification to be amortized over many (re)uses. However, reuse also introduces new challenges:

• How can a model user be confident that a planned re-use of the model is within the range of uses intended by the model creator?
• How can one characterize the uncertainty of a model that is reused (possibly with some adaptations to a new context)?

- How can one characterize the uncertainty of simulation models obtained through composition of multiple models?
- How can one accelerate the process of adapting and reusing models for different purposes? What are the fundamental limitations of technologies for model reuse?

16.5.6 Workshop Outcomes

The output produced by the workshop is a report documenting the observations and findings produced by its participants. (National Science Foundation, 2016). In addition, broad dissemination of the report is anticipated, for example, through publications of summaries of the report as well as the report itself in professional journals and other publication venues. Longer term, further dissemination, discussion, and elaboration of workshop results are expected through follow on meetings and other activities associated with conferences in order to engage the broader M&S community. The NM&SC which initiated discussions leading to this workshop and sponsored various related events leading up to this workshop, will assist in disseminating workshop results, and working with groups such as the M&S Congressional Caucus in the United States to build broad support for this initiative.

16.6 Summary

NM&SC remains the capstone organization to promote and leverage M&S to strengthen the United States' work force. Its mission is to create a coalition of "unified national community of individuals and organizations" around the M&S discipline and to be the national principal advocate. The purpose of the coalition is to serve the needs of all components of the M&S community of practice and to recognize it as a profession, as an industry, and as a catalyst within the marketplace. The coalition provides a forum for dialog across industry, government, academia, and professional societies and support for the discipline in the areas of technology/research and development, education/professional development, industrial development, and business practice through a campaign of public awareness.

This coalition understands that the discipline of M&S, as both an art and a science, will continue to grow and expand throughout many professional and workforce domains: entertainment, medicine, vocational education, public television, real estate sales, airplane accident investigations, and so on. NM&SC also understands that the discipline and profession will continue to make possible major industrial advances in areas such as electronic technologies, big data visualization, advanced manufacturing, animation and, perhaps more significantly, as an expanse to the "creative mind" of the simulation industry. As a science (which includes many fields of engineering) M&S will be used to conceptualize, frame, model, analyze, and solve problems, some of which would be impossible to solve without the capability. Indeed, as a problem solving discipline it may very well be the "calculus" of the next century.

Following its foundation charter, the intent for the NM&SC is to support these trends by building a strong coalition of professionals and promoting workforce development. These initiatives include establishing an over-arching national research agenda, mapping to visualize M&S geographical activities and helping organizations identify potential collaborators, and direct efforts to identify and develop M&S NAICS codes. NM&SC is the M&S communities of interest's direct link to Congress and remains its central channel of communication in order to simplify, expedite, and improve the national consideration of the many policies, regulations, problems, opportunities, and questions of broad application involved within the M&S enterprise.

References

Bergeron, B. (2006) *Developing Serious Games*, Charles River Associates, Hingham, MA.

Goldsman, Nance, Wilson (2010) A brief history of simulation revisited. In *Proceedings of the 2010 Winter Simulation Conference, Baltimore, MD*, pp. 567–574.

House Resolution 487 (2007) In the House of Representatives, US, July 16.

Markovits, G., Markovits, D., and Teter, J., (2005) Accelerate Innovation. Defense AT&L: September 1–4.

National Science Foundation (2016) Research challenges in modeling & simulation for engineering complex systems. *National Research Agenda Workshop, January 13–14.*

Taylor, S.J., Khan, A., Morse, K.L., Tolk, A., Yilmaz, L., Zander L., and Mosterman P.J. (2015) Grand challenges for modeling and simulation: simulation everywhere–from cyberinfrastructure to clouds to citizens. *SIMULATION: Transactions of the Society for Modeling and Simulation International*, 91 (7), 648–665.

Tolk, A., Balci, O., Combs, C.D., Fujimoto, R., Macal, C.M., Nelson, B.L., and Zimmerman, P. (2015) Do we need a national research agenda for modeling and simulation? *Proceedings of the 2015 Winter Simulation Conference, IEEE Press*, pp. 2571–2585.

Index

The Profession of Modeling and Simulation: Discipline, Ethics, Education, Vocation, Societies, and Economics, First Edition. Andreas Tolk and Tuncer Ören.
© 2017 John Wiley & Sons, Inc. Published 2017 by John Wiley & Sons, Inc.